The Chemistry of a County

Bryan Hanley

The Chemistry of a County

Norfolk as a Paradigm for the Impact of Chemistry over Time

Springer

Bryan Hanley
Department of Preventative Dentistry
Academic Center for Dentistry Amsterdam
Norwich, UK

ISBN 978-3-032-21490-4 ISBN 978-3-032-21491-1 (eBook)
https://doi.org/10.1007/978-3-032-21491-1

This Springer imprint is published by the registered company Springer Nature Switzerland AG
The registered company address is: Gewerbestrasse 11, 6330 Cham, Switzerland

Preface

The national and international importance of science and innovation has been widely described. Thinking in terms of economic and commercial success has helped to drive the ever-increasing specialisation of science and, while multidisciplinarity is a frequently described aspiration, the translation (and application) of scientific ideas and concepts across boundaries is relatively rare and faces considerable barriers. Projects are frequently broken down into sub-projects with high levels of specific technical expertise as a required input for each of them. Chemistry is no different from other disciplines in this respect. This is, in part, explained by the interface between chemistry and adjacent disciplines such as biology and the need to translate chemical concepts into useful information for other scientists. This creates a twofold problem which can be expressed in terms of the accompanying scientific disciplines of biology and physics. The first challenge is that the application of chemistry can provide information that is of value to chemists but may not add to the sum of knowledge regarding the adjacent discipline. For example, the application of chemistry principles may enable the isolation of specific enzymes and the measurement of the rate of the reaction they catalyse as an isolated entity but this information may not provide much information regarding the biological system that the enzyme is isolated from; so, reaction rate measurement without context is not particularly valuable. This problem is actually a signifier of a more basic and philosophical challenge. Chemistry is the investigation of molecular interactions. It is therefore a reductionist exercise. The descriptions of atoms and molecules (and how they interact) is the driving force behind the concepts of the existence of atoms, molecules and bonds. In order to make sense of these concepts (and, in particular, how they interact), it is necessary to break systems into their component parts which can then be measured (analysed) separately. Biological systems are defined by the complex interactions of their component parts and this is, on the whole, opaque to chemical reductionism. The second challenge is illustrated by the comparison of chemistry with physics. Physics deals with all aspects of nature on both the macroscopic and submicroscopic levels. As such it is concerned with matter, motion and energy and in the development of mathematical laws that explain the interactions between these. It is, therefore, the ultimate reductionist scientific discipline.

Chemistry is the ultimate empirical and transformative science. While physics deals with things as they are and uses mathematics to describe them unambiguously, chemistry is concerned with implied structures and behaviours. Even atoms, molecules and bonds are convenient models to explain behaviours rather than mathematically provable truths. The definitions provided by physics do not explain the chemistry (transformative properties) of a system. In summary, one scientific discipline is not 'superior' to another, they simply express different optics of a system.

Against this background of scientific heterogeneity, chemists have to make sense of the world and develop ways of transforming it to provide intellectual and economic value. This cannot be achieved on a global basis but by taking one particular area (rather than a sub-discipline) as an example, it is possible to examine the flow of chemistry through time and to develop an illustrative understanding of how that flow is interdependent. In order to develop this idea, it is necessary to concentrate on one particular region (Norfolk) and to examine the impact of chemistry at three levels—individually, institutionally and societally. These do not exist as separate and distinct concepts but they facilitate each other and are dominant on different timescales and have different dependencies and impacts.

In order to illustrate differences and interdependencies, the book chapters are grouped into four themes as follows:

- Theme One: The Essence of Chemistry and Its Relationship with and Exploitation of the Natural World (Chaps. 1–3)
- Theme Two: Chemistry and Commercialisation (Chaps. 4 and 5)
- Theme Three: Chemistry and Manufacturing (Chaps. 6 and 7)
- Theme Four: Chemistry and Products from Nature. Food Processing for Value (Chaps. 8 and 9)

Each of these themes will illustrate aspects of the growth and (to a lesser extent) eventual decline of Norfolk and how chemistry provides a unifying theme between these paradigms.

Norwich, UK Bryan Hanley

Competing Interests The author has no competing interests to declare that are relevant to the content of this manuscript.

Contents

Editor Biography

Bryan Hanley has worked both in academia and in industry in the UK, the USA and Europe. He is a chemist by training and holds degrees from the University of St Andrews and the University of Edinburgh. He has authored over 80 scientific papers and has authored or edited five books. He was a Visiting Professor at the University of Ulster (2013–2017).He is also the President of the International Life Sciences Institute, Europe, and was formerly the President of the Industry and Technology Division Council of the Royal Society of Chemistry (UK) and Chairman of the Food Group of the Royal Society of Chemistry.

Part I
The Essence of Chemistry and Its Relationship with and Exploitation of the Natural World

Chapter 1
The Essence of Chemistry

Abstract The essence of chemistry lies in the concepts of reductionism and transformation. Within this, there is a collision between attempts to reduce chemistry to measurable entities and the complexity of the world in general and living systems in particular. Chemistry also lies at the intersection of science, technology and industry and so is defined by its usefulness. The combination of reductionism and usefulness was a theme of 'Frankenstein' by Mary Shelley in which a creature was made from the component parts of a body. This also raises the question of Vitalism and whether living systems have some unique property. This was largely disproven by the synthesis of urea (a component also isolated from living systems) by Wohler. Such investigations also popularised the concept of analysis and synthesis as linked approaches to proving the structure of natural products. This has been joined, in the twentieth century, by the addition of the concept of bio-synthesis (the creation of molecules by living systems). In addition to reductionism, chemistry is also the science of transformation. While many living systems operate within certain physiological boundaries, some of them (extremophiles) are able to function outside of these boundaries. The adaptation of chemistry by individuals, institutions and society provides a conceptual thread that encompasses products from nature and from the application of manufacturing processes and examples of these can be seen at different stages in the development of Norfolk.

Keywords Reductionism · Transformation · Internationalism · Alchemy · Vitalism · Entropy

Chemistry encompasses both reductionist and transformative aspects. These approaches are used to not simply develop an understanding of the world by separating the different components into recognisable facets and finding how these interact, but also to try to create systems that are able to enhance and improve, in a structured way, what is available in nature. Both of these are part of the paradigm whereby the potential impact of chemistry is increased and they also demonstrate how chemistry can change paradigms and ways of considering substances by understanding component parts (and their relationship to each other) and by changing

B. Hanley, *The Chemistry of a County*,
https://doi.org/10.1007/978-3-032-21491-1_1

what is available and hence modifying the starting point for any investigation. In this way chemistry is both a descriptive and consequential discipline. This is perhaps most obvious in the use of natural products as the starting point for the synthesis of novel medicinal agents [1] however the development of novel alloys makes use of chemistry—sometimes coupled to advanced techniques such as machine learning [2]. In the development of the Periodic Table and the expression of the concept of affinities, chemists first of all tried to reduce compounds to their component parts which could not be further sub-divided (atomic theory). Subsequently there was a realisation that not all elements had an equal degree of attractiveness for all other elements and these differences were referred to as 'affinities'. The very earliest studies on affinities by Geoffroy in the eighteenth century sought to classify individual elements based upon their capability to form salts and to displace other elements. The first affinity table, created by Geoffroy in 1718 comprised 16 columns and 9 rows. The bonding species were listed in the first row and the species to which they could bind in successive rows in order of decreasing affinity. This also introduced the concept of displacement reactions in which an element with greater affinity could displace one of lower affinity. Geoffroy's table was the first to introduce the concept of reactivity and this played an essential role in the development of chemistry and the understanding of molecular interactions and bonding. The consequences of the affinity paradigm continues to be appreciated—particularly in the context of the thermodynamic properties of elements [3] and the development of model systems.

It is integral to chemical thought that substances can be transformed. As we shall see, this can happen on a molecular level with eventual consequences at a whole organism level. The consistency of the link between microscopic changes at a molecular level and the macroscopic properties of substances is, as will be discussed later, an important essence of the philosophy of chemistry. In some cases (for example, crop rotation, which will be considered in Chap. 2), the molecular understanding does not develop immediately and, while the macro level consequences of some actions are clear, the underpinning molecular interactions are only able to be understood much later. It is not the case that the hierarchy of understanding invariably begins on a molecular level and that the consequential scale up to a systems level (macroscopic) effect is observed and understood later. In some cases, knowledge is acquired in a piecemeal fashion and illumination becomes apparent retrospectively. In other cases, the consequences of a given molecular event may not be predictable in the context of the complex macromolecular system. This conflict between reductionism to a molecular level and the impact on a complex, multifunctional entity is one of the persistent challenges for chemistry. Nonetheless, retrospective analysis based on the acquisition of new knowledge may lead to further advances based on a combination of current knowledge and the revision of a historical interpretation. In Norfolk, the development of chemistry over time and its application on a systemic and molecular level can be seen to define a number of threads of thought with common themes. These are investigated in the following chapters with successive themes of agriculture, commercial exploitation and manufacturing from systemic and molecular standpoints. These comprise the three thematic foci of the book as a whole.

In addition to reactivity, the transformative nature of chemistry is also reflected in and a consequence of the development of social habits and thoughts. For example, in the book 'Frankenstein' by Mary Shelley, which was first published in 1818, the scientist, Dr. Victor Frankenstein, adopted a very reductionist approach to the creation of his new being. He began with the premise that a creature was made up of several constituent parts. He then created his new organism by combining parts from other constituent organisms and added life by harnessing the power of electricity (which had been described as intrinsic animal electricity by Galvani in 1771 [4]). In some ways, this reflects a very literal interpretation of the analysis/synthesis paradigm developed by Descartes in his scientific and philosophical investigations [5] and used by chemists studying the nature of natural substances by first of all decomposing them into recognisable components then proving the structure by joining the components together to recreate the natural substance. In essence, Shelley defines a human creature as being composed of a number of distinct physical entities that can be separated then recombined to create something that is new but recognisably derived from its component parts and given the essence of life by the addition of electricity. This can be seen as a refutation of the concept of 'vitalism' which was disproven, on a molecular level, by Wohler's synthesis of urea in 1828 [6]. In order to create something that results from an additive transformation of existing parts, the present day equivalent of a creature such as that created by Dr. Frankenstein might either be a purely biological chimera [7] or a biological/machine hybrid (a cyborg) [8]. In molecular terms, the analogy is with the synthesis of a compound that is found in nature or something that is only able to be created in a laboratory. The distinction between natural and chemist (or laboratory)-driven approaches to natural products will be examined in more detail later and will include considerations of biosynthetic, synthetic and semi-synthetic approaches at a molecular rather than a macromolecular level.

Chemistry, like most scientific endeavours, is a trans-national effort. As noted above, contributions from England, France, Germany and Italy all added to the empirical underpinning of the philosophical grounding of chemistry. The development of scientific thought was a collaborative effort. While Newton developed equations that described the movements and interactions of bodies on an observable scale in his Principia in 1687, chemistry, since it was primarily concerned with molecular interaction that, in most cases could not be observed or measured directly was less intensively studied. In addition, the separation between chemistry and alchemy was an ongoing challenge. Theories of atomism and 'affinities' helped to define chemistry in the context of testable hypotheses and Chairs of Chemistry were established at Cambridge in 1702 and in Edinburgh in 1713. Although the Chemistry Chair was established at Cambridge, the importance of chemistry as a respectable discipline was not particularly appreciated in England and it was not considered to be a suitable field of study for 'gentleman scholars' and future members of the Church. Chemistry was considered to be a craft rather than a philosophically reputable branch of science. Indeed, there has been considerable debate about whether chemistry (and biology) can be 'reduced' to physics and thereby gain scientific credibility. This debate will be considered later. Arguments have been made both for the interpretation of chemistry as an inferior branch of physics and as a philosophically respectable field of study in its own right [9] and the idea that chemistry can be thought of purely in terms of equations derived from physics seems to be as irrational as the idea that biology can be explained purely in terms of chemical reactions—if only we understood the thermodynamics of living systems in enough detail. In contrast, to the situation in England, there was a considerable interest in chemistry as a distinct discipline in Europe (particularly at Leiden). In Edinburgh, the study of chemistry was closely linked to Medicine and to the practical (and demonstratable) aspects of chemistry. This is perhaps the key to the distinction between chemistry and physics. The principles of reductionism and transformation will be examined in more detail in Sect. 1.1; however it should be noted that classical physics and Newtonian physics describes the properties of macroscopic entities in terms of measurable relationships and these properties can be tested by carrying out measurements. It has, therefore, a core of empiricism (proof by testing) however it differs from chemistry since the proof is primarily concerned with the relationships between entities on a macroscopic level (quantum physics is a different matter and will be dealt with later) and not on their molecular interactions to produce something that is not simply the additive blend of its constituents. Chemistry, like physics, is also driven by testing and empiricism but at the core of chemistry is the concept of molecular interactions and the chemical bond. Whether chemists adopt a molecular orbital theory for the chemical bond or think of it in terms of the valence theories espoused by Pauling, their concepts are not capable of being subjected to direct inspection. The empiricism of chemistry involves something more than the macromolecular measurement of properties carried out by physicists. Chemistry involves change (or transformation) in a way that provides a result that can be predicted based upon models and hypotheses. The basis of the result is a consequence of the nature and the properties of chemical bonds. The influences of Europe and of the importance of empiricism and accurate measurement were demonstrated by the fact that first two professors of chemistry at Edinburgh trained for a while at Leiden and were influenced by the pre-eminent chemist of the time, Herman Boerhaave. Prior to the establishment of chemistry in Edinburgh, with its distinctly European orientation, Norfolk, through its interactions with the Hanseatic League and trading links, was already looking towards Europe rather than to the insularity of London, Oxford and Cambridge. Continental building styles (particularly Dutch) and commercial considerations (textile making, in particular) were very evident in Norfolk and some aspects of these will be discussed further in Chap. 6.

While chemistry is concerned with all the elements of the Periodic Table, the chemistry of carbon—due to its particular tetravalent nature and its crucial role in living systems—provides levels of complexity that have contributed considerably to the growth of chemistry. In order to understand the importance and relevance of chemistry to the development of, in particular, science and industry in Norfolk, some of the key attributes of organic chemistry should be explained. This will not be a comprehensive account of organic chemistry but rather a consideration of some philosophical aspects of the subject both as a reductionist and transformative discipline. By highlighting some of the key principles of chemistry, it will be possible to demonstrate the linearity of principles that underpin different aspects of chemistry and the unified nature of the discipline.

1.1 Reductionism and Transformation in Chemistry

The concepts of empiricism, reductionism and transformation in chemistry has already been introduced. The alchemical tradition of chemistry refers to carrying out various actions that are designed to purify substances and to develop their 'essences'. The seven operations of alchemy were Calcination, Dissolution, Separation, Conjunction, Fermentation, Distillation and Coagulation and these were designed to cleanse or purify substances. These will be discussed in more detail later. Traditionally alchemy has embraced both reductionism and transformation. Substances are 'transmuted' into other substances (hence the belief that base metal can be converted into gold) and even Paracelsus was, at heart, a transmutational alchemist [10]. However alchemical operations also extended to living systems and the promise of eternal life was also part of the alchemical tradition [11]. The language of alchemy (and to an extent the thinking behind it) remains part of modern science [12], particularly in the study of complex biological systems. More recently alchemical free energy (AFE) has been described as a concept. This is a simulation method in which one state of a system changes into a different state. The AFE method of simulation calculates the free energy difference between the states (but not the path taken) [13]. This simulation method is used to assess drug potency and binding to biological macromolecules but it also has other applications [14] and seeks to bridge the gap between molecules and materials—between molecular and macroscopic properties.

The principle that underpins alchemy is that of balance and, in particular, the concept that combinations of elements including, mercury and sulphur, can be used to create a material that will have multiple properties and could cure all disease and offer the possibility of eternal life [15]. In a sense, therefore the practices associated with alchemy can be found at the core of thinking about health and disease including Ehrlich's concept of a biologically specific 'Magic Bullet' [16] to combat disease and current thinking about personalised medicine [17]. One aspect of the practice of alchemy which continues to resonate with modern chemistry is the development of hypotheses that can then be tested by empirical methods [18]. Indeed, the concepts of affinities as explained by Geoffroy and mentioned previously can be seen to be part of the lineage of combining elements to create something new that was practiced by alchemists. The emphasis on prolonging life and converting base metal into gold became the most important drivers for alchemy (as noted previously) and this, in turn, facilitated the development of scientific culs de sac leading to practices such as homeopathy which were not able to be tested in a coherent mechanistic way and which, therefore, were not part of a multi-step progression resulting in a significant advance. The divergence of alchemy and chemistry in the seventeenth century was an inevitable consequence of the increased understanding of reactions and the eventual acceptance of atomic theory (corpuscularism) after much scepticism led, in part, by the thinking of Aristotle [19]. The source of some of the scepticism at that time relates to the difference between chemistry and physics during that period. While the motions of objects and the splitting of light by a prism could be demonstrated, atomism was not a testable hypothesis [20]. In addition the division of the observable (macromolecular) and the postulated but unquantifiable (microscopic and molecular) worlds was promoted by Lavoisier (and others). As technical advances allowed the development of the accurate measurement of different types of atoms in substances to give empirical formulae then atomism grew in popularity and the evidence base became more established however this still resulted in the measurement of the bulk properties of (supposedly identical) atoms and not the individual atoms themselves. Although atomism was supported by the development of measurements of the fixed proportions of the constituent atoms in a compound by weight, this did not disprove the theory of simple affinities [21]. Indeed, it was only when more complex (non-binary) mixtures were investigated that the atomic theory was found to be more coherent than the dual nature of the affinity system (compounds consisted of combinations of pairs of substances).

Another aspect of reductionism in science that applies to chemistry, and which was previously mentioned in passing, is the idea that chemistry (and, indeed, other science disciplines such as biology) can be 'reduced' to physics [22] and, according to this theory, it is only the complexity of the systems that prevents this from being carried out. The implication is that, once sufficiently robust, powerful and comprehensive computing systems have been developed, there will be a unity of sciences based on the principles of physics and everything will be able to explained in terms of mathematical equations. Alternative arguments have been explored and the feasibility of reductionism of scientific disciplines to a core set of principles remains a topic for debate [23]. This will be reconsidered at different points in the narrative from different perspectives. Even such a fundamental chemical concept as the Periodic Table has been examined from the perspective of physics [24]. A key concept of chemistry, however, that distinguishes it from physics is the creation of novel substances with new properties (both inherent and external) as an objective in itself. Philosophers, including Diderot, have examined the question of the creation of new materials (chemically charged materialism) [25] to explain that, while they are composed of different elements, they do not have properties that can be directly derived from the additive properties of their constituent parts. In addition, chemistry is in a constant state of flux due to the interconversions and transformations of chemical substances. The empirical basis of chemistry—both from a synthetic and an analytical standpoint—means that the transformative culture of experimentation is a key aspect of chemistry and distinguishes it from physics. In a practical sense, the elements that compose a material have properties that are different to those of the elements in isolation. In this way, the elements are affected by the compound they are a part of as well as being unique entities. The same type of argument applies in principle to both micro and macro properties. Diderot considered that living systems were not composed of several series of chemical reactions (the reduction of biology to chemistry) but rather, are characterised by the divergence between living and non-living matter [26] and living systems may have properties which its material parts do not possess [27]. While on the one hand, this can be seen as a defence of the uniqueness of living systems which is close to vitalism, it can also be used to help to define chemical reactions. If living systems are unique both because of their components and how these components act in the context of the living entity then, on a molecular level, individual elements are altered by the environment of the substance they are a constituent part of, and chemical transformations of molecules are affected by the environment of the organic system they in turn are a part of. This emphasis on the impact and effect of environment suggests that the reductionism that leads to the measurement of single components of a biological system in isolation (for example, enzyme-catalysed reaction rates or protein crystal structures) are misleading snapshots of how specific component parts of biological systems might operate in isolation but these can only have very limited biological relevance since they ignore context. A similar principle applies to the development of chemistry in different environments. The direction a particular chemical advance will proceed in will be in part dependent upon three different levels of context—the individual and their level of knowledge and experience, the history and background of facilitating organisations such as companies, research institutes and universities and finally the collective awareness and maturity of thought of society as a whole. Aspects of this will be observed when considering Norfolk and the impact and consequences of chemistry on the county.

Since the foundations of chemistry as an empirical science persist in some of the practices of alchemy, it is relevant to examine these associations (which were mentioned previously) in more specific detail and to draw an analogy with later chemistry approaches. The first stage in alchemy is calcination which is the combustion of material. This is also the basis of the divination of the fixed proportions of elements in a substance as developed by Lavoisier [28]. The second stage described in alchemy is dissolution. In the practice of natural product chemistry, fractional dissolution to separate components is a common process and is used, for example, in the isolation of substances with different physicochemical properties such as alkaloids and acids. The third stage in alchemy is separation in which the soluble component after dissolution is separated from the insoluble. The fourth stage is the combination of the separated components to produce a new substance. This is analogous to the analysis/synthesis procedure which was developed as a way of proposing and then confirming a chemical structure and is still used [29]. Fermentation is a way of testing the new substance and was incorporated into Galenic thinking in alchemical methods [30] and it is followed by distillation (purification) and coagulation (crystallisation)—both of which are techniques used in contemporary chemistry. Alchemical tradition is more than a series of empirical steps however, and the nature of the techniques used is a guide to a way of thinking that underpins chemistry and, in particular, the investigation of the chemistry of natural systems. The steps outlined above include both reductionist and transformational steps and also the creation of new substances with novel properties. These operations will become apparent (in different ways) as the thread of chemistry in Norfolk is examined and illustrated in more detail.

As noted previously, the concept of the constituents of a substance being present in fixed proportions is an essential part of the atomism theory developed in the progression of chemistry and its divergence from alchemy. This theory did not, however, enable chemists to carry out all the manipulations and transformations that they wished. For example, W H Perkin used such a knowledge-based approach in his attempt to synthesise quinine. Perkin knew that the empirical formula for quinine was $C_{20}H_{24}N_2O_2$, however the structural formula was unknowable at that time (even the existence of benzene rings was not known until publications in 1865 and 1866—some 10 years after Perkin's attempt to synthesise quinine [31]. Perkin then hypothesised that if he took 2 molecules of allyl toluidine ($C_{10}H_{13}N$—$C_{20}H_{26}N_2$) and added oxygen atoms (x3) from the known oxidising agent, potassium dichromate, he would have been able to synthesise quinine and water. This reaction failed to give the desired product but instead, in a great piece of serendipity, Perkin produced the first aniline dye. The method developed by Perkin required the use of a number of techniques that would have been familiar to alchemists including filtration (separation), dissolution, combination, distillation and crystallisation.

1.2 Reactions and Entropy

The importance of reactions and the consequent formation and breaking of bonds has already been noted as a defining characteristic of chemistry. In the seventeenth century, the Cartesian mechanical philosophy was invoked to explain the chemical properties of compounds. The underpinning assumption was that all chemical properties could be reduced to combinations of the mechanistic properties of the fundamental particles. This position was contrary to the thinking of the prevailing chemical school (including Robert Boyle) who proposed that, while fundamental particles could be used to explain fundamental, mechanical properties, there was also a set of distinctly chemical phenomena which could not be explained in this way and required the adoption of ideas about chemical properties [32]. A necessary consequence of the adoption of a theory about the nature of chemistry is that the properties of compounds cannot be explained solely by the additive consideration of their component parts. In addition to an upward causation which describes the properties of molecules in terms of atoms (components), there is also a downward causation from complete chemical entities to their component parts. In other words (and in agreement with the previous discussion concerning the importance of environment) reductionism suggests that the properties of a molecule are dependent upon its constituent parts but the constituents are also affected by the environment of the molecule as a whole. A simple example illustrates this concept and the importance of defining the most appropriate unit as a contributor to the overall properties of a molecule. If the properties of an organic molecule were solely defined by the proportions of its component parts (the empirical formula) then, in organic chemistry, molecules with the same empirical formula would have identical properties (Fig. 1.1). This is clearly not the case for either simple or more complex structures as illustrated for dimethyl ether & ethanol and the natural products carvacrol and thymol. The molecular formula of each pair of molecules is the same but their properties are different.

Reactions and transformations are essential to chemistry and, as will be discussed in later chapters, to the application of chemistry as a useful tool to create wealth and knowledge. The concept of experimental and transformative chemistry is based, as mentioned previously in the discussion about Diderot, in the alchemical tradition of chemistry since many of the processes that are employed by chemists have their origins in alchemy. As such chemistry, at least partially, circumvents Newtonian ideas about the properties of materials. Nonetheless, thermodynamics is important to obtaining a better understanding of chemistry and to its application on a macromolecular scale (and hence for its utility in daily life) and various attempts have been made to model molecular interaction and even, in a limited way, living systems. The problem in creating a comprehensive model for chemistry and chemical entities is, as alluded to previously, not one of complexity but rather a consequence of the nature of chemical interactions. Entropy calculations for reactions have however been modelled and various strategies developed [33] to use the results in examining molecular interactions. The application of such methods to complex biological systems is much more challenging, although maximum entropy models have been used to compare the experimental and the simulated properties of biological systems [34]. While the free energy of binding in biomolecular reactions can be calculated, entropy calculations are much more challenging even in bimolecular systems [35] and this creates a significant challenge even though bimolecular interactions may make a significant contribution to overall biological effects. One application of modelling that has been regularly reviewed and continues to show promise [36] is the use of molecular modelling in drug design. These are necessarily limited and simplistic. Such modelling, therefore, does not obviate the need for experimental investigations, but they can suggest new approaches and help to prioritise

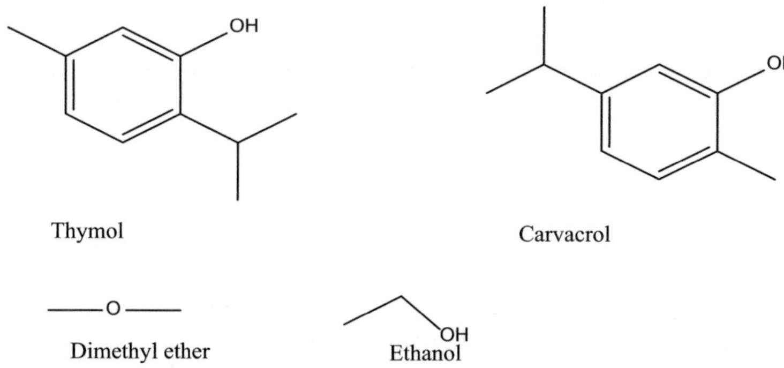

Fig. 1.1 Structural isomers with different properties

potential leads. Another area of interest in the development of entropic models refers specifically to bioactive molecules rather than the overall entropy of biological systems which was noted previously. For example, studies of the antimicrobial peptide, thiocillin have been carried out [37]. Thiocillin contains a macrocyclic peptide structure. Replacement of each component with an alternative has indicated that a rigid, low entropy form of the molecule has the greatest activity. The rigid form of the molecule is limited in the conformations it can adopt and this facilitates tight and specific binding to the biological target (the interface between ribosomal protein L11 and the 23S rRNA in Gram positive bacteria) and hence facilitates optimal biological activity.

As noted previously (and illustrated in the example of thiocillin), entropy change is an important measurable consequence of the interaction of molecules with systems. One of the important tools for investigating the entropy of systems at a molecular level by inference from their macroscale properties is statistical physics [38]. Such significant tools are likely to be applied to increasingly complex systems as computational power increases however it should be noted that, as mentioned previously, direct inspection of chemical bonds at a molecular level remains unachievable and extrapolation (or simulation) is the likeliest route to developing an understanding and the intrinsic properties of chemistry mean that reductionism to a set of mathematical formulae is not conceptually possible. This theme will also be explored elsewhere.

In the context of entropy and natural systems, there remains a tendency towards an increase in entropy with time and this is relevant to the study of growth and evolution in irreversible systems [39]. This is part of the nature of isolated and closed systems [40]. This increase in entropy has been postulated to occur at the molecular and organ level—including in the brain [41]. On a superficial level, it may seem that the growth of living systems results in a reduction in entropy, certainly from a chemical standpoint since nutrients are used to create ordered macromolecules including proteins, nucleic acids, fats and lipids and cellular structures however the release of carbon dioxide by living systems and the generation of heat (defined as the kinetic energy of the atoms and molecules in a system [42]) means that, if the surroundings (i.e. the whole system) are taken into account, the overall entropy increases. There are other energetic considerations that means that living systems obey the Second Law of Thermodynamics and the overall Free Energy is negative and the entropy is positive for any process [43]. The energetic status of living systems have been studied however these investigations have generally employed the premise that the individual components of a living system must obey thermodynamic principles—even if they operate in a non-equilibrium mode [44]. The application of reductionism to the thermodynamic properties of the components of a living system is an important principle. One particular aspect of living systems, in the context of the energetics of its components, is the use of enzymes to lower the activation energy of a particular individual reaction or set of reactions. One of the practical consequences of the acceleration of a reaction catalysed by an enzymes is the increase in the turnover rate (the rate of conversion of a given substrate to a product). Without the involvement of an enzyme the rate of many reactions would be insufficient to maintain the complex interdependences of living systems. Enzymes are, therefore, crucial for the operation of living systems [45] and a number of reactions that are essential to life would not occur under physiological conditions in the absence of enzymes. However the introduction of an enzyme catalyst may complicate simple thermodynamic considerations—particularly when interrelated reactions are involved—and make the reductionist study of what happens in living systems more difficult to disentangle. For example, the Krebs cycle (a key series of metabolic reactions) can operate in different ways and, depending on inputs and outputs, may accumulate different intermediate metabolites [46], for example in cancer cells. For uncatalyzed reactions, on the other hand, the reaction rate increases exponentially with increased temperature [47] and this enables reaction rates under specified conditions to be both calculated and predicted. In the case of enzyme-catalysed reactions, the thermal properties of the enzyme means that the effects of increasing the reaction temperature are not necessarily completely predictable [48] and are not simply a consequence of the denaturation of the enzyme structure and loss of catalytic capability when the temperature exceeds a certain level. Other conditions may affect enzymes. For example, the use of enzymes in low water systems (either natural enzymes or those derived from extremophile organisms [49]) where conformational rigidity may be imposed (with a lowering of the degrees of freedom and hence the entropy of the enzyme molecule) can enhance thermal stability [50]. Such studies may also more closely mimic the *in vivo* conditions under which an enzyme operates than a buffered aqueous solution (see below). In the case of membrane-bound enzymes, for example, their properties and characterisation has been more challenging to study than those of water-soluble enzymes which can be isolated and investigated under *ex vivo* conditions [51]. Considerable efforts have been made, adopting a reductionist approach, to classify enzymes under controlled (not necessarily biomimetic) conditions and this is generally carried out in a buffered aqueous solution. While optimisation of the *in vitro* conditions for enzyme-catalysed biotransformations to take place is important from a biotechnological application standpoint, this approach may have more limited applicability in understanding how living systems operate since, as noted previously, it does not include any consideration of the biological environment in which the enzyme occurs and operates. This is not solely important in the investigation of the contribution of enzyme-catalysed reactions to complex living systems. In the area of biotechnology, for example, manipulation of an enzyme by immobilisation may be a useful step in the practical use of enzymes to carry out valuable biotransformations [52] and has been the subject of considerable efforts. The existence of organisms that are able to operate in extreme physiological environments (as noted previously) implies (based on the paradigm of micro to macro coherence) that enzyme-catalysed reactions are also able to operate in these extreme conditions. Extremophilic organisms can be classified in different ways (depending upon their environmental growth conditions) as halophiles, thermophiles, psychrophiles, acidophiles, alkaliphiles, and xerophiles based on their ability to inhabit and thrive under specific conditions, ranging from high salinity and high or low temperature to acidic, alkaline or dry conditions. One way that extremophile organisms counter the effects of these environments on a molecular level is through the production of heat shock proteins (HSPs) which are able to stabilise some protein structures [53]. The production of HSPs is also a normal response of non-extremophiles to environmental stress however increased expression levels are fond in some extremophiles in their stress response [54]. Adaptations of extremophiles to non-optimal conditions have been studied on a genomic level [55] and this provides evidence for specific adaptive responses at an individual protein level rather than a single set of adaptations (such as the production of anti-freeze proteins by a range of species [56]) that are used to confer some overall stability to a range of extreme environments [57] and a range of strategies have been found to operate [58] in different organisms. This has led to the examination of extremophile organisms as a source of novel molecules and as model organisms for astrobiology [59]. The use of enzymes from extremophiles in industrial processes has led to the 'mining' of extreme environments for the enzymes from viable organisms [60] and this has also resulted in a better understanding of the biological processes that facilitate survival under extreme conditions [61]. It is worth noting that, in terms of human habitation, the majority of the planet represents an 'extreme' environment and what is regarded as benign and normal occupies the minority of the available space. Being able to better use extreme environments is likely to become increasingly important. The biotechnological advantages of the components of extremophile species for the *in situ* or *ex vivo* synthesis of novel molecules can be particularly useful [62] however there are also instances of problems being caused on a complete organism level by the survival of pathogenic species in food after heat processing [63]. While much of the interest in extremophiles has been a consequence of the interest in (and discovery of) adapted prokayotic organisms (particularly bacteria) and their acquired properties, the possibility of carrying out genome remodelling to increase the viability of some species to specific extreme conditions has also been noted as a strategy to deal with environmental pollutants [64]. Such a strategy may also be a potential route to developing eukaryotic species able to survive and grow in a variety of environments [65] thereby increasing the amount of available (cultivatable) land mass. In addition to modification at an organism level, specific directed mutation has been used to improve the thermal properties of specific, commercially important enzymes including lipases [66] and others [67]. A further iteration of enzyme manipulation which has particular potential biotechnological application is the production of fused enzymes with multiple functionality [68]. Another application which is directly related to the functional properties of extremophiles is the use of HSP fusion proteins to enhance immunogenicity [69]. Such an approach might also increase the thermal stability of expressed fusion proteins and improve functionality through enhanced protein folding.

1.3 Nature and Laboratory Derived Chemistry

Chemistry in the laboratory is, as Diderot recognised, an evolving and changing landscape. This act of creation and transformation is part of what defines chemistry. It should not be described as being in a state of homeostasis or even equilibrium. Rather, each iteration opens some and closes some possibilities for what is possible next. The processes for the synthesis of molecules in nature and in the laboratory are conceptually different and are dependent both on the types of process that are available and on the purpose to which the molecule will be put. This is particularly evident in the development of schemes to synthesise molecules. An example is given in Fig. 1.2 in the synthesis of the small molecule, 3-nitropropanoic acid.

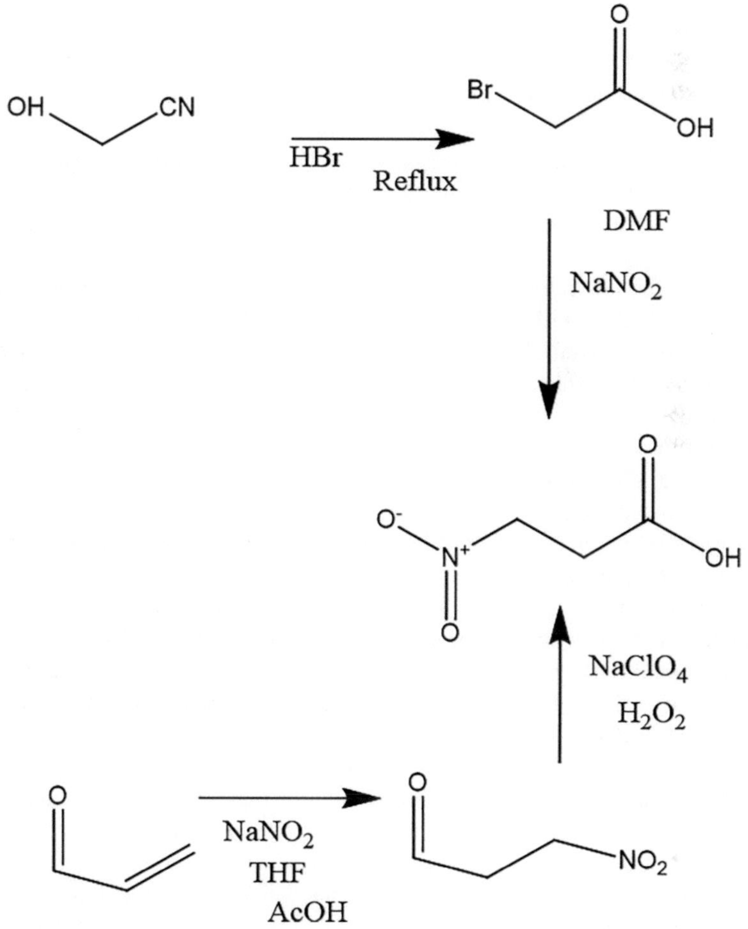

Fig. 1.2 Putative schemes for the synthesis of 3-nitropropanoic acid

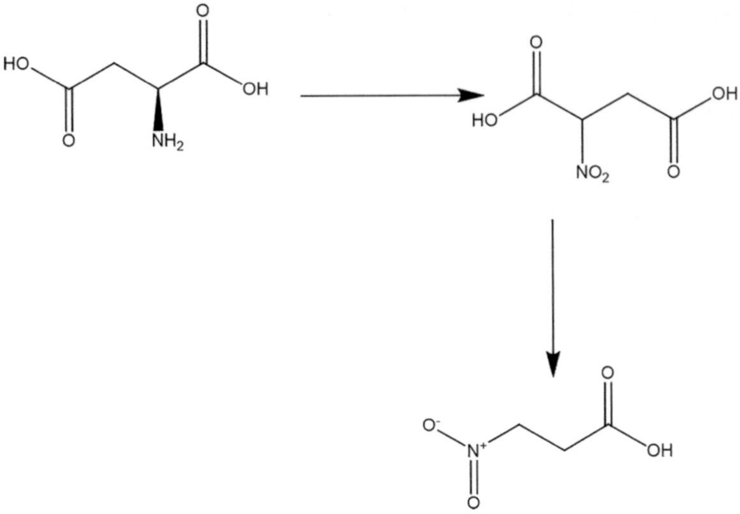

Fig. 1.3 The fungal biosynthesis of 3-nitropropanoic acid

This small (3-carbon) molecule can be assembled in a number of different ways using different components, starting materials and reagents. In contrast the biosynthesis of the same molecule by microorganisms follows one distinct pathway (Fig. 1.3).

The aim of the laboratory synthesis is generally to prepare the target molecule in the highest yield and with the greatest efficiency. The biosynthetic route is influenced, to a greater extent, by the availability of certain biologically available starting materials and that of enzymes that are able to carry out specific biotransformations.

The difference between the objectives of chemistry in nature and in the laboratory helps to explain how different approaches have developed. In the laboratory, there is no clear distinction between the processes for the production of different substances. The function of the production of the molecule is the production of the molecule itself. In nature, on the other hand, the production of different molecules can be a requirement for survival (primary metabolism and maintenance of structural integrity) or for some other purpose (secondary metabolism). The development of chemical biodiversity in nature can be driven directly by favourable changes that eventually affect both phenotype (structural characteristics) and primary metabolism. More effective energy generation (a better yield of ATP in a given time) may be considered to be a favourable outcome of a random mutation in a gene for an enzyme involved in primary metabolism. Primary metabolism can be considered, therefore, to be mainly involved with the processes that are essential for an organism to remain viable. Increased efficiency of energy production or of the molecules required to maintain the structure can result in stimulation of growth, replication and optimal survival. In the case of secondary metabolites, it is perhaps more difficult to define a single advantage that is the consequence of the production of a given molecule since it will rely not only on increased 'fitness' of the parent organism but also on its environment and factors such as the competition for space and resources. Nonetheless, studies have been carried out on secondary natural products to try to define benefits that accrue from the synthesis of molecules that have no direct consequence in terms of improved physiology and, indeed, represent an investment in terms of raw materials diverted from primary metabolism and energy being used. The reductionist chemistry approach has prioritised the mechanism of secondary product formation (enzyme specificity/promiscuity, gene clusters etc.) followed by a better understanding of functional benefits [70]. The question of whether the diversity of natural product biosynthesis is target or diversity driven has been posed [71]. This distinction is of greater consequence for the producing organism than, perhaps, for the organism (often humans) that find a use for the secondary metabolite. In many cases, the major use of compounds is unintended and examples of this will be shown in the discussion of the chemistry that evolved in Norfolk in subsequent chapters. It has been suggested that, over time, biosynthetic enzymes evolved to produce a particular compound however the promiscuity of these (and, indeed, of other enzymes) has resulted in the production of clusters of related compounds as a consequence of molecular changes in initial substrates and post-biosynthetic modification. An example of this can be seen in glucosinolate biosynthesis where chain elongation and other modifications of a precursor amino acid can occur prior to the construction of the core glucosinolate structure (thioglucose and sulphonated oxime moieties). Such modifications can also occur later—for example the post-biosynthetic modification of the indole glucosinolate, glucobrassicin, leads to the production of ring-substituted compounds (Fig. 1.4).

One of the most interesting developments in the production of new structures from nature has been the development of techniques to modify the products that able to be synthesised by natural systems. One approach is to change one or more of the genes in a pathway—traditional genetic modification [72]. This is a valuable tool but can be cumbersome to carry out, can give rise to unexpected products and fine tuning to give a specific product in high enough yield to make the process viable is not invariably possible. A second possibility is to make use of the modular nature of the enzymes of some biosynthetic pathways. An example of this approach uses the modular nature of the biosynthetic polyketide synthases. By manipulating the order of the modules, novel products can be created with a range of properties [73]. While such products make use of biological systems, it is important to realise that, conceptually and philosophically, they belong to the product-oriented, chemistry laboratory lineage described above. They are simply using biological systems to reach that goal rather than for the purpose for which the biosynthetic pathways were originally designed. Such approaches have been described as being 'evolution-based' [74] however it is not clear that the driving principles actually owe as much to evolution as to practical ingenuity. More generally, the conversion of biological systems into, in essence, collections of reagents with a specific purpose in the synthesis of particular molecules can be considered to be the basis of synthetic biology [75] which crucially brings together engineering, biology, genetics and chemistry to design and redesign biological systems to create known or novel molecules or uses

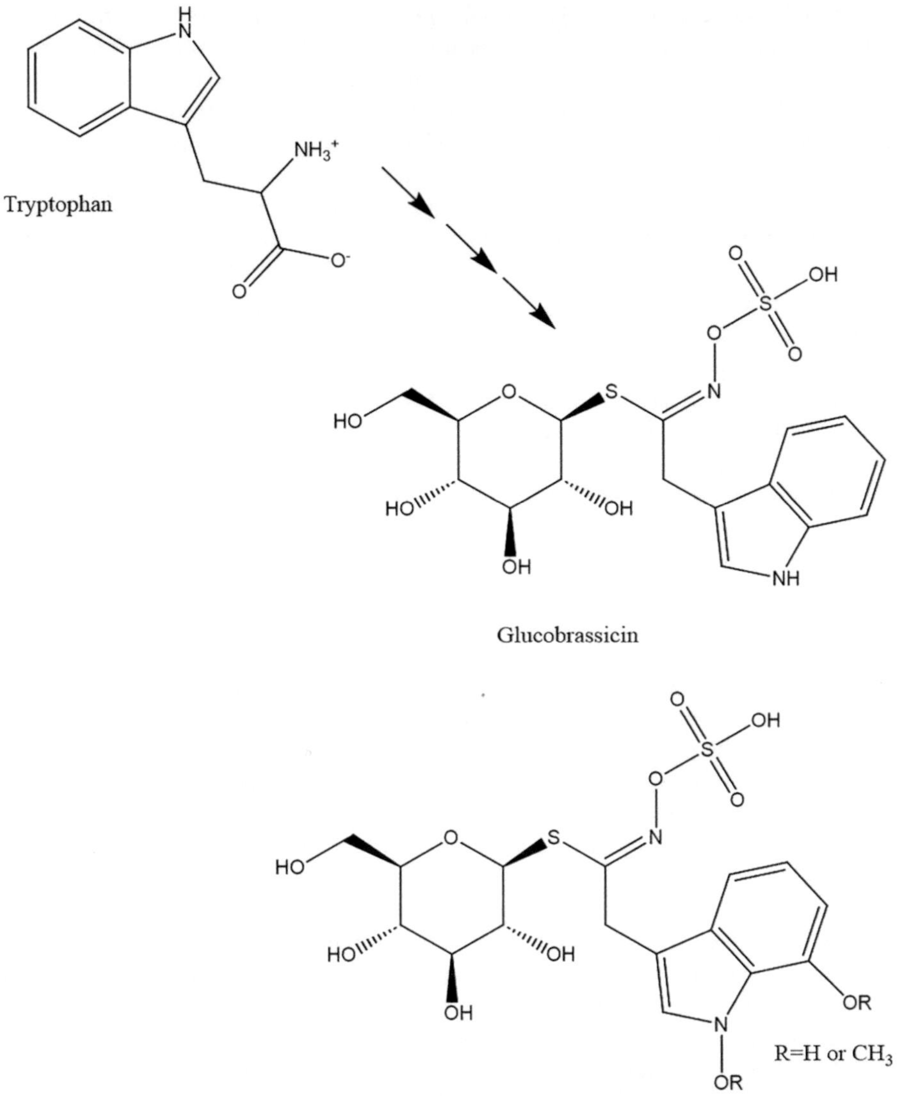

Fig. 1.4 Biosynthesis of ring substituted indolyl-3-methyl glucosinolates (glucobrassicins)

for the system. One of the most significant approaches to the adaptation of biosynthetic pathways has been the development of directed, *in vitro* evolution [76]. The directed evolution paradigm in essence adopts the principles of evolution but on a greatly accelerated timeframe. Importantly, directed evolution requires selection of desirable outcomes and rapid progression towards these [77] however, in addition to the modification of existing enzyme reactions, the directed evolution approach has enabled 'new to nature' enzyme-catalysed biotransformations to be achieved [78]. Strictly biomimetic evolution is a less explored option [79], not least because chemists are primarily concerned with the functional properties of molecules that can be measured and are of value to society or to chemists themselves rather than those which confer an advantage for the producing species and are only serendipitously of value to an external user. The theme of the value of chemistry to different stakeholders is one which will recur in future chapters. This is intertwined with the recognition of potential value at the individual, organisational or societal level. The contrast between the external value of the products of chemistry and the evolutionary drivers for the optimisation of chemistry, as shall be demonstrated, underpins advances in agriculture, industry, art and science and serves to quantify the perceived success of scientific and technological progress.

1.4 The Development of the Analysis/Synthesis/Biosynthesis Paradigm

Chemistry evolved as an experimental and empirical science from the techniques and knowledge built up by alchemy. At the core of chemistry were the twin objectives of determining the nature of substances and the development of methods to create these and other entities. This fuelled the ambition of chemists to try to break substances down to give known components (the original basis of analytical chemistry) and to transform materials into other substances. In the case of the latter, the driving force (which also underpins much of modern chemistry), was twofold—the creation of enhanced value products and the development of benefits and usefulness. These two objectives lay at the core of attempts to convert (transmute) base metal into gold and to cure disease and find the secret (elixir) of eternal life. In principle metal working and knowledge of how to make 'gold-like' alloys were more achievable and more measurable aims than curing all disease so the focus of much of medieval alchemy was towards base metal transformation. This led to the development of apparatus to heat, distil and concentrate materials—some of which later formed the basis of chemical laboratories. The link between the transmutation of metals and the elixir of life was thought to be some essence which enabled the manipulation of life and matter. Indeed, Newton believed in a 'spirit' that existed in living systems [80] and distinguished between nature (vegetable spirit) and simple mechanical processes such as those advocated by Boyle who developed paradigms around the concepts of experimental reasoning thereby promoting mechanical philosophy and empirical methodology. While some of these ideas may seem somewhat fanciful, the concept of a special property for anything derived from a living system (vitalism) was firmly established and at least some of the scientific method developed by Newton and others had an influence on later thinkers such as Darwin [81]. While the demise of vitalism and the removal of the division between living and non-living systems from the point of view of their chemical content can be traced (at least retrospectively) to the synthesis of urea by Wohler from inorganic starting materials (potassium cyanate and ammonium chloride) in 1828 [82], the 'vital force' hypothesis remained important in medicine throughout the nineteenth century and was arguably not abandoned until Claude Bernard wrote his treatise 'An Introduction to the Study of Experimental Medicine' in 1927 [83].

It is against the background of the derivation of chemical principles and ideas from alchemy that the development of the main paradigms of chemistry should be considered. While inorganic chemistry was initially studied the most, this was at least in part due to the quest for transmutation and the practical application of metallurgy [84] which was carried out long before the advent of chemistry as a discipline. Heating and dissolution were two of the key elements that underpinned alchemy and this, together with purification, were important techniques and concepts when experimenting with metals, minerals and alloys. These are shown in reviews of the history of chemistry [85]. Inorganic chemistry also had a practical application in the development of colours from ancient cave art using earth colours and ochres [86] to the synthesis of Prussian Blue (Ferric hexacyanoferrate, $Fe_4[Fe(CN)_6]_3 \cdot xH_2O$) [87] and Ultramarine (sodium aluminosilicate) [88]. While alchemy was to an extent a way of thinking about the composition of the world and how it could be manipulated, the practical application of alchemical methods and experimentation and their recording by figures as eminent as Newton in the 1690s [89] through grew and provided a way for alchemists to create value [90]. The idea of transmutation (rather like that of vitalism) persisted and it continued to be practiced even after scientific opinion had judged it to be unachievable [91]. It was only when the corpuscular theories of Boyle [92] and the affinity concepts promoted by Geoffroy [93] became widely accepted and supported by experimental data that transmutation was finally laid to rest. However it remained a practice of practical, experimental chemistry even when it was intellectually unsustainable. Interestingly, species transmutation remained an established way of thought throughout the nineteenth century [94]. The idea of chemical transmutation proved difficult to completely eliminate from popular opinion [95] and a small number of researchers continue to work on biological transmutation from the standpoint of cold fusion [96] however such studies are contrary to the Law of Mass Conservation and therefore have little credibility among the wider scientific community.

The progression of organic chemistry was, in some ways, more closely allied to the Analysis/Synthesis/Biosynthesis paradigm developed in the nineteenth century. This was in part due to the composition and constituents of organic compounds. Using this approach the nature of substances is determined by breaking them down into smaller, identifiable units which can be characterised and then proving the structure by reassembling it from the constituent units. In the case of inorganic chemistry, the smallest units are the individual elements that make up the compound and much can be learned or inferred from empirical data on the proportions of the different constituents that are present [97]. In the case of organic chemistry, identification of the proportions of the constituent atoms led to the realisation that a large number of molecules with different structures (and very different properties) were formed predominantly from carbon and hydrogen with smaller proportions of oxygen and a few other atoms (N, S, P and a small number of others). This means that empirical formulae in which the ratios of the different atoms that are present are deduced are of comparatively little value in determining the structure of an isolated compound. As noted previously, several molecules can have the same empirical formula but have very different structures and properties. It also means (as Perkin discovered when he tried to synthesise quinine [98]) that, using simple combinations of equivalent empirical formulas in an additive fashion, it would not be possible to synthesise complex organic molecules. What was required was analysis at functional group rather than atomic level. This level of analysis of substances was, at first, carried out by the degradative actions of heat or other chemicals (reagents). Indeed, in addition to helping to identify pertinent structural characteristics, identification of functional groups helped in the purification of certain types of molecules (for example alkaloids with an ionisable N atom and organic acids). The key distinction between synthesis and biosynthesis (as noted previously) is the usefulness of the compounds and how this influences the methodologies that have been developed. The biosynthesis of compounds is generally designed to provide an ecological advantage for the producing species. This can, inadvertently, also give rise to molecules with physiological effects that are of benefit to other species [99] but this is not the primary function of the biosynthetic method. Laboratory synthesis of molecules is often subject to Quantitative Structure Activity Relationship (QSAR) testing to optimise a given activity that is of value to the chemist. This concept has come full circle (as demonstrated previously) with the development of synthetic biology in which biological systems are manipulated (and sometimes modified) to facilitate the production of molecules that are not primarily of value to the biological host but of benefit to the manipulating chemist. The concept of 'usefulness' will be considered further in Sect. 1.5 and in other chapters. Another traditional driver for synthetic chemistry in the laboratory is the 'synthesis as proof of structure following analysis' which was proposed by Descartes [5] based upon claims of a metaphysical nature [100]. In the absence of a clear understanding of the nature of (particularly organic) molecules, structure determination by synthesis was an abstract concept rather than a way of proving a well formulated hypothesis. A significant advance in thought and conceptual understanding arose from the work of Lavoisier who, by combustion, was able to determine the empirical formulae of organic compounds (see above) and, by decomposition and synthesis, the components of water. As noted previously, the development of ideas of chemical types enabled selective isolation of some classes of compounds including acids (for example lactic, oxalic, tartaric and others) and alkaloids (alkali-like substances (a term first used by Meissner in 1819)) [101, 102]. Despite the work by Lavoisier, German natural product chemists and others, the synthesis (and even the structural identification) of molecules lagged behind their isolation and purification from nature. The great expansion of chemistry driven by the concept of the creation of something of use has provided the driver for many millions of compounds to be synthesised or isolated for application in areas such as drug discovery [103] and other functional properties [104]. Advances in genome mining and the discovery of biosynthetic gene clusters (even if the final products have not been found and characterised) has enabled synthetic biology to facilitate the discovery of many more natural products than were previously envisaged through purely isolative chemical methods [105] and this has revitalised the study of the discovery of useful compounds from natural sources. It was not until the second half of the nineteenth century that the principles of the structures of organic compounds began to be widely accepted. Indeed it was in 1865 that Kekule published the correct structure for benzene [106]. It was not until an older generation of chemists (including, significantly, Liebig and Berzelius) had passed that the concepts of chemical structures were widely accepted. This was some 60 years after Dalton developed the concept of synthesis and analysis in the context of atoms in 1808. These ideas and their application in the laboratory and to living systems were necessary prerequisites for the development of chemistry as a transformative and useful discipline.

1.5 The Usefulness of Chemistry

The concept of usefulness in chemistry derives from the idea that chemists are able to create things (substances) that were not available previously. While the development of a sort of mechanical chemistry can be traced to Lavoisier and others, this was found to be insufficient. The core of chemistry is both analytical (mechanical) and synthetic (transformative). Chemistry does not exist as a series of explanatory ideas but rather as a discipline in which bonds are created and broken and substances with properties are created and these are based on the concepts of molecular interaction—which are not directly observable but are consequential. This means that chemistry creates an ever-changing landscape and is as much about substances as it is about theories. This gives chemistry a position as a scientific discipline that sits at the intersection of science, technology and industry. When considering the impact of chemistry, it is important to consider the contributions of all three areas. These are not solely sequential. As Perkin discovered, a useful outcome can predate an understanding of the details of that outcome. This means that it is not necessary to complete the accumulation of knowledge before making practical use of chemistry. The other factor is context. The products of chemistry (ideas and substances) and their use depends on the context and how chemistry is being used. For example, if an assay for anti-malarial activity had been available to Perkin, the direction of his research may have been influenced by this rather than by trying to find a use for the compound he had synthesised in his quest for quinine. Perkin might have become a leading medicinal chemist rather than a maker of dyestuffs. A second aspect of chemistry that underlines its usefulness is its ubiquity. The idea of trying to define systems based on their component chemical species is not new. Indeed the concept of living systems and being composed of the changes they undergo has been proposed [107] and the concept of self-assembling molecular machines is one that has captured the imagination of chemists and sparked considerable research [108]. By considering the growth and practical applications of chemistry, it is possible to arrive at some idea of some areas of potential value and utility and this will be considered in future chapters. Finally, as noted previously, chemistry and the development of chemistry ideas is influenced by the environment (individual, organisational and societal) and this is one reason why the thread of chemistry runs through Norfolk and why it provides a paradigm for the influence of chemistry over time.

References

1. Singh K, Gupta JK, Chanchal DK, et al. Natural products as drug leads: exploring their potential in drug discovery and development. Naunyn Schmiedeberg's Arch Pharmacol. 2025;398:4673–87. https://doi.org/10.1007/s00210-024-03622-6.
2. Zinkovich V, Sotskov V, Shapeev A, et al. Exhaustive search for novel multicomponent alloys with brute force and machine learning. npj Comput Mater. 2024;10:269. https://doi.org/10.1038/s41524-024-01452-x.
3. Swart M, Bickelhaupt FM. Proton affinities of anionic bases: trends across the periodic table, structural effects, and DFT validation. J Chem Theory Comput. 2006;2(2):281–7. https://doi.org/10.1021/ct0502460.
4. Piccolino M. Luigi Galvani and animal electricity: two centuries after the foundation of electrophysiology. Trends Neurosci. 1997;20(10):443–8. https://doi.org/10.1016/s0166-2236(97)01101-6.
5. Duncan H. Descartes and the method of analysis and synthesis. In: Brown JR, Mittelstrass J, editors. An intimate relation. Boston studies in the philosophy of science, vol. 116. Dordrecht: Springer; 1989. https://doi.org/10.1007/978-94-009-2327-0_4.
6. Ramberg PJ. The death of vitalism and the birth of organic chemistry: Wohler's urea synthesis and the disciplinary identity of organic chemistry. Ambix. 2000;47(3):170–95. https://doi.org/10.1179/amb.2000.47.3.170.
7. Madan K. Natural human chimeras: a review. Eur J Med Genet. 2020;63(9):103971. https://doi.org/10.1016/j.ejmg.2020.103971.
8. Papakonstantinou E, Mitsis T, Dragoumani K, Bacopoulou F, Megalooikonomou V, Chrousos GP, Vlachakis D. The medical cyborg concept. EMBnet J. 2022;27:e1005. https://doi.org/10.14806/ej.27.0.1005.
9. Chang H. Reductionism and the relation between chemistry and physics. In: Arabatzis T, Renn J, Simões A, editors. Relocating the history of science. Boston studies in the philosophy and history of science, vol. 312. Cham: Springer; 2015. https://doi.org/10.1007/978-3-319-14553-2_13.
10. Sparling A. Paracelsus, a transmutational alchemist. Ambix. 2020;67(1):62–87. https://doi.org/10.1080/00026980.2020.1720358.
11. Sumrall L. Alchemical medicine and distillation. In: Sgarbi M, editor. Encyclopedia of renaissance philosophy. Cham: Springer; 2018. https://doi.org/10.1007/978-3-319-02848-4_1086-1.
12. Moss LG. Creating new β cells: cellular transmutation by genomic alchemy. J Clin Invest. 2013;123(3):1007–10. https://doi.org/10.1172/JCI68348.
13. Straatsma TP, McCammon JA. Computational alchemy. Annu Rev Phys Chem. 1992;43:407–35. https://doi.org/10.1146/annurev.pc.43.100192.002203.
14. Saravanan K, Kitchin JR, von Lilienfeld OA, Keith JA. Alchemical predictions for computational catalysis: potential and limitations. J Phys Chem Lett. 2017;8(20):5002–7. https://doi.org/10.1021/acs.jpclett.7b01974.
15. Kundu D, Chauhan RPS. The medical use of alchemy: a historical exploration of the intersection of medicine and transmutation. J Integr Med Ther. 2024;7(1):1.

16. Strebhardt K, Ullrich A. Paul Ehrlich's magic bullet concept: 100 years of progress. Nat Rev Cancer. 2008;8(6):473–80. https://doi.org/10.1038/nrc2394.
17. Verna R. From alchemy to personalised medicine: the journey of laboratory medicine. J Clin Pathol. 2023;76(5):301–7. https://doi.org/10.1136/jcp-2022-208492.
18. Newman WR. What have we learned from the recent historiography of alchemy? Isis. 2011;102(2):313–21. https://doi.org/10.1086/660140.
19. Gregory JC. An aspect of the history of atomism. Sci Progr. 1927;22(86):293–304.
20. Banchetti-Robino MP, Villani G. John Dalton and chemical atomism. In: From the atom to living systems: a chemical and philosophical journey into modern and contemporary science. New York: Oxford Academic; 2023.
21. Usselman MC, Brown TA. Atomic theory and multiple combining proportions: the search for whole number ratios. Ann Sci. 2015 Apr;72(2):153–69. https://doi.org/10.1080/00033790.2015.1007522.
22. Bunge M. Is chemistry a branch of physics? Z Allg Wissensch. 1982;13:209–23. https://doi.org/10.1007/BF01801556.
23. Lombardi O. Linking chemistry with physics: arguments and counterarguments. Found Chem. 2014;16:181–92. https://doi.org/10.1007/s10698-013-9197-x.
24. Pyykkö P. The physics behind chemistry and the periodic table. Chem Rev. 2012;112(1):371–84. https://doi.org/10.1021/cr200042e.
25. Wolfe C. Diderot's vital materialism. Ambix. 2024;71(3):342–59. https://doi.org/10.1080/00026980.2024.2379961.
26. Baertschi B. Diderot, Cabanis and Lamarck on psycho-physical causality. Hist Philos Life Sci. 2005;27(3–4):451–63.
27. Kaitaro T. Can matter mark the hours? Eighteenth-century vitalist materialism and functional properties. Sci Context. 2008;21(4):581–92. https://doi.org/10.1017/S0269889708001968.
28. Banchetti-Robino MP, Villani G. Lavoisier and the quantification of chemistry. In: From the atom to living systems: a chemical and philosophical journey into modern and contemporary science. New York: Oxford Academic; 2023. https://doi.org/10.1093/oso/9780197598900.003.0005.
29. Paul D, Kundu A, Saha S, Goswami RK. Total synthesis: the structural confirmation of natural products. Chem Commun. 2021;57(27):3307–22. https://doi.org/10.1039/d1cc00241d.
30. Schmechel C. Leaven of dough, ferment of gold: the Breadmaking analogy in medieval metallic transmutation. Ambix. 2024;71(3):243–70. https://doi.org/10.1080/00026980.2024.2396702.
31. Rocke AJ. Kekulé's benzene theory and the appraisal of scientific theories. In: Donovan A, Laudan L, Laudan R, editors. Scrutinizing science. Synthese library, vol. 193. Dordrecht: Springer; 1988. https://doi.org/10.1007/978-94-009-2855-8_7).
32. Chalmers A. Robert Boyle's corpuscular chemistry: atomism before its time. In: Scerri E, Fisher G, editors. Essays in the philosophy of chemistry. New York: Oxford Academic; 2020. https://doi.org/10.1093/oso/9780190494599.003.0006).
33. Shin W, Yang ZJ. Computational strategies for entropy modeling in chemical processes. Chem Asian J. 2023;18(9):e202300117. https://doi.org/10.1002/asia.202300117.
34. Boomsma W, Ferkinghoff-Borg J, Lindorff-Larsen K. Combining experiments and simulations using the maximum entropy principle. PLoS Comput Biol. 2014;10(2):e1003406. https://doi.org/10.1371/journal.pcbi.1003406.
35. Kassem S, Ahmed M, El-Sheikh S, Barakat KH. Entropy in bimolecular simulations: a comprehensive review of atomic fluctuations-based methods. J Mol Graph Model. 2015;62:105–17. https://doi.org/10.1016/j.jmgm.2015.09.010.
36. Wade RC, Salo-Ahen OMH. Molecular modeling in drug design. Molecules. 2019;24(2):321. https://doi.org/10.3390/molecules24020321.
37. Tran HL, Lexa KW, Julien O, Young TS, Walsh CT, Jacobson MP, Wells JA. Structure-activity relationship and molecular mechanics reveal the importance of ring entropy in the

biosynthesis and activity of a natural product. J Am Chem Soc. 2017;139(7):2541–4. https://doi.org/10.1021/jacs.6b10792.
38. Pachter JA, Yang YJ, Dill KA. Entropy, irreversibility and inference at the foundations of statistical physics. Nat Rev Phys. 2024;6:382–93. https://doi.org/10.1038/s42254-024-00720-5.
39. Martyushev LM. The significance of the entropic measure of time in natural sciences. Entropy. 2025;27(4):425. https://doi.org/10.3390/e27040425.
40. Kostic MM. The second law and entropy misconceptions demystified. Entropy. 2020;22(6):648. https://doi.org/10.3390/e22060648.
41. Ji S, Zhang Y, Chen N, Liu X, Li Y, Shao X, Yang Z, Yao Z, Hu B. Shared increased entropy of brain signals across patients with different mental illnesses: a coordinate-based activation likelihood estimation meta-analysis. Brain Imaging Behav. 2022;16(1):336–43. https://doi.org/10.1007/s11682-021-00507-7.
42. Cassidy D, Holton G, Rutherford J. Heat-a matter of motion. In: Understanding physics. Undergraduate texts in contemporary physics. New York: Springer; 2002. https://doi.org/10.1007/978-1-4757-7698-0_7.
43. Sauerheber RD. Thermodynamics and entropy in natural and artificial systems. Am Res J Chem. 2018;2(1):1–26.
44. Kim CS. Free energy and inference in living systems. Interf Focus. 2023;13(3):20220041. https://doi.org/10.1098/rsfs.2022.0041.
45. Lewis T, Stone WL. Biochemistry, proteins enzymes. In: StatPearls. Treasure Island: StatPearls Publishing; 2025.
46. Franco R, Serrano-Marín J. Robustness of the Krebs cycle under physiological conditions and in cancer: new clues for evaluating metabolism-modifying drug therapies. Biomedicine. 2022;10(5):1199. https://doi.org/10.3390/biomedicines10051199.
47. Arrhenius S. Über die Reaktionsgeschwindigkeit bei der inversion von Rohrzucker durch Säuren. Z Phys Chem. 1889;4:226–48.
48. van der Kamp MW, Prentice EJ, Kraakman KL, Connolly M, Mulholland AJ, Arcus VL. Dynamical origins of heat capacity changes in enzyme-catalysed reactions. Nat Commun. 2018;9(1):1177. https://doi.org/10.1038/s41467-018-03597-y.
49. Karan R, Capes MD, Dassarma S. Function and biotechnology of extremophilic enzymes in low water activity. Aquat Biosyst. 2012;8(1):4. https://doi.org/10.1186/2046-9063-8-4.
50. Tuena de Gómez-Puyou M, Gómez-Puyou A. Enzymes in low water systems. Crit Rev Biochem Mol Biol. 1998;33(1):53–89. https://doi.org/10.1080/10409239891204170.
51. González Flecha FL. Kinetic stability of membrane proteins. Biophys Rev. 2017;9(5):563–72. https://doi.org/10.1007/s12551-017-0324-0.
52. Datta S, Christena LR, Rajaram YR. Enzyme immobilization: an overview on techniques and support materials. 3 Biotech. 2013;3(1):1–9. https://doi.org/10.1007/s13205-012-0071-7.
53. Sato Y, Okano K, Honda K. Effects of small heat shock proteins from thermotolerant bacteria on the stress resistance of Escherichia coli to temperature, pH, and hyperosmolarity. Extremophiles. 2024;28:12. https://doi.org/10.1007/s00792-023-01326-y.
54. Sato H, Nakasone K, Yoshida T, Kato C, Maruyama T. Increases of heat shock proteins and their mRNAs at high hydrostatic pressure in a deep-sea piezophilic bacterium, Shewanella violacea. Extremophiles. 2015;19(4):751–62. https://doi.org/10.1007/s00792-015-0751-4.
55. Kumar A, Alam A, Tripathi D, Rani M, Khatoon H, Pandey S, Ehtesham NZ, Hasnain SE. Protein adaptations in extremophiles: an insight into extremophilic connection of mycobacterial proteome. Semin Cell Dev Biol. 2018;84:147–57. https://doi.org/10.1016/j.semcdb.2018.01.003.
56. Cheung RCF, Ng TB, Wong JH. Antifreeze proteins from diverse organisms and their applications: an overview. Curr Protein Pept Sci. 2017;18(3):262–83. https://doi.org/10.2174/1389203717666161013095027.
57. Reed CJ, Lewis H, Trejo E, Winston V, Evilia C. Protein adaptations in archaeal extremophiles. Archaea. 2013;2013:373275. https://doi.org/10.1155/2013/373275.

58. Angelakis GN, Psarologaki C, Pirintsos S, Kotzabasis K. Extremophiles and extremophilic behaviour-new insights and perspectives. Life. 2024;14(11):1425. https://doi.org/10.3390/life14111425.
59. Rawat M, Chauhan M, Pandey A. Extremophiles and their expanding biotechnological applications. Arch Microbiol. 2024;206(6):247. https://doi.org/10.1007/s00203-024-03981-x.
60. Demirjian DC, Morís-Varas F, Cassidy CS. Enzymes from extremophiles. Curr Opin Chem Biol. 2001;5(2):144–51. https://doi.org/10.1016/s1367-5931(00)00183-6.
61. Aulitto M, Gallo G. Extremophiles breakthrough: hot topics and current issues in their isolation, identification, and biotechnological applications. Life. 2025;15(3):434. https://doi.org/10.3390/life15030434.
62. Coker JA. Extremophiles and biotechnology: current uses and prospects. F1000Res. 2016;5:396. https://doi.org/10.12688/f1000research.7432.1.
63. Guragain M, Smith GE, Bosilevac JM. Methods for screening and isolating extremely heat-resistant Escherichia coli from meat sources. Life. 2024;14:1123. https://doi.org/10.3390/life14091123.
64. Gallo G, Puopolo R, Carbonaro M, Maresca E, Fiorentino G. Extremophiles, a nifty tool to face environmental pollution: from exploitation of metabolism to genome engineering. Int J Environ Res Public Health. 2021;18(10):5228. https://doi.org/10.3390/ijerph18105228.
65. Rappaport HB, Oliverio AM. Lessons from extremophiles: functional adaptations and genomic innovations across the eukaryotic tree of life. Genome Biol Evol. 2024;16(8):evae160. https://doi.org/10.1093/gbe/evae160.
66. Sharma M, Kumar R, Singh R, et al. Thirty-degree shift in optimum temperature of a thermophilic lipase by a single-point mutation: effect of serine to threonine mutation on structural flexibility. Mol Cell Biochem. 2017;430:21–30.
67. Silva C, Martins M, Jing S, Fu J, Cavaco-Paulo A. Practical insights on enzyme stabilization. Crit Rev Biotechnol. 2018;38(3):335–50. https://doi.org/10.1080/07388551.2017.1355294.
68. Elleuche S. Bringing functions together with fusion enzymes–from nature's inventions to biotechnological applications. Appl Microbiol Biotechnol. 2015;99(4):1545–56. https://doi.org/10.1007/s00253-014-6315-1.
69. Li D, Li H, Zhang P, Wu X, Wei H, Wang L, Wan M, Deng P, Zhang Y, Wang J, Liu Y, Yu Y, Wang L. Heat shock fusion protein induces both specific and nonspecific anti-tumor immunity. Eur J Immunol. 2006;36(5):1324–36. https://doi.org/10.1002/eji.200535490.
70. Chevrette MG, Gutiérrez-García K, Selem-Mojica N, Aguilar-Martínez C, Yañez-Olvera A, Ramos-Aboites HE, Hoskisson PA, Barona-Gómez F. Evolutionary dynamics of natural product biosynthesis in bacteria. Nat Prod Rep. 2020;37(4):566–99. https://doi.org/10.1039/c9np00048h.
71. Noda-Garcia L, Tawfik DS. Enzyme evolution in natural products biosynthesis: target- or diversity-oriented? Curr Opin Chem Biol. 2020;59:147–54. https://doi.org/10.1016/j.cbpa.2020.05.011.
72. Chen J, Wu Q, Hawas UW, et al. Genetic regulation and manipulation for natural product discovery. Appl Microbiol Biotechnol. 2016;100:2953–65. https://doi.org/10.1007/s00253-016-7357-3.
73. Huang Z, Xie S, Liu RZ, et al. Plug-and-play engineering of modular polyketide synthases. Nat Chem Biol. 2025. https://doi.org/10.1038/s41589-025-01878-4.
74. Grininger M. Design principles imprinted by evolution. Nat Chem Biol. 2025. https://doi.org/10.1038/s41589-025-01880-w.
75. Ornelas MY, Cournoyer JE, Bram S, Mehta AP. Evolution and synthetic biology. Curr Opin Microbiol. 2023;76:102394. https://doi.org/10.1016/j.mib.2023.102394.
76. Schmidt-Dannert C, Umeno D, Arnold FH. Molecular breeding of carotenoid biosynthetic pathways. Nat Biotechnol. 2000;18(7):750–3. https://doi.org/10.1038/77319.
77. Fasan R, Jennifer Kan SB, Zhao H. A continuing career in biocatalysis: Frances H. Arnold. ACS Catal. 2019;9(11):9775–88. https://doi.org/10.1021/acscatal.9b02737.

78. Tyagi V, Fasan R. Myoglobin-catalyzed olefination of aldehydes. Angew Chem Int Ed. 2016;55:2512–6.
79. Barona-Gómez F, Chevrette MG, Hoskisson PA. On the evolution of natural product biosynthesis. Adv Microb Physiol. 2023;83:309–49. https://doi.org/10.1016/bs.ampbs.2023.05.001.
80. Newman WR. Toward a general theory of vegetability and mechanism. In: Newton the alchemist. Princeton: Princeton Scholarship; 2018.
81. Bellon R. There is grandeur in this view of Newton: Charles Darwin, Isaac Newton and Victorian conceptions of scientific virtue. Endeavour. 2014;38(3-4):222–34. https://doi.org/10.1016/j.endeavour.2014.10.008.
82. Kinne-Saffran E, Kinne RK. Vitalism and synthesis of urea. From Friedrich Wöhler to Hansa. Krebs Am J Nephrol. 1999;19(2):290–4. https://doi.org/10.1159/000013463.
83. Normandin S. Claude Bernard and an introduction to the study of experimental medicine: "physical vitalism," dialectic, and epistemology. J Hist Med Allied Sci. 2007;62(4):495–528. https://doi.org/10.1093/jhmas/jrm015.
84. Smith CS. The interaction of science and practice in the history of metallurgy. Technol Cult. 1961;2(4):357–67.
85. Brobjer TH. History of chemistry. In: Atlas of the history of modern science 1500–2020. Cham: Springer; 2025. https://doi.org/10.1007/978-3-031-76627-5_5.
86. Velliky EC, Porr M, Conard NJ. Ochre and pigment use at Hohle Fels cave: results of the first systematic review of ochre and ochre-related artefacts from the upper palaeolithic in Germany. PLoS One. 2018;13(12):e0209874. https://doi.org/10.1371/journal.pone.0209874.
87. Parma MG, Beraldo H, Mendes IC. Syntheses of Prussian blue pigment following 18th-century methodologies: factors influencing product purity and syntheses yields. ACS Omega. 2025;10(11):11375–85. https://doi.org/10.1021/acsomega.4c11328.
88. Mertens J. The history of artificial ultramarine (1787–1844): science. Ind Secrecy Ambix. 2004;51(3):219–44. https://doi.org/10.1179/amb.2004.51.3.219.
89. Allen MS. Revisiting Isaac Newton's index chemicus: a response to Richard S. Westfall. Ambix. 2019;66(1):72–81. https://doi.org/10.1080/00026980.2019.1573523.
90. Nummedal TE. Words and works in the history of alchemy. Isis. 2011;102(2):330–7. https://doi.org/10.1086/660142.
91. Principe LM. The end of alchemy? The repudiation and persistence of chrysopoeia at the Académie Royale des sciences in the eighteenth century. Osiris. 2014;29:96–116. https://doi.org/10.1086/678099.
92. Chalmers AF. Boyle and the origins of modern chemistry: Newman tried in the fire. Stud Hist Phil Sci. 2010;41(1):1–10. https://doi.org/10.1016/j.shpsa.2009.12.001.
93. Joly B. Etienne-François Geoffroy (1672-1731), a chemist on the Frontiers. Osiris. 2014;29:117–31. https://doi.org/10.1086/678100.
94. Rouyan A. Reforming uncultivated minds: the species transmutation debate and American science of life in the antebellum agricultural press, 1820-1859. Stud Hist Phil Biol Biomed Sci. 2019;76:101170. https://doi.org/10.1016/j.shpsc.2019.01.006.
95. Rutherford L. The transmutation of the atom. Sci Mon. 1934;38(1):15–23.
96. Biberian J-P. Biological transmutations. J Condens Matter Nucl Sci. 2019;28(1):21–7. https://doi.org/10.70923/001c.72482.
97. Yam VW-W. Inorganic chemistry: a prestigious history and a bright future. Angew Chem Int Ed. 2015;54:8304–5. https://doi.org/10.1002/anie.201504512.
98. Ball P. Perkin, the mauve maker. Nature. 2006;440:429. https://doi.org/10.1038/440429a.
99. Chopra B, Dhingra AK. Natural products: a lead for drug discovery and development. Phytother Res. 2021;35(9):4660–702. https://doi.org/10.1002/ptr.7099.
100. Cunning D. Analysis versus synthesis. In: Nolan L, editor. The Cambridge Descartes lexicon. Cambridge University Press; 2015. p. 7–12.
101. Meissner W. Über Pflanzenalkalien: II. Über ein neues Pflanzenalkali (Alkaloid) [About Plant Alkalis: II. About a New Plant Alkali (Alkaloid)]. J Chem Phys. 1819;25:379–81.
102. Hosztafi S. The discovery of alkaloids. Pharmazie. 1997;52(7):546–50.

103. Singh SB, Culberson JC. In natural product chemistry for drug discovery. London: The Royal Society of Chemistry; 2009. p. 28–43.
104. Huang X, Yang J, Ho CT, Ke Q, Kou X. Functional flavor agents: enhancing health benefits and consumer preferences. Crit Rev Food Sci Nutr. 2025;65(32):8188–216. https://doi.org/10.1080/10408398.2025.2494297.
105. Kunakom S, Eustáquio AS. Natural products and synthetic biology: where we are and where we need to go. mSystems. 2019;4(3):e00113–9. https://doi.org/10.1128/mSystems.00113-19.
106. Rocke AJ. It began with a daydream: the 150th anniversary of the Kekulé benzene structure. Angew Chem Int Ed Engl. 2015;54(1):46–50. https://doi.org/10.1002/anie.201408034.
107. Whitesides GM. Reinventing chemistry. Angew Chem Int Ed. 2015;54:3196–209. https://doi.org/10.1002/anie.201410884.
108. Amadi EV, Venkataraman A, Papadopoulos C. Nanoscale self-assembly: concepts, applications and challenges. Nanotechnology. 2022;33(13) https://doi.org/10.1088/1361-6528/ac3f54.

Chapter 2
The Essence of Chemistry. Chemistry and Agriculture

Abstract The application of the principles of crop rotation can be considered to provide benefits in terms of nutrient replenishment (for example, manure from grazing animals and the use of legumes as a rotation crop. It can also help to prevent the build up of disease as happened to the Irish potato crop leading to the nineteenth century famine. A particular form of crop rotation which became known as the 'Norfolk rotation' (although it was developed in Belgium) made use of a fodder crop as part of the rotation cycle. Much of the interest in crop rotation has focused on the macro properties of crops and, in particular, on the correlation between specific crop rotations and crop yields. On a micro level (including effects on the soil microbiome) specific effects can be causally linked to some crop rotations including the production of allelochemicals and the effect of these (and of, for example, quorum sensing molecules) on the rhizosphere. The communication between Norfolk and Europe that led to the adoption of crop rotation was also an important aspect of the exchange of people and technology that led to wealth creation in Norfolk and resulted in Norwich being the second biggest (behind London) city in England.

Keywords Agriculture · Correlation · Causality · Crop rotation · Microbiome · Allelopathy

Theme one begins the exploration of chemistry in terms of its effect on agriculture and raw materials and the conceptualisation of chemistry as an empirical and experimental discipline. The development of chemistry in agriculture began with largely observational studies—growing crops in succession for example leading to crop rotation which, together with the increased use of farming implements such as the plough [1] and the decline in family farming and its replacement (particularly in the South and East of England) by larger, more commercially viable units [2] resulted in significant improvements in agricultural practice. The benefits of these changes included improved yield, less loss of yield due to the accumulation of phyto-specific diseases in soil and, importantly, less wastage as the residues after harvest were ploughed back into the soil to provide additional nutrients. Many of these advances were imported from Europe (particularly from Scandinavia, Belgium and Holland)

B. Hanley, *The Chemistry of a County*,
https://doi.org/10.1007/978-3-032-21491-1_2

and then refined in Norfolk to suit local conditions. As will be shown later, (Theme Three Chap. 6) the importation of technology that was of benefit to England was not limited to agricultural concepts and knowledge but also included the arrival and settling of large groups of individuals and their expertise. Crop breeding and the concept of information about specific characteristics of varieties of plants being stored in discrete packages (later identified as genes) also originated in Europe (Mendel, Austria) [3] and the progression from macro events (crop growth) to the planned manipulation of plant characteristics (plant breeding and crossing) to molecular understanding of genes and the principles of conversion of genetic information into phenotypic characteristics and interference with the process on a molecular level leading to specific macromolecular outcomes (genetic modification and gene editing) shows the presence of a clear lineage with subsequent increasing specificity and control and increased reductionism (from agronomic systems to individual plants to genes). The awareness of the complexity of agricultural systems and the development of more precise techniques for generating and manipulating information (genomics and all the other omics technologies) has, to an extent, reversed the reductionist knowledge flow and increased the understanding that it is not simply individual molecules and interactions that are important but rather it is the nature of the interrelationships between and within systems and how they operate, that is of greatest importance and has the most significant impact.

Living systems do not operate in a state of actual or perceived equilibrium or homeostasis but rather, through a series of transitions. This is expressed in Kant's view of chemistry as being the science that studies the changes of matter into new forms [4]. Chemistry then reflects the ever-changing characteristics of natural systems, and it is the ultimate empirical (applied) scientific endeavour [5]. The progress of chemistry is, then, dependent both on the creation of viable hypotheses and also having the methods to test their validity. The knowledge base thus created is subject to modification as newer methods arise and are applied.

Having considered the main attributes of chemistry as an empirical discipline and considered it as a science of transitions which fits the paradigms of Kant, Theme one will now develop a consideration of the impact of chemistry on raw material production through an analysis of the development of agriculture in Norfolk. By treating agriculture as an activity that can be influenced by chemistry, it is possible to apply the same principles in different circumstances as the sophistication and power of chemistry is supplanted by greater knowledge and the growth of a better understanding of molecular sciences. This begins with a description of the use of crop rotation to affect the macroscopic properties of growing crops and next sequentially develops the key paradigms involved in breeding and strain selection then, ultimately, arrives at the manipulation of biosynthetic pathways to affect plant phenotype through genotypic manipulation. The optimisation of the macroscopic properties of particular target crops through control of the growing environment and the agronomic conditions, the optimisation of variety selection and (eventually) the control of molecular interactions within the crop itself is a common theme which not only resulted in improved agronomic characteristics for the target crop but also provided the means to extend the knowledge base to enable further optimisation to be attained. This is demonstrated particularly in the case of the production of polyketides. The increased understanding of how secondary metabolite production and optimisation is controlled in a soil microorganism has led to the increase in the development of the knowledge base for molecular manipulation of genes, biosynthetic pathways and products beyond the microorganism. While advances in molecular biology are not the sole preserve of Norfolk, the thread of chemistry from agronomy to plant to genes demonstrates a clear lineage and flow. Later themes (commercialisation and manufacturing) will also show a genesis and persistence of ideas that will have a range of different consequences as a result of optimisation and refinement. Perhaps the most significant point is that an awareness that an external agent can be used to optimise a crop and that this involves the use of different types of chemistry is a key aspect of how individuals, organisations and society are able to take advantage of opportunities. The flow of ideas over time creates different options but the capability to realise these requires the development of specific ways of thinking and working.

The development of crop rotation systems was part of a series of agricultural innovations that were introduced before experimental procedures were available or in place to examine the molecular mechanisms that were being affected. In a sense, therefore, crop rotation bears some similarity to experimental chemistry in the eighteenth and nineteenth centuries when the limited knowledge and techniques available meant that the consequences of the manipulation of molecules could only be measured by looking for macromolecular (whole organism) changes. The impacts of changes in agronomic practice on crop yields were able to be measured even if the details of why the benefits resulted from rotation were observed were not able to be understood. The essence of crop rotation is the practice of growing different crops in different places in successive sets of cultivation. This has two main consequences—first of all, it breaks the cycle of development of certain pests and diseases and secondly it enables the replenishment of nutrients thereby promoting the growth of healthy, high yielding crops. The impact of crop rotation can be considered in terms of both a macro and a microscopic scale. One of the consequences of growing the same crop successively was the Irish potato famine which occurred in the middle of the nineteenth century caused by the fungus *Phytopthera infestans* [6]. In the eighteenth century, much of the population of Ireland were subsistence tenant farmers. After paying their rent, the only crop that could provide sufficient food to prevent starvation was the potato and one variety in particular was almost exclusively grown. Potato blight originated in Mexico [7] and was first recorded in the USA in Philadelphia and New York City, in early 1843 and spread from there to the rest of the USA and Canada before crossing the Atlantic (probably in a consignment of potatoes) and becoming established on mainland Europe. It is unclear if crop rotation (which was widely practiced in Europe) helped to mitigate the effects of potato blight but the narrow genetic base of the single variety grown in most of Ireland (Irish Lumper) contributed to the substantial loss in crop yield. In addition the Irish population was particularly dependent upon the potato as a source of nutrients and it grew particularly well in that climate. The loss of such a significant proportion of nutrients increased the susceptibility of the population to a range of infectious diseases as well as the direct effects of starvation. The effectiveness in growth and yield of the Irish Lumper potato variety (although it was widely criticised for its poor taste) prior to the spread of potato blight (to which it was particularly susceptible) helps to explain the relatively high nutritional status of the Irish population before the advent of the Famine although poverty was widespread [8].

2.1 Crop Rotation: Main Features

Crop rotation has been practiced for thousands of years [9]. One of the first manifestations of a birotation system involved rotating a legume crop with a cereal crop. This has the advantage that the legume crop is able to 'fix' atmospheric nitrogen and to thereby increase the fertility of the soil. A third rotation involved leaving an area fallow for a year. More commonly in England, in medieval times, a two-stage rotation was carried out with a fallow year being followed by a growing year [10]. If animals were allowed to graze on the fallow area, they would then further enrich the soil with manure. The rotation methods that were practiced culminated in the development of the 'Norfolk' method, using a fodder crop as a 'break' rather than a fallow period, which was first used in Belgium and then adopted and adapted in Norfolk. The details and consequences of this method will be discussed in more detail in Sect. 2.4. The environment in Norfolk was conducive to the adoption of new agronomic practices giving higher yields even if the precise details of how they worked was unclear.

Crop rotation has had a significant history and while the details of why it was adopted are unclear, a consequence of growing the same crop in successive years (monoculture) can be both a gradual loss in yield and an increase in disease occurrence and, as was seen in the case of *Phytopthera infestans* in Ireland and monoculture agricultural practices have also been linked to a loss in biodiversity [11]. While observation on a macro level played an important role, it has also become increasingly clear that a number of molecular mechanisms may be operating—particularly at a soil microbiome level [12]. Crop rotation is, however, in essence, an experimental, empirical approach in which the result of a particular course (or courses) of action can be observed even if the molecular rationale remains unclear or is revealed later when detailed analysis at a microbiological level can be carried out [13]. The development of various agronomic practices (particularly the use of fertilisers and crop protection products) was thought to obviate the need for crop rotation (see below) but, in places where monoculture is employed, this may have a negative effect on plant and animal biodiversity [14] and the particular advantages of crop rotation have been realised in a number of studies [15]. Having established the positive consequences of crop rotation, the next stage is to develop a better understanding of the factors (apart from global effects on biodiversity which have already been mentioned) that are a consequence of crop rotation and to determine which of them is most significant. In this way crop rotation can be optimised under different conditions and in different environments and the most significant benefits determined and linked to particular processes and causal effects. The benefits of crop rotation, for example, may be a consequence of the protection of soil structure and the beneficial retention of organic matter in the growing medium. However other factors which contribute to the additional benefits of crop rotation based on historical data have been noted [16]. Differences in the crop rotation regimens and in the results in terms both of ecological diversity and yield have been noted and these imply that, perhaps unsurprisingly, different rotation methodologies may be important under different growing conditions [17]. In addition to interventions followed by observational investigation, a detailed analysis of different types of crop rotations has been carried out and this has helped to define the scientific basis for the practice and those factors which are most significant [18]. The impact of crop rotation on the micro scale properties of growing systems (particularly the growth and diversity of microbial species) will be examined later (Sect. 2.3) however it is noteworthy that microbial diversity shows an increase when crop rotation is carried out [19] and this may have a number of agronomic consequences. The general principle of crop rotation is that diverse (phylogenetically unrelated) species (each with their own preferred population of microbial species) should be successively planted in a particular site in order to get the most beneficial effect. While planting the same species successively appeared to lead to a reduction in yield however, beyond the species level, there was no effect of phylogenetic distance on crop performance [20]. It was noted, however, that soil microbial communities were more similar in soil in which closely related species were successively planted. There are a number of possible reasons for this including the presence of different plant-derived secondary metabolites in root exudates [21]. The differences in performance were not apparently attributable to nutrient depletion. The study was not conducted for long enough (only 2 years were monitored) to determine if there was a longer-term effect on plant disease due to a build up of disease vectors. The possibility of phytochemical influence on plant performance is possible and this may introduce a further stress on repeatedly unrotated crops [22]. There are suggestions that some phytochemicals may affect plant growth directly rather than through the soil microbial community. For example, the plant *Arabidopsis thaliana,* which contains isothiocyanates derived from glucosinolates is inhibited by a range of these compounds [23] and this might suggest an effect on plant growth if it is grown without rotation. Further work is needed to confirm any such hypothesis using a range of plants and their resident phytochemical constituents. The development of rational agriculture was one aspect of the control of nature. It would, however, be a mistake to assume that the only driver was improved yield and better agronomic practices. Rotation cropping was also a way of making use of farm labour at throughout the year rather than having a single annual sowing and harvesting cycle. In addition, obtaining consistent comparable data on the effects of different types of crop rotation can be challenging to assemble—for example, evidence for the direct economic benefits of crop rotation were, in some cases, clear only once it ceased. For example, the development of synthetic fertilisers aided by the invention and scale-up of the Haber-Bosch process, which converts nitrogen into ammonia using an iron catalyst [24], encouraged farmers in the 1950s and 1960s to consider the abandonment of crop rotation by preserving the nutrient balance in the soil through the use of inorganic fertiliser [25] and preventing disease and loss due to insect predation by using crop protection products [26]. It has subsequently been found, however, that inorganic fertilisation alone leads to a loss of diversity [27] and there has been suggestions (based upon epidemiological studies) that there are possible associations between exposure to inorganic fertilizers and solid organ tumours and hematological malignancies and organic fertilizers with infections and diarrhoea [28]. The evidence base is, however both weak and heterogenous and further studies have been suggested as a requirement. Nonetheless rotation has been considered to provide a significant advantage and so it was reintroduced as a significant contribution to agronomy [15]. Part of the advantage was a consequence of improvements to the rhizosphere and these will be considered in more detail in Sect. 2.3. Increasing crop yields and the agricultural impacts of climate change means that rapid assessments of the impact of certain types of crop rotation on farming on a commercial scale are necessary in order that between season optimisation of practices can be carried out and models have been developed [29] which can facilitate this process. While data analysis can provide valuable practical advice, it is very much an observational (correlative) tool, and it is not hypothesis driven. This means that it does not normally allow for the development of predictive tools that create mechanism-based optimisation. In addition, retrospective data on crop rotation practices can only provide a limited degree of certainty about the effect of rotation on the crop [30]—even, as noted above, if rotation is discontinued for whatever reason—and the impact on other factors such as environmental quality means that using maximisation of yield over a single growing season as the sole key measure of effectiveness is no longer sufficient [31]. Apart from indices of soil health such as available nitrogen, most investigations have still used crop yield as the ultimate determinant of the success of different agronomic practices [32] albeit over longer time periods than a single growing season. Comparison of the effect of crop rotation on different production methods (for example organic vs conventional) have been carried out [17] and some studies have attempted to determine the impact of existing crop rotation practices on future, climate influenced, conditions [33] however the suggested interventions to maintain both yields and soil quality have been general and tentative. In general, however, more diverse crop rotation is thought to be beneficial in terms of yield and important sustainability indices [34].

One approach to identifying the way that chemistry (and in particular small secondary metabolite molecules) specifically affects whole crop systems in the context of rotation is to consider allelopathy. Allelopathy is a growth communication system between organisms (particularly plants, although other forms are known and cross-kingdom effects have also been observed [35]). If a particular plant secretes chemicals (most often non-nutritive secondary metabolites) that affect the growth and agronomy of crops sown afterwards then this allelopathic effect can be considered one of the consequences of crop rotation. It can, in some ways, be considered to be qualitatively similar to the cross-species quorum sensing system used by microbes (see Sect. 2.3) since it is manifest in an effect on the growth of one species by a chemical produced by another species. The effects of allelopathy can be positive or negative for any given plant and is modulated by the production of these specific chemicals. Initially allelopathy was considered to be a plant: plant interaction [36] however the concept of microbial allelopathy has been accepted [37]. Plant allelochemicals comprise a number of different structures (Fig. 2.1) and operate on a molecular level at a range of target sites.

In addition to stimulating the growth of particular crops, the use of specific crop rotations can also inhibit the growth of weeds and minimise damage caused by plant destructive factors such as nematodes and other parasites [38]. In the case of stimulation of growth, the biological effects of allelochemicals can be targeted towards a range of growth-related mechanisms and some of these have been summarised [39]. In addition to an effect on systems that influence plant growth and development directly, allelochemicals may also modulate plant hormonal status, impact on various enzyme systems and affect fundamental aspects of physiology such as respiration and photosynthesis. While a link between specific chemicals and certain plant species has been suggested (for example benzoxazinones in wheat and rye [40]), further mechanistically-based investigations are required to build a more complete picture of the different molecular targets of allelochemicals. In addition to stimulatory effects on growth, allelochemicals can also suppress competing species. Prevention of weed growth is an important part of crop husbandry and can lead to a reduction in the synthetic control products that need to be applied [41]. The effect

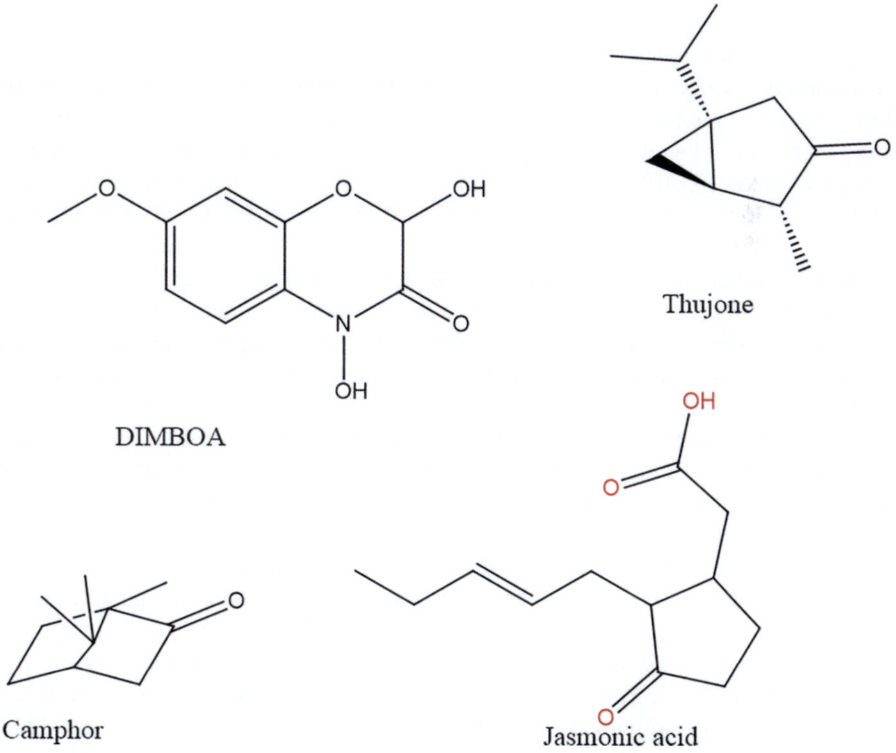

Fig. 2.1 Examples of plant allelochemical structures

of allelochemicals associated with crop rotation with the yield of subsequent crops is suggestive of an important role for plant and microbial secondary metabolites that impacts upon primary metabolism and growth and hence on crop yield. By understanding how to manipulate secondary metabolite production, it may be possible to optimise the production of particular allelochemicals and thereby increase crop yields from crops grown in subsequent years and under particular environmental conditions [42]. One of the simplest ways of examining potential allelopathic effects a whole plant (macro) level is to study the impact of the root exudates of various plants on a target crop [43]. It is also possible to postulate an impact upon specific targets as well as an overall effect on plant growth. One of the postulated specific mechanisms of action of some allelochemicals is the result of an effect on the redox status of the growing plant and on enzymes associated with this [44]. The consequence of this interaction can be hypothesised to be either positive or negative. It can result in the inhibition of the removal of reactive oxygen species (ROS) and a consequential inhibitory effect on plant growth through disruption of plant oxidative homeostasis or, alternatively, an increase in the resistance to abiotic stress with a subsequent positive effect on plant growth [45, 46]. One of the problems with developing a mechanistic basis for the study of allelopathic natural products is the difficulty in finding an appropriate test system (since different species will vary in their response) and in determining the effects of mixtures of chemicals such as will be found in crop growing environments. Allelopathy has also been suggested (because of the size of effects) to the success of alien plants and to the genetic convergence of some species [47]. This might imply that similar biosynthetic mechanisms might operate on a molecular level and the genetic relationships between the biosynthesis of allelopathic chemicals from different species has been investigated [48] Such studies have great predictive potential for the determination of the consequences for the macro properties of plant interactions. The use of other screening methods such as the inhibitory circle method have been proposed [49] however challenges both in the screening method chosen and its application to an actual agronomic situation has been noted. Changes in transcriptome and phytohormone production in response to allelopathic chemicals has been noted [50] and the use of gene transcription screens is, at least potentially, an effective method for measuring the effects of allelochemicals on a molecular level—even if the consequences for plant growth are not necessarily easily correlated. Other approaches have been noted [51] and, as noted previously, although detailed analysis of all allelochemicals is not available, the study of the interaction with biosynthetic gene clusters remains an active area of interest. Just as the effects of knowledge of chemistry can impact at various levels (individual, institutional and societal) and be immediate or may persist over a longer timeframe, the effects of allelochemicals can be seen at various levels (molecular, individual plants and environmental/agronomic characteristics) and in both immediate and longer timeframes. A further example of a molecular approach to the study of allelochemicals is the use of bioassay-guided isolation and purification methodology and it can be seen to be analogous to the approaches taken in the study of medicinal (pharmacological) principles from living systems [52]. Once again, for molecular analysis, metabolomic and transcriptomic analysis of species at a baseline level is information that will be required if the effects of specific chemicals is to be correctly assessed [53]. The link between allelopathy and abiotic stress response was noted previously and such studies also require the acquisition of baseline relational information [54]. In addition to plants, allelopathy has also been observed in algae [55] and extension into the animal and insect kingdoms has been observed and may be of benefit in agricultural pest management [56]. Allelopathy is part of the communication and effector systems of living systems and further investigation using a combination of reductive techniques (studies at the molecular level) and multivariate analysis of the interactions of complex living systems will be necessary. Another potential area of investigation is the possibility of allosteric effects mediated by allelopathic chemicals. If a chemical can alter the reactivity of an enzyme or the binding characteristics of a receptor by binding at a site distant from the active site, then this allosteric interaction may alter the characteristics or growth of the affected species. Hundreds of natural products have been found to be allosteric modulators of proteins.

For example both magnolol and honokiol (Fig. 2.2), which are present in magnolia bark, are positive allosteric modulators of $GABA_A$ receptors [57]. This may seem

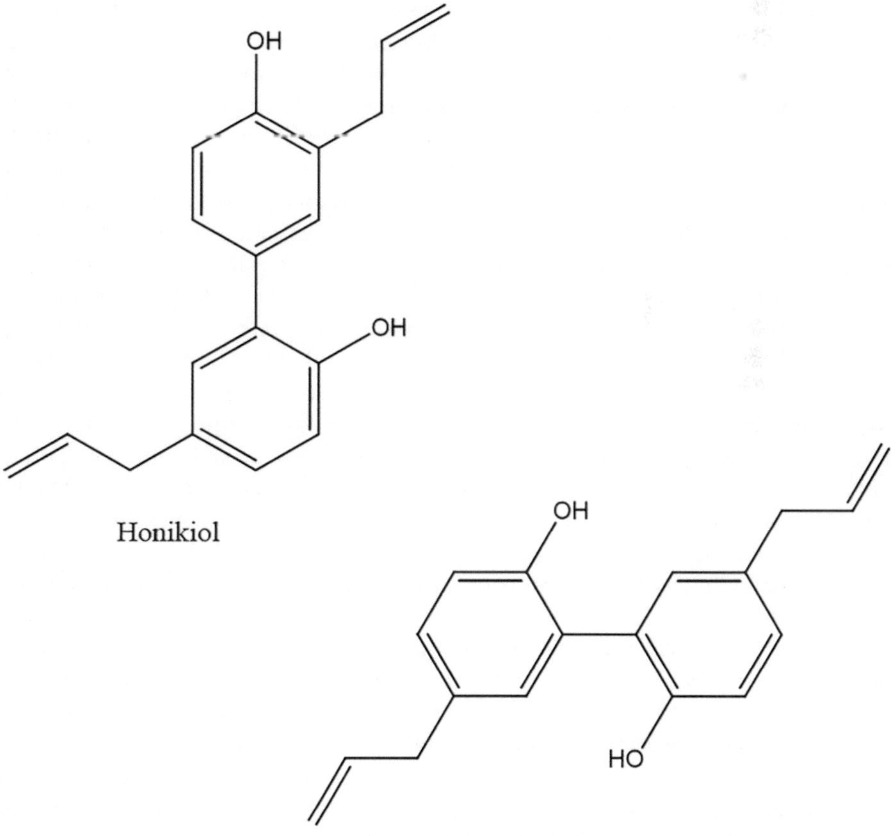

Honikiol

Magnolol

Fig. 2.2 Structures of magnolol and honokiol

like a rather esoteric observation but GABA is known to have a functional role in plant development and stress response [58] and a GABA receptor has been found in plants [59] and may have a role in signal transduction. While this does not prove a biologically significant allosteric modulation effect for magnolol or honokiol, both have antifungal activity that may be beneficial for plant growth and could, therefore, be allelopathic in addition to being allosteric modulators [60]. In addition the GABA receptor in plants is implicated in stress response including fungal attack [61] and therefore the production of magnolol or honokiol may be part of a multifunctional allosteric/allelopathic response which acts to inhibit the spread of fungi in plants.

2.2 Crop Rotation: Macro Scale Impacts

Having considered some general aspects of the macro and potential molecular impacts of crop rotation, macro and micro scale impacts will now be considered in more specific detail. There are different ways of measuring the impact of crop rotation. Perhaps the simplest methods are by measuring the observational effect of rotation (compared to the absence of rotation) on the end products—the yield of the crop and the incidence of disease. While this approach is, in principle, a set of quite uncomplicated measurable results, it is not necessarily quantifiable, comparable between different sets of conditions and free of confounders. An alternative, macro scale impact is on the soil structure and health. This can include the measurement of the presence (or retention) of macronutrients such as essential components including minerals, water and essential components such as phosphates, nitrates and other minerals. The promotion of efficient and effective nutrient cycling is an important aspect of this. One of the challenges of measuring impacts was alluded to previously—considerable amounts of data have to be analysed and interpreted. Data collection of itself is a significant task and the relative importance of different measurable components has to be determined. In addition, synergies between different characteristics may also affect outcomes both directly and indirectly. This is not simply a question of looking for components of the same (or similar) type but also the interplay between, for example, chemicals and soil structure (and the consequences of these interactions), the amount of heat and level of water in the system. Big data modelling and long-term data collection are both essential components of macro scale assessment of the impact of crop rotation.

Another large-scale but indirect consequence of crop rotation is its effect on a range of environmental factors. This can be measured in the context of improved greenhouse gas capture (secondary output) along with enhanced crop yield (primary output) and improvements to soil health [34]. Not all crop rotations have the same impact either on yield or on environmental factors and it has been observed that even a single alteration such as the rotation of different species can produce variable effects. In addition, crop rotation may not simply be an annual event and time course can also have a significant effect. Patterns in crop rotations and their consequences have been calculated from land use data [30] and these were used to predict crop rotations—with varying degrees of accuracy. Crop rotation remains, historically, a largely empirical exercise however analysis and a better understanding of the characteristics of different rotations may enable the optimisation of rotation patterns to provide the maximum benefit for both primary and secondary outcomes. Such studies will remain, however, largely correlative in nature since there are generally too many variables and also unknown factors that may impact upon measured outcomes.

As noted previously, the correlation of specific types of crop rotation or of different components in the crop rotation process with increases in outcome measures can be difficult to assign unambiguously. One approach is to consider the impact of crop rotation on a surrogate marker of agronomic health and hence of yield. One such marker which was briefly mentioned previously, is soil structure and this has been widely considered to be a beneficial characteristic [62] which is also related to crop growth conditions. It is often tempting to attempt to develop a causal relationship between crop rotation, a particular agronomic characteristic (for example the efficiency of carbon utilisation) and a particular aspect of soil such as the existence of a particular microbial community [63]. The validity of such causal relationships under all conditions and with confounding effects which may or may not be known, can be difficult to prove. The evidence base for these and similar effects (which can be classified under the term 'micro' events) will be considered in more detail in Sect. 2.4. As noted previously, crop rotation may also have an impact on (and be affected by) ongoing climate change—a distinctly macro effect. While a significant amount of agricultural production is based upon the successive growth of monocultures of commercial crops and the use of fertilisers and crop protection products, this may not ultimately be the most sustainable approach [64] and the adoption of crop rotation as an integral part of land management may become an important consideration if effective methods can be both developed and their impact measured in a standardised and consistent way [65]. Two of the most important macronutrients for ensuring the growth of crops and production of high yields are biologically available forms of phosphorous and nitrogen [66]. Both are affected by each other and by crop rotation and this in turn affects nutrient cycling and soil fertility and hence yield. The measurement of the crude levels of important soil macronutrients is important however it should be noted that the bioavailability of nitrogen, phosphorous and potassium, as noted above, are also dependent upon the chemical form in which they exist. A number of techniques, including 31-P nuclear magnetic resonance (NMR) spectroscopy [67] have been used to assess the proportions of the differentially bioavailable forms of soil macronutrients under various standardised conditions [68] and measurement of nutrient ratios in rotational schemes has been explored [69], however the accuracy of such simple ratio measurements has been questioned (see below). In addition to having an effect upon macronutrient bioavailability, crop rotation can also help to minimise the leaching of inorganic N and P from agricultural soils which is, of itself, a significant environmental issue. Real time detection of nutrient ratios in soil remains challenging [70] and while the nitrogen to phosphorous ratio (N:P) is widely used to assess soil fertility, it is also subject to measurement uncertainty [71] and caution is advised in its widespread use although non-invasive methods including electrical conductivity [72] and techniques based on NMR (see above) and near infra red spectroscopy have been developed [73]. Rapid, simple and cost effective methods are still required to be further developed. One of the remaining challenges is obtaining sampling data over a large enough area to be meaningful. Satellite imagery to measure field nutrient levels is possible [73] however such remote techniques are best used in conjunction with field data.

2.3 Crop Rotation: Micro Scale Impacts Including Nutrients and Secondary Metabolites

While the effect of crop rotation can be measured in the context of its effects on macronutrient content, its consequences on the plants (improved crop yield) and/or its environmental impact (soil structural improvement) [74] or even the resilience of the ecosystem to the consequences of growing crops under increasingly extreme conditions as a result, for example, of climate change, the importance of molecular scale events should also be considered. Some aspects of this were previously discussed under the heading of allelopathic effects however micronutrients and secondary metabolites can, as will be discussed, affect both the healthy growth of species and their resistance to challenging growth conditions [75]. In addition to the macronutrients noted above (Sect. 2.2), crop rotation also has an impact upon micronutrients and on minor components (secondary natural products) and these, in turn can affect the growth of crops, resistance to disease and the soil microbiome (see Sect. 2.4 for a more detailed discussion). The effect of crop rotation on micronutrients can be considered from two standpoints—the provision of plant-essential micronutrients to improve the agronomic characteristics of the crop and the enhanced micronutrient profile of the crop for the consuming organism. In addition, the bioaccumulation of harmful elements (for example, cadmium [76]) may have a negative nutritional effect. Determining the best plant micronutrient status to give optimal plant growth and optimal subsequent nutrition is of prime importance [77]. The combination of crop rotation and the application of mineral and other fertilisers can help to optimise plant growth [78] due to a combination of macro and micronutrient effects. There is also a perceived overlap between the adoption of practices that facilitate sustainable and healthy plant growth (often referred to as regenerative agriculture) and the optimal nutritional profile of an edible crop [79]. While much of the emphasis of the role of macronutrients in plant (and consumer) health has remained known although it continues to be studied (particularly in the context of processed food [80]), the increasing drive for health has created opportunities in the area of optimisation of micronutrients in particular [81] however, unlike macronutrients, there is an optimal beneficial window of opportunity for micronutrients and this must be taken into account. Overlaid upon this is the realisation that biofortification of plants for positive nutritional outcomes is a viable option [82] and could improve nutritional status and help to alleviate some aspects of chronic, non-communicable diseases (NCDs). Overall, the potential for biofortification in a number of areas to improve health through dietary means has been recognised [83] however many of these efforts have concentrated on genetic-based breeding programmes [84] rather than on agronomic practices such as crop rotation. Other approaches that consider more the potential role of agronomic practices have been reviewed [85].

An alternative to fortification of crops with macro and micronutrients to improve crop yields and nutritional value is the use of bioactive secondary natural products as part of a crop rotation strategy. While micro and macronutrients can be defined either in terms of chemical identity (P, N, K, Ca etc.) or in the context of their impact on biological processes (for example, the necessity of Mg to the photosynthetic system), secondary natural products are more difficult to define in those simple terms. First of all because they have a large number of complex structures which, as we have previously noted, may be empirically identical but are structurally (and functionally) different and secondly because their impact on biological processes may be both minor and partial and they may only be a part of the effector system for a given process. For these reasons, the analysis of the importance (and even presence) of secondary natural products relies more on limited molecular rather than atomic reductionism. A given molecule must be separated from others then the amounts present determined. In this way, its identity progresses from being a molecule in a biological system to being a collection of data which is produced from its characteristics (solubility, volatility, size, polarity etc.). Having been separated and identified, the chemical entity which has been the source of characterisation-based data however, has been taken from its biological environment, it no longer has biological relevance and its function is lost. In order to determine the biological significance, the chemical entity has to be reintroduced into a suitable biological system and its impact (a whole new set of data) determined. Just as the analysis and identification of a molecule deprives it of biological relevance, so the introduction of the molecule into the biological system deprives it of clear and unambiguous chemical data. Many physicochemical attributes are no longer identifiable and even the identity or structure of the molecule itself may change. Much of the investigation of chemicals in a biological world is to do with obtaining imperfect data from partially understood interactions and attempting to correlate that with biological relevance. The identity/relevance dichotomy lies at the core of chemistry in the world. The better a molecule is defined and characterised, the less it can be understood as part of a chemical/biological system. In the study of secondary natural products, their role lies primarily (and by definition) in the impact they have on the secondary characteristics of plants such as disease resistance rather than their effect on primary metabolism, growth and yield. Many secondary metabolites have an ill-defined biological role and so, on the whole, studies designed to find ways to maximise their production are comparatively rare. Despite this apparent lack of understanding of their core role, natural plant growth inhibitors are known [86] and many of them are classified as allelochemicals and were discussed previously. In some cases, natural chemicals found in soil can have multifunctional purposes in agronomy. For example, residues of allyl isothiocyanate, which may be present in soil that has been used to grow *Brassica* vegetables, seems to show a range of positive activities against plant pathogens [87] however it also inhibits the growth of several plants by inhibiting actin-dependant intracellular transport [88]. These dual functionalities add further complication to the design of the most advantageous crop rotation systems. In addition to multiple functions in their components, plants also contain a range of anti-microbial compounds although the effect of these on agronomically important microbial species has not been extensively tested. Similarly the effect of both plant and microbial natural products on plant growth is known but not intensively studied [89]. Perhaps the simplest route would be to assay compounds for effects (agonist and antagonist) on plant pathways associated with modulation by known plant hormones. It may then be possible to determine which aspects of crop rotation have the most significant effects and also have the greatest potential to synergise with other crop rotation possibilities [74].

2.4 Crop Rotation: The Rhizosphere and the Microbiome

While the molecular (micro) scale impact of crop rotation has been discussed in terms of micronutrients and secondary metabolites, one of the principal impacts of crop rotation is on microbial species in the soil. While crop rotation improves crop yield and increases the soil quality (see above) at least some of this improvement is a consequence of an improvement in the soil microbiome and the rhizosphere.

The rhizosphere is defined both in terms of its location and its functional properties. It is the narrow region of soil whose properties (and, in particular its microbiological population) is influenced and largely defined by the compounds produced by the resident plant population and, in particular, by secretions from their roots. The term 'rhizosphere' was first used by Hiltner in 1904 [90, 91] and was underpinned by his belief that the root areas of plants contained a range of microbes (both beneficial and deleterious) which were adapted to that environment and furthermore that the rhizosphere could be manipulated to enhance the benefits and minimise the negative aspects of pathogenesis. The importance of the rhizosphere and its component parts for plant health and growth has been reviewed and explained [92] and this forms the basis of significant amounts of research. While the rhizosphere is now considered to be a complex mixture of interdependent organisms and includes plants, microbes, fungi and chemicals, there are some aspects that are more able to be interpreted in terms of distinct mechanisms than others. One of these is the role of quorum sensing molecules [93] and the sensitivity of some microbial species to quorum sensing mechanisms. In addition to affecting microbial growth, quorum sensing also impacts upon macronutrient properties such as nitrogen cycling in the rhizosphere [94] and the production of plant bioactive agents such as indole acetic acid. The microbiome of the rhizosphere as a whole is controlled by quorum sensing and other complex signalling pathways [95] and unravelling these individual and interlinked pathways is a major challenge.

On a molecular level, then, manipulation of the microbiome of the rhizosphere through understanding microbial and plant signalling is a potential lever to manipulate crops and to develop advantageous endpoints. There is an increased understanding of how specific crop rotations can affect the rhizosphere microbiome and hence both improve crop yield and increase resistance to disease [96] however the link remains principally associative rather than mechanistic. Such approaches are only possible with the adoption of high throughput sequencing technology. It is possible to carry out analysis at three levels—the identification of distinct microbial populations and the associated diversity of species, utilisable plant nutrients and the yield and fitness of the crop itself [97]. In another strand of the link between the rhizosphere, its microbial community and plants, the correlation between crop rotation, root biomass production and the soil rhizosphere has been noted [98] and further work, building on the body of knowledge based on the empirical and observational studies of crop rotation, will provide the basis for the creation of predictive models of crop rotation and the development of responsive biomarkers of soil health. While AI-based models for crop rotation have been developed [99], there are limitations to their accuracy and many of them use bioavailable macronutrient data as the basis for their modelling. The incorporation of new data and the combination of different types of models are likely to produce more powerful predictive tools [100] and increase the strength of the association between specific factors (for example the soil microbiome) and plant growth endpoints *via* crop rotation. There are, however, some confounders (for example, seasonal climate variation) which are challenging to incorporate into any crop rotation, agronomic model [101] and there is, therefore, a limit to the extent to which predictive modelling can be successful.

In summary, the quantification of the effects of crop rotation on the yield of products from specific rotation practices as a consequence of impacts upon the soil microbiome and the rhizosphere is challenging. There is a substantial body of empirical evidence, built up over many years, which provide information on which combinations of rotations are most effective for a given endpoint and untangling the key influencing factors will enable predictive models to be developed and to then be verified through the use of appropriate interventions. This will enable improvements to be made to current practices and to the effective introduction of novel crops or the more efficient production of existing useful products. One analogy to the effects of crop rotation on the soil microbiome is the human gut microbiome and this congruency has been noted [102]. The similarity has been extended to compare the use of antibiotics in humans to the use of pest control agents in agriculture. The concept of a healthy human microbiome and a healthy agricultural microbiome being, not simply analogous but also interdependent, is one that is, at least superficially, attractive. One point of divergence is the immune system. In plants, the innate immune system helps in resistance to disease (including pathogens—see below) [103] while in humans (and other mammals) both an innate and inducible immune system are present and this dual mechanism is a key feature in the resistance against external challenges [104]. The microbiome is postulated to have an effect on the performance of the human immune system [105]. The plant innate immune system has been considered to be primarily a non-inducible response to pathogens, it has

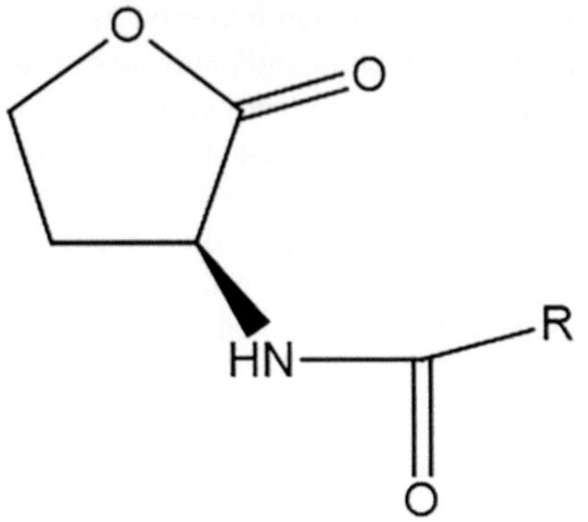

N-Acyl Homoserine Lactone (ACL)

Fig. 2.3 The structure of the homoserine lactone QS molecule

also been postulated to play a role in shaping the plant-associated microbiome and this in turn provides an additional defence against pathogenic attack [106] and provide, in synergy with plant-derived nutrients, an extra level of protection [107]. An alternative level of pathogen protection is suggested by the discovery of quorum sensing inhibitors (QSIs) and this may be a mechanism to control pathogens and to protect species that are able to produce QSIs [108] and this may have particular relevance to plant/microbe interactions [109].

The presence of QSIs in mammalian systems (based on induction of a lactonase enzyme system that cleaves one of the major classes of quorum sensing molecules—homoserine lactones (Fig. 2.3)) was initially thought to be a positive development however QSI-resistant strains of pathogenic bacteria have been found [110] and this may represent a significant challenge to strategies to combat anti-microbial resistance and to modulate the control of the soil microbiome by plant-derived components.

2.5 Crop Rotation in Norfolk: Actions and Consequences

As noted previously, the idea of crop rotation and giving fields time to recover fertility is not new. Indeed the annual Nile delta floods can be considered a rapid form of nutrient replenishment. For most places, however, crop rotation involved leaving about half the agricultural land to lie fallow at a time. It was frequently the case that land was grazed on by ruminants during the fallow period however because the climate did not allow fallow land to be productive in winter, livestock was slaughtered in autumn and preserved over winter. One advantage of the fallow system was that soil replenishment was supplemented by manure from the grazing animals. Over time a three-field system was developed in which legumes were grown (and because they facilitated the microbial fixing of nitrogen thereby improving soil fertility, the fallow period could be reduced). This improved the fertility and production of food. The passing of the Enclosures Act in 1773 increased the control that landowners had over their acquired land resources. This Act enabled landowners to remove common land by enclosing it and this land, which was run under a central control, could be rented out to tenants who then had to follow the agricultural practices laid down by the landowner. The various enclosure measures greatly limited peasant strip farming and allowed for the creation of large, essentially homogenous estates. This allowed farming practices to change and greatly increased productivity. It also facilitated the cultivation of land by fewer people and this helped to provide urban labour for the Industrial Revolution. While changes in agriculture created wealth for the relatively few, it also allowed for the substantial increase in population required to operate the new factories. It dislocated the link between food production and consumption. A key figure at that time was the Professor of History and Political Economy, Thomas Malthus. The publication of 'An Essay on the Principle of Population' by Malthus in 1798 was an important contribution to the debate over populations and food production. In this tract, Malthus proposed the idea that populations grow by geometric progression. What this means is that each successive term in a population sequence following the first can be found by multiplying the previous one with a common ratio, which is a fixed, non-zero number. For instance, in the sequence 2, 6, 18, 54, 162, the common ratio is 3. He furthermore suggested that food production increases by an arithmetic progression in which successive numbers are increased additively by a constant amount and this means for example, in the sequence 2, 4, 6, 8, 10, the constant amount is 2. The consequence of this is that the population will grow faster than the food supply and this will inevitably lead to starvation and death. The answer to the Malthusian catastrophe lies in finding a way of either reducing the rate of growth of the population (natural disasters and wars were suggested methods) or increasing food productivity. The Enclosures Act and the introduction of crop rotation facilitated the latter.

While the Enclosures Acts enabled farming to be carried out at scale, the development of crop rotation was the practical realisation of the potential of the larger units of land. It was part of the Agricultural Revolution that grew to prominence in mainland Europe in the eighteenth century. In particular, the development of new rotation systems with root crops such as potatoes and turnips allowed the fallow period to be eliminated. There is, however, a degree of revisionism attached to the history of crop rotation. While the four-course rotation was advocated by Coke, Townshend and others in Norfolk (see below) to the extent that it became known as the 'Norfolk Rotation', it was not simply due to the introduction of turnips and clover as rotation crops and the removal of the less productive fallow period. Indeed, turnips were grown in England as early as 1638—long before the introduction of the Norfolk rotation—however the combination of the changes in agricultural practices and the adoption of the Enclosures Acts were sufficiently synergistic to lead to an Agricultural Revolution and this in turn enabled the population increases observed in the eighteenth and nineteenth centuries. By 1750, the population of England was 5.5 million. This grew to nine million by 1800 and to 16 million by 1850—all sustainable by the increase in agricultural production. A further consequence of the growth of root crops was their use to feed livestock over the winter period until forage grass was available in the Spring and this, together with advances in breeding of livestock, greatly increased the availability of an affordable animal food supply. These new crops were incorporated into farming practice at scale by Lord Charles Townshend at his estate at Raynham in Norfolk and Thomas Coke, the first Earl of Leicester at Holkham (also in Norfolk) in the early part of the nineteenth century. The four-crop rotation had been introduced in Flanders (Waasland) in the sixteenth century and was imported into Norfolk by Townshend. The area of Waasland was boggy and so the development of new agricultural methods was a necessity. There are analogies with the draining of the Fens in the seventeenth century which was also a consequence of Flemish/Dutch influence (Cornelius Vermuyden) and is further considered in Chap. 4. The adoption of the crop rotation system was not the first time that a technology was developed in Flanders and brought to Norfolk (see Chap. 6 for a discussion of the Flemish weavers and the technology they brought into the county). This cross-fertilisation of new, scientifically based farming methods from the near continent was a major factor in the development of wealth in Norfolk and paved the way for the importation and adoption of future developments, some of which will be introduced in succeeding chapters. The importance of the use of turnips and clover as break crops has been referred to previously. In the case of the latter, an important aspect was the capability of the crop to facilitate nitrogen fixing thereby to increase the levels of this important macronutrient [111]. The use of crops (legumes or clover) to enable bioavailable nitrogen to be created was a major aspect of agriculture until the large-scale production of ammonia by the Haber-Bosch process [112]. The increase in agricultural production was particularly beneficial for Norfolk. The crop rotation technology that was imported and then applied in Norfolk led to a substantial surplus being created such that, by 1794, more grain was being exported from Norfolk to Holland than from the whole of the rest of England. Many of the large estates of Norfolk were created as a consequence of the wealth that had been created through the importation of technology. Although trains eventually reached Norfolk and Norwich was connected to London (*via* Cambridge in 1845 and directly in 1849), this was relatively late. Train services had linked London with Birmingham, Manchester, and Liverpool by 1838. Norfolk and Norwich, perhaps in part as a consequence of the development of the Hanseatic League, maintained strong trading links with Europe. The distance from Norwich to Amsterdam is the same as from Norwich to London and, while in medieval times, Amsterdam was just 1 day sailing away, travel to London took up to 4 days. In contrast, to Europe, links from Norwich to the industrial cities in the north of England (Liverpool, Manchester, Birmingham) were much less well developed and were not seen as a priority by railway companies. In addition, the canal network did not extend to Norfolk although Norwich was a sea port with access to the North Sea *via* the rivers Wensum and Yare to Great Yarmouth. The orientation of Norwich towards Europe and its relative isolation from the rest of England had economic consequences and, while Norwich was the second city in England (after London) until the industrial revolution in the 18th and 19th centuries, the agricultural industrial base became less important for wealth generation even though initiatives from Europe (in particular those with a visible strand of chemistry) continued to benefit the local economy and to stimulate knowledge and technology exchange. This will be examined further in some of the following chapters.

References

1. Payne FG. The British plough: some stages in its development. Agric Hist Rev. 1957;5(2):74–84.
2. Shaw-Taylor L. The rise of agrarian capitalism and the decline of family farming in England. Econ Hist Rev. 2012;65(1):26–60. https://doi.org/10.1111/j.1468-0289.2010.00585.x.
3. Sussmilch FC, Ross JJ, Reid JB. Mendel: from genes to genome. Plant Physiol. 2022;190(4):2103–14. https://doi.org/10.1093/plphys/kiac424.
4. McNulty MB. What is chemistry, for Kant? Kant Yearb. 2017;9(1):85–112. https://doi.org/10.1515/kantyb-2017-0005.
5. Carrier M. Kant's theory of matter and his views on chemistry. In: Watkins E, editor. Kant and the sciences. New York: Oxford Academic; 2006. https://doi.org/10.1093/0195133056.003.0011.
6. Powderly WG. How infection shaped history: lessons from the Irish famine. Trans Am Clin Climatol Assoc. 2019;130:127–35.
7. Yoshida K, Schuenemann VJ, Cano LM, Pais M, Mishra B, Sharma R, Lanz C, Martin FN, Kamoun S, Krause J, Thines M, Weigel D, Burbano HA. The rise and fall of the Phytophthora infestans lineage that triggered the Irish potato famine. Elife. 2013;28(2):e00731. https://doi.org/10.7554/eLife.00731.
8. Mokyr J. Irish history with the potato. Irish Econ Soc Hist. 1981;8:8–29.
9. White KD. Fallowing, crop rotation, and crop yields in Roman times. Agric Hist. 1970;44(3):281–90.
10. Hamerow H, et al. Crop rotation and seasonal sowing. In: Feeding medieval England: a long 'agricultural revolution. Oxford: Oxford Academic; 2025. p. 700–1300. https://doi.org/10.1093/9780191988905.003.0004.
11. de Vareilles A, Woodbridge J, Pelling R, Fyfe R, Smith D, Campbell G, Smith W, Carruthers W, Adams S, le Hégarat K, Allot L. The development of arable cultivation in the south-east of England and its relationship with vegetation cover: a honeymoon period for biodiversity? The Holocene. 2023;33(11):1389–406. https://doi.org/10.1177/09596836231185836.
12. Malik AI, Bell R, Zang H, et al. Exploring the plant and soil mechanisms by which crop rotations benefit farming systems. Plant Soil. 2025;507:1–9. https://doi.org/10.1007/s11104-024-06994-z.
13. Li C, Shi L, Wang K, Liu B, Liao J, An Z, Chang SX. Crop rotation differentially increases soil bacterial and fungal diversities in global croplands: a meta-analysis. Nat Commun. 2025. https://doi.org/10.1038/s41467-025-66823-4.
14. Tissier ML, Kletty F, Handrich Y, Habold C. Monocultural sowing in mesocosms decreases the species richness of weeds and invertebrates and critically reduces the fitness of the endangered European hamster. Oecologia. 2018;186(2):589–99. https://doi.org/10.1007/s00442-017-4025-y.
15. Bullock DG. Crop rotation. Crit Rev Plant Sci. 1992;11(4):309–26. https://doi.org/10.1080/07352689209382349.
16. Knox OGG, Leake AR, Walker RL, Edwards AC, Watson CA. Revisiting the multiple benefits of historical crop rotations within contemporary UK agricultural systems. J Sustain Agric. 2011;35(2):163–79. https://doi.org/10.1080/10440046.2011.539128.
17. Barbieri P, Pellerin S, Nesme T. Comparing crop rotations between organic and conventional farming. Sci Rep. 2017;7:13761. https://doi.org/10.1038/s41598-017-14271-6.

18. Wang B, Liu J, Liu Q, Sun J, Zhao Y, Liu J, Gao W, Chen Y, Sui P. Knowledge domain and research progress in the field of crop rotation from 2000 to 2020: a scientometric review. Environ Sci Pollut Res Int. 2023;30(37):86598–617. https://doi.org/10.1007/s11356-023-28266-6.
19. Tiemann LK, Grandy AS, Atkinson EE, Marin-Spiotta E, McDaniel MD. Crop rotational diversity enhances belowground communities and functions in an agroecosystem. Ecol Lett. 2015;18(8):761–71. https://doi.org/10.1111/ele.12453.
20. Kaplan I, Bokulich NA, Caporaso JG, Enders LS, Ghanem W, Ingerslew KS. Phylogenetic farming: can evolutionary history predict crop rotation via the soil microbiome? Evol Appl. 2020;13(8):1984–99. https://doi.org/10.1111/eva.12956.
21. Ku YS, Cheng SS, Luk CY, Leung HS, Chan TY, Lam HM. Deciphering metabolite signalling between plant roots and soil pathogens to design resistance. BMC Plant Biol. 2025;25(1):308. https://doi.org/10.1186/s12870-025-06321-3.
22. Fais A, Era B. Phytochemical composition and biological activity. Plants. 2024;13(3):331. https://doi.org/10.3390/plants13030331.
23. Urbancsok J, Bones AM, Kissen R. Glucosinolate-derived Isothiocyanates inhibit Arabidopsis growth and the potency depends on their side chain structure. Int J Mol Sci. 2017;18(11):2372. https://doi.org/10.3390/ijms18112372.
24. Leigh GJ. Haber-Bosch and other industrial processes. In: Smith BE, Richards RL, Newton WE, editors. Catalysts for nitrogen fixation. Nitrogen fixation: origins, applications, and research Progress, vol. 1. Dordrecht: Springer; 2004. https://doi.org/10.1007/978-1-4020-3611-8_2.
25. Parveen A, et al. Potential of inorganic fertilizers for sustainable development in agriculture. In: Mahmood Q, editor. Sustainable plant nutrition under contaminated environments. Sustainable plant nutrition in a changing world. Cham: Springer; 2022. https://doi.org/10.1007/978-3-030-91499-8_3.
26. Sparks TC, Bryant RJ. Crop protection compounds - trends and perspective. Pest Manag Sci. 2021;77(8):3608–16. https://doi.org/10.1002/ps.6293.
27. Shi TS, Collins SL, Yu K, et al. A global meta-analysis on the effects of organic and inorganic fertilization on grasslands and croplands. Nat Commun. 2024;15:3411. https://doi.org/10.1038/s41467-024-47829-w.
28. Tagkas CF, Rizos EC, Markozannes G, Karalexi MA, Wairegi L, Ntzani EE. Fertilizers and human health-a systematic review of the epidemiological evidence. Toxics. 2024;12(10):694. https://doi.org/10.3390/toxics12100694.
29. Meisner MH, Rosenheim JA. Ecoinformatics reveals effects of crop rotational histories on cotton yield. PLoS One. 2014;9(1):e85710. https://doi.org/10.1371/journal.pone.0085710.
30. Upcott EV, Henrys PA, Redhead JW, Jarvis SG, Pywell RF. A new approach to characterising and predicting crop rotations using national-scale annual crop maps. Sci Total Environ. 2023;860:160471. https://doi.org/10.1016/j.scitotenv.2022.160471.
31. Lemaire G, Gastal F, Franzluebbers A, Chabbi A. Grassland-cropping rotations: an avenue for agricultural diversification to reconcile high production with environmental quality. Environ Manag. 2015;56(5):1065–77. https://doi.org/10.1007/s00267-015-0561-6.
32. Zhang L, Yuan J, Zhang M, et al. Long term effects of crop rotation and fertilization on crop yield stability in Southeast China. Sci Rep. 2022;12:14234. https://doi.org/10.1038/s41598-022-17675-1.
33. Liu R, Liu H, Wang T, Wang T, Lu Z, Yuan X, Song Z, Li R. Sustainability of maize–soybean rotation for future climate change scenarios in northeast china. J Agro Crop Sci. 2025;211:e70033. https://doi.org/10.1111/jac.70033.
34. Yang X, Xiong J, Du T, Ju X, Gan Y, Li S, Xia L, Shen Y, Pacenka S, Steenhuis TS, Siddique KHM, Kang S, Butterbach-Bahl K. Diversifying crop rotation increases food production, reduces net greenhouse gas emissions and improves soil health. Nat Commun. 2024;15(1):198. https://doi.org/10.1038/s41467-023-44464-9.
35. Schandry N, Becker C. Allelopathic plants: models for studying plant-Interkingdom interactions. Trends Plant Sci. 2020;25(2):176–85. https://doi.org/10.1016/j.tplants.2019.11.004.

36. Molisch H. Der Einfluss Einer Pflanze Auf Die Andere-allelopathie. Jena: Fischer; 1937.
37. Rice EL. Allelopathy. 2nd ed. New York: Academic; 1984.
38. Li Y, Wang B, Chang Y, Yang Y, Yao C, Huang X, Zhang J, Cai Z, Zhao J. Reductive soil disinfestation effectively alleviates the replant failure of Sanqi ginseng through allelochemical degradation and pathogen suppression. Appl Microbiol Biotechnol. 2019;103(8):3581–95. https://doi.org/10.1007/s00253-019-09676-4.
39. Cheng F, Cheng Z. Research Progress on the use of plant allelopathy in agriculture and the physiological and ecological mechanisms of allelopathy. Front Plant Sci. 2015;17(6):1020. https://doi.org/10.3389/fpls.2015.01020.
40. Krogh SS, Mensz SJ, Nielsen ST, Mortensen AG, Christophersen C, Fomsgaard IS. Fate of benzoxazinone allelochemicals in soil after incorporation of wheat and rye sprouts. J Agric Food Chem. 2006;54(4):1064–74. https://doi.org/10.1021/jf051147i.
41. Khamare Y, Chen J, Marble SC. Allelopathy and its application as a weed management tool: a review. Front Plant Sci. 2022;28(13):1034649. https://doi.org/10.3389/fpls.2022.1034649.
42. Kumar N, Singh H, Giri K, et al. Physiological and molecular insights into the allelopathic effects on agroecosystems under changing environmental conditions. Physiol Mol Biol Plants. 2024;30:417–33. https://doi.org/10.1007/s12298-024-01440-x.
43. Zhang X, Shi S, Li X, Li C, Li Q. Potential effect of root exudates from ten crops on promoting stress tolerance in alfalfa (Medicago sativa) seedlings. Life. 2025;15(4):600. https://doi.org/10.3390/life15040600.
44. Zeng RS, Luo SM, Shi YH, Shi MB, Tu XY. Physiological and biochemical mechanism of allelopathy of secalonic acid F on higher plants. Agron J. 2001;93:72–9.
45. Rama DS, Prasad MNV. Ferulic acid mediated changes in oxidative enzymes of maize seedlings: implications in growth. Biol Plant. 1996;38:387–95.
46. Maqbool N, Wahid A, Farooq M, Cheema ZA. Allelopathy and abiotic stress interaction in crop plants. In: Cheema Z, Farooq M, Wahid A, editors. Allelopathy. Cham: Springer; 2013. p. 451–68.
47. Zhang Z, Liu Y, Yuan L, Weber E, van Kleunen M. Effect of allelopathy on plant performance: a meta-analysis. Ecol Lett. 2021;24:348–62. https://doi.org/10.1111/ele.13627.
48. Sultana MH, Liu F, Alamin M, Mao L, Jia L, Chen H, Wu D, Wang Y, Fu F, Wu S, Wang W, Ye C, Zhu QH, Qiu J, Fan L. Gene modules co-regulated with biosynthetic gene clusters for allelopathy between rice and barnyardgrass. Int J Mol Sci. 2019;20(16):3846. https://doi.org/10.3390/ijms20163846.
49. Li JY, Guo XK, Zhang Q, Liu CH, Lin ZH, Yu ZM, Wu H, He HB. A novel screening method for rice allelopathic potential: the inhibitory-circle method. Weed Res. 2015;55:441–8.
50. Zhou T, Li Q, Huang X, Chen C. Analysis transcriptome and phytohormone changes associated with the allelopathic effects of ginseng hairy roots induced by different-polarity ginsenoside components. Molecules. 2024;29(8):1877. https://doi.org/10.3390/molecules29081877.
51. Kong C-H, Li Z, Li F-L, Xia X-X, Wang P. Chemically mediated plant–plant interactions: allelopathy and allelobiosis. Plants. 2024;13(5):626. https://doi.org/10.3390/plants13050626.
52. Kato-Noguchi H. Isolation and identification of allelochemicals and their activities and functions. J Pestic Sci. 2024;49(1):1–14. https://doi.org/10.1584/jpestics.D23-052.
53. Jiang M, Yan Y, Zhou B, et al. Metabolomic and transcriptomic analyses highlight metabolic regulatory networks of salvia miltiorrhiza in response to replant disease. BMC Plant Biol. 2024;24:575. https://doi.org/10.1186/s12870-024-05291-2.
54. Tan QW, Lim PK, Chen Z, et al. Cross-stress gene expression atlas of Marchantia polymorpha reveals the hierarchy and regulatory principles of abiotic stress responses. Nat Commun. 2023;14:986. https://doi.org/10.1038/s41467-023-36517-w.
55. Semmouri I, Janssen CR, Asselman J. Allelopathy in macroalgae: ecological principles, research opportunities and pitfalls reviewed. J Appl Phycol. 2024;36:441–58. https://doi.org/10.1007/s10811-023-03110-z.

56. Farooq M, Jabran K, Cheema ZA, Wahid A, Siddique KH. The role of allelopathy in agricultural pest management. Pest Manag Sci. 2011;67(5):493–506. https://doi.org/10.1002/ps.2091.
57. Alexeev M, Grosenbaugh DK, Mott DD, Fisher JL. The natural products magnolol and honokiol are positive allosteric modulators of both synaptic and extra-synaptic GABA(a) receptors. Neuropharmacology. 2012;62(8):2507–14. https://doi.org/10.1016/j.neuropharm.2012.03.002.
58. Li L, Dou N, Zhang H, Wu C. The versatile GABA in plants. Plant Signal Behav. 2021;16(3):1862565. https://doi.org/10.1080/15592324.2020.1862565.
59. Žárský V. Signal transduction: GABA receptor found in plants. Nat Plants. 2015;1:15115. https://doi.org/10.1038/nplants.2015.115.
60. Chen YH, Lu MH, Guo DS, Zhai YY, Miao D, Yue JY, Yuan CH, Zhao MM, An DR. Antifungal effect of Magnolol and Honokiol from Magnolia officinalis on Alternaria alternata causing tobacco Brown spot. Molecules. 2019;24(11):2140. https://doi.org/10.3390/molecules24112140.
61. Ramos-Ruiz R, Martinez F, Knauf-Beiter G. The effects of GABA in plants. Cogent Food Agric. 2019;5(1) https://doi.org/10.1080/23311932.2019.1670553.
62. Ball BC, Bingham I, Rees RM, Watson CA, Litterick A. The role of crop rotations in determining soil structure and crop growth conditions. Can J Soil Sci. 2005;85(5):557–77. https://doi.org/10.4141/S04-078.
63. Meng X, Wang B, Zhang X, Liu C, Ji J, Hao X, Yang B, Wang W, Xu D, Zhang S, Wang X, Cao M, Wang Y. Long-term crop rotation revealed the relationship between soil organic carbon physical fraction and bacterial community at aggregate scales. Microorganisms. 2025;13(3):496. https://doi.org/10.3390/microorganisms13030496.
64. Azarbad H, Conventional vs. Organic agriculture-which one promotes better yields and microbial resilience in rapidly changing climates? Front Microbiol. 2022;9(13):903500. https://doi.org/10.3389/fmicb.2022.903500.
65. Teixeira EI, de Ruiter J, Ausseil AG, Daigneault A, Johnstone P, Holmes A, Tait A, Ewert F. Adapting crop rotations to climate change in regional impact modelling assessments. Sci Total Environ. 2018;616-617:785–95. https://doi.org/10.1016/j.scitotenv.2017.10.247.
66. Tian J, Dunfield K, Condron L. Biological cycling of nitrogen and phosphorus in soils. Plant Soil. 2024;498:1–4. https://doi.org/10.1007/s11104-024-06601-1.
67. Li M, Mazzei P, Cozzolino V, et al. Optimized procedure for the determination of P species in soil by liquid-state 31P-NMR spectroscopy. Chem Biol Technol Agric. 2015;2:7. https://doi.org/10.1186/s40538-014-0027-8.
68. Harmsen J. Measuring bioavailability: from a scientific approach to standard methods. J Environ Qual. 2007;36(5):1420–8. https://doi.org/10.2134/jeq2006.0492.
69. Bansal S, Yin X, Savoy HJ, Jagadamma S, Lee J, Sykes V. Long-term influence of phosphorus fertilization on organic carbon and nitrogen in soil aggregates under no-till corn-wheat-soybean rotations. Agron J. 2020;112:2519–34. https://doi.org/10.1002/agj2.20200.
70. Ameer S, Ibrahim H, Kulsoom FNU, et al. Real-time detection and measurements of nitrogen, phosphorous & potassium from soil samples: a comprehensive review. J Soils Sediments. 2024;24:2565–83. https://doi.org/10.1007/s11368-024-03827-5.
71. Yan Z, Tian D, Han W, Tang Z, Fang J. An assessment on the uncertainty of the nitrogen to phosphorus ratio as a threshold for nutrient limitation in plants. Ann Bot. 2017;120(6):937–42. https://doi.org/10.1093/aob/mcx106.
72. Reddykapa V, et al. Real-time estimation of nitrogen, phosphorus, and potassium. In: Soil 2022 IEEE Delhi section conference (DELCON). New Delhi: IEEE; 2022. p. 1–6. https://doi.org/10.1109/DELCON54057.2022.9753548.
73. He Y, Song HY, Pereira AG, Gómez AH. Measurement and analysis of soil nitrogen and organic matter content using near-infrared spectroscopy techniques. J Zhejiang Univ Sci B. 2005;6(11):1081–6. https://doi.org/10.1631/jzus.2005.B1081.

74. Wen Y, Zang H. Crop rotation boosts yields and soil quality. Nat Food. 2025;6:130–1. https://doi.org/10.1038/s43016-024-01098-4.
75. Jan AU, Hadi F, Ditta A, Suleman M, Ullah M. Zinc-induced anti-oxidative defense and osmotic adjustments to enhance drought stress tolerance in sunflower (Helianthus annuus L.). Environ Exp Bot. 2022;193:104682. https://doi.org/10.1016/j.envexpbot.2021.104682.
76. Yang Y, Nan Z, Zhao Z. Bioaccumulation and translocation of cadmium in wheat (Triticum aestivum L.) and maize (Zea mays L.) from the polluted oasis soil of Northwestern China. Chem Speciat Bioavailab. 2014;26:43–51.
77. Assunção AGL, Cakmak I, Clemens S, González-Guerrero M, Nawrocki A, Thomine S. Micronutrient homeostasis in plants for more sustainable agriculture and healthier human nutrition. J Exp Bot. 2022;73(6):1789–99. https://doi.org/10.1093/jxb/erac014.
78. Mukhametov A, Ansabayeva A, Efimov O, Kamerova A. Influence of crop rotation, the treatment of crop residues, and the application of nitrogen fertilizers on soil properties and maize yield. Soil Sci Soc Am J. 2024;88:2227–37. https://doi.org/10.1002/saj2.20760.
79. Manzeke-Kangara MG, Joy EJM, Lark RM, Redfern S, Eilander A, Broadley MR. Do agronomic approaches aligned to regenerative agriculture improve the micronutrient concentrations of edible portions of crops? A scoping review of evidence. Front Nutr. 2023;12(10):1078667. https://doi.org/10.3389/fnut.2023.1078667.
80. Venn BJ. Macronutrients and human health for the 21st century. Nutrients. 2020;12(8):2363. https://doi.org/10.3390/nu12082363.
81. Chrysargyris A, Höfte M, Tzortzakis N, Petropoulos SA, Di Gioia F. Editorial: micronutrients: the borderline between their beneficial role and toxicity in plants. Front Plant Sci. 2022;11(13):840624. https://doi.org/10.3389/fpls.2022.840624.
82. Di Gioia F, Petropoulos SA, Ozores-Hampton M, Morgan K, Rosskopf EN. Zinc and iron agronomic biofortification of Brassicaceae microgreens. Agronomy. 2019;9:110677. https://doi.org/10.3390/agronomy9110677.
83. Ofori KF, Antoniello S, English MM, Aryee ANA. Improving nutrition through biofortification-a systematic review. Front Nutr. 2022;9(9):1043655. https://doi.org/10.3389/fnut.2022.1043655.
84. Zulfiqar U, Khokhar A, Maqsood MF, Shahbaz M, Naz N, Sara M, Maqsood S, Sahar S, Hussain S, Ahmad M. Genetic biofortification: advancing crop nutrition to tackle hidden hunger. Funct Integr Genomics. 2024;24(2):34. https://doi.org/10.1007/s10142-024-01308-z.
85. Azeem A, Khan S, Zia Ul Haq M, et al. Biofortification strategies for enhancing crop nutritional value: a review of methods, challenges, and future directions. Discov Plants. 2025;2:185. https://doi.org/10.1007/s44372-025-00276-3.
86. Kefeli VI, Kalevitch MV. Natural growth inhibitors and phytohormones in the intact plants and isolated cells, organs and tissues. In: Borsari B, editor. Natural growth inhibitors and phytohormones in plants and environment. Dordrecht: Springer; 2003. https://doi.org/10.1007/978-94-017-0315-4_9.
87. Ren Z, Li Y, Fang W, Yan D, Huang B, Zhu J, Wang X, Wang X, Wang Q, Guo M, Cao A. Evaluation of allyl isothiocyanate as a soil fumigant against soil-borne diseases in commercial tomato (Lycopersicon esculentum mill.) production in China. Pest Manag Sci. 2018. https://doi.org/10.1002/ps.4911.
88. Sporsheim B, Øverby A, Bones AM. Allyl isothiocyanate inhibits actin-dependent intracellular transport in Arabidopsis thaliana. Int J Mol Sci. 2015;16(12):29134–47. https://doi.org/10.3390/ijms161226154.
89. Gross D, Parthier B. Novel natural substances acting in plant growth regulation. J Plant Growth Regul. 1994;13:93–114. https://doi.org/10.1007/BF00210953.
90. Hiltner L. Ueber neuere Erfahrungen und Probleme auf dem Gebiete der Bodenbakteriologie und unter besonderer BerUcksichtigung der Grundungung und Brache. Arb Deut Landw Gesell. 1904;98:59–78.

91. Hartmann A, Rothballer M, Schmid M. Lorenz Hiltner, a pioneer in rhizosphere microbial ecology and soil bacteriology research. Plant Soil. 2008;312:7–14. https://doi.org/10.1007/s11104-007-9514-z.
92. McNear DH Jr. The rhizosphere - roots, soil and everything in between. Nat Educ Knowl. 2013;4(3):1.
93. Müller H, Westendorf C, Leitner E, Chernin L, Riedel K, Schmidt S, Eberl L, Berg G. Quorum-sensing effects in the antagonistic rhizosphere bacterium Serratia plymuthica HRO-C48. FEMS Microbiol Ecol. 2009;67(3):468–78. https://doi.org/10.1111/j.1574-6941.2008.00635.x.
94. DeAngelis KM, Lindow SE, Firestone MK. Bacterial quorum sensing and nitrogen cycling in rhizosphere soil. FEMS Microbiol Ecol. 2008;66(2):197–207. https://doi.org/10.1111/j.1574-6941.2008.00550.x.
95. Venturi V, Keel C. Signaling in the rhizosphere. Trends Plant Sci. 2016;21(3):187–98. https://doi.org/10.1016/j.tplants.2016.01.005.
96. Hong S, Yuan X, Yang J, Yang Y, Jv H, Li R, Jia Z, Ruan Y. Selection of rhizosphere communities of diverse rotation crops reveals unique core microbiome associated with reduced banana fusarium wilt disease. New Phytol. 2023;238(5):2194–209. https://doi.org/10.1111/nph.18816.
97. Jin L, Lyu J, Jin N, Xie J, Wu Y, Zhang G, Feng Z, Tang Z, Liu Z, Luo S, Yu J. Effects of different vegetable rotations on the rhizosphere bacterial community and tomato growth in a continuous tomato cropping substrate. PLoS One. 2021;16(9):e0257432. https://doi.org/10.1371/journal.pone.0257432.
98. Chen S, Yao F, Mi G, Wang L, Wu H, Wang Y. Crop rotation increases root biomass and promotes the correlation of soil dissolved carbon with the microbial community in the rhizosphere. Front Bioeng Biotechnol. 2022;6(10):1081647. https://doi.org/10.3389/fbioe.2022.1081647.
99. Wang X, Tang Q, Yang K. Improving crop rotation classification using a random forest model incorporating spatial heterogeneity. Geocarto Int. 2024;39(1) https://doi.org/10.1080/10106049.2024.2384473.
100. Song Y, et al. Potato crop performance is predicted by tuber microbiome. Nat Microbiol. 2025;10:8–9. https://doi.org/10.1038/s41564-024-01893-6.
101. Rodriguez D, de Voil P, Hudson D, Brown JN, Hayman P, Marrou H, Meinke H. Predicting optimum crop designs using crop models and seasonal climate forecasts. Sci Rep. 2018;8(1):2231. https://doi.org/10.1038/s41598-018-20628-2.
102. Hirt H. Healthy soils for healthy plants for healthy humans: how beneficial microbes in the soil, food and gut are interconnected and how agriculture can contribute to human health. EMBO Rep. 2020;21(8):e51069. https://doi.org/10.15252/embr.202051069.
103. Jones J, Dangl J. The plant immune system. Nature. 2006;444:323–9. https://doi.org/10.1038/nature05286.
104. Nicholson LB. The immune system. Essays Biochem. 2016;60(3):275–301. https://doi.org/10.1042/EBC20160017.
105. Lambring CB, Siraj S, Patel K, Sankpal UT, Mathew S, Basha R. Impact of the microbiome on the immune system. Crit Rev Immunol. 2019;39(5):313–28. https://doi.org/10.1615/CritRevImmunol.2019033233.
106. Teixeira PJP, Colaianni NR, Fitzpatrick CR, Dangl JL. Beyond pathogens: microbiota interactions with the plant immune system. Curr Opin Microbiol. 2019;49:7–17. https://doi.org/10.1016/j.mib.2019.08.003.
107. Vannier N, Agler M, Hacquard S. Microbiota-mediated disease resistance in plants. PLoS Pathog. 2019;15(6):e1007740. https://doi.org/10.1371/journal.ppat.1007740.
108. Bhardwaj AK, Vinothkumar K, Rajpara N. Bacterial quorum sensing inhibitors: attractive alternatives for control of infectious pathogens showing multiple drug resistance. Recent Pat Antiinfect Drug Discov. 2013;8(1):68–83. https://doi.org/10.2174/1574891x11308010012.

109. Majdura J, Jankiewicz U, Gałązka A, Orzechowski S. The role of quorum sensing molecules in bacterial-plant interactions. Meta. 2023;13(1):114. https://doi.org/10.3390/metabo13010114.
110. Koul S, Prakash J, Mishra A, Kalia VC. Potential emergence of multi-quorum sensing inhibitor resistant (MQSIR) bacteria. Indian J Microbiol. 2016;56(1):1–18. https://doi.org/10.1007/s12088-015-0558-0.
111. Griffith GS, Cresswell A, Jones S, Allen DK. The nitrogen handling characteristics of white clover (Trifolium repens L.) cultivars and a perennial ryegrass (Lolium perenne L.) cultivar. J Exp Bot. 2000;51(352):1879–92. https://doi.org/10.1093/jexbot/51.352.1879.
112. Guo J, Chen P. Ammonia history in the making. Nat Catal. 2021;4:734–5. https://doi.org/10.1038/s41929-021-00676-0.

Chapter 3
The Essence of Chemistry. Chemistry, Microorganisms and Plants

Abstract What is the nature of matter and the chemical bond. The chemistry of living systems at a macro and micro level. Analysis to give empirical formulae and the existence of isomeric forms of compounds. Molecules as distinct entities with particular properties and not simply blends of the properties of its constituents. Optimal reaction rates in chemical (reductionist) and biological (complex) systems. The failure of Perkin to synthesise quinine. The development of colours from natural sources and by synthesis. The use of gene transfer and fermentation to make colours. Plant breeding and the relationship between genotypic and phenotypic variation in peas. The corpuscular nature of genetics and the absence of inherited. Average properties. Genetic modification and gene editing using CRISPR. The manipulation of genes to produce different polyketides. Soil microorganisms and diseases of plants and the link back to crop rotation. The use of natural and synthetic fertilisers.

Keywords Atoms · Phlogiston · Vitalism · Isomers · Alkaloids · Phenotype · Genotype · CRISPR

In order to further develop the theme of agriculture and factors that affect it, the importance of molecular interactions (which were touched on in Chap. 2) should be further developed. While the development of practices like crop rotation and improved agronomy are measured at a macro level (ultimately improvement of crop yields), some attempts have been made to understand the most significant molecular interactions and consequences in order to develop predictive tools [1]. Chemistry is concerned with the consequences of the interactions of molecules. While these interactions may have consequences at a systems level, the path from micro (molecular) events to macro effects is often unable to be precisely defined. The molecular interactions are models that are derived from observations, predictions, hypotheses and empirical testing. The theories of chemical bonds are considered to be at the centre of understanding of the molecular interactions that underpin chemistry [2] and, while sophisticated models have been developed, they are models that explain observed behaviour rather than being an actual truth. There is a considerable

B. Hanley, *The Chemistry of a County*,
https://doi.org/10.1007/978-3-032-21491-1_3

history—and progression of ideas—associated with chemical bond theory [3] and further modifications are likely. In essence, a chemical bond is the realisation of the combination of atoms and molecules to give entities that are sufficiently stable, under the specific conditions that apply, to be considered to be a separate molecule. The understanding of chemical bonds also requires that the entities being bonded together (atoms or molecules) should have particular characteristics. While the ancient Greeks developed the conceptual idea of atoms as being the basic, indivisible components which made up substances, they also (perhaps more importantly) also identified the concept of space between atoms. This may not seem immediately to be of significance, it is, in fact, crucial to the concept of bond formation. If matter is continuous then new molecules can only be formed by the substitution (replacement) of one atomic entity (leaving a space) with another. The existence of atoms implies that voids (spaces between atoms and molecules) must also exist [4] and this is perhaps the most significant aspect of atomic theory but it remains controversial [5] and was opposed by Plato, Aristotle and religious bodies at the time of the Renaissance and the Enlightenment. The lack of direct observation of atoms posed problems for scientists who held to the empirical nature of science. Furthermore, the development of elements of thought concerning Dark Matter and the theoretical and empirical basis for its existence [6].

While the development of chemical bond theory progressed through the realisation that some atoms and molecules had greater affinities for each other than others (hence introducing the concept of bond strength), the nature and importance of the chemical bond (and hence of chemical reactions) in living systems began to be realised [7]. One of the most significant advances was the realisation (and subsequent proof) that molecules derived from living systems did not contain some unique essence (vitalism). The synthesis of urea (which had been isolated from urine and characterised by Boerhaave et al. [8]) from inorganic starting materials by Wohler was a significant step in the demise of vitalism [9]. The disproving of the theory of vitalism is comparable in impact to the discrediting of the idea of phlogiston by Lavoisier and others. Both phlogiston and an essence in living systems were shown to have no physicochemical (or philosophical) basis. The idea that living systems were composed of sets of chemical reactions became more popular and this fitted with theories regarding the mechanical nature of living systems. Such paradigms underpin current concepts of biomechanics [10] and synthetic biology [11] however the complexity of an individual living entity (and the molecules it produces) remains a considerable challenge to both chemists and biologists.

The development of better and more precise methods of measuring the elemental components of molecules, together with ways of breaking molecules apart into recognisable components, enabled chemists to understand more about the composition of molecules, however this advance was of only limited use when faced with the complexity of living systems and the scale at which that complexity is manifest. The reductionist approach of measuring the proportions of the components of molecules was appropriate when considering the relatively small number of inorganic molecules and the limited numbers of different inorganic structures with the same elemental composition.

In the case of living systems, they were composed largely of organic molecules that consisted of atoms of carbon, hydrogen and oxygen and small amounts of other elements (nitrogen, phosphorous, sulphur etc.). The core elements were combined in different ways to give a plethora of molecules with the same (or similar) relative proportions of elements but very different properties (Fig. 3.1).

The relative proportions of elements in a molecule gives the *empirical formula* and it is reflective of components rather than properties. This is an important philosophical and physicochemical distinction. The atoms in a molecule do not reflect the collective properties of their origin. The particular attributes of a molecule are not a blend of the properties of its constituent elements but those of a transformed molecule with different properties to its constituents. This is particularly so in the case of organic molecules. For such systems, the empirical formula contains little, if any, indication of the functional properties of the molecule. In order to obtain such functional information, the reductionist level of analysis has to operate at a higher level of complexity. The molecule has to be examined in the context of its actively functional groups rather than its elemental composition. One of the first functional groups to be targeted by chemists was the carboxylic acid moiety since it enabled selective dissolution and precipitation of compounds containing that characteristic group. In the eighteenth century, Scheele isolated citric and tartaric acids [12] and the manipulation of organic molecules from natural living sources became possible. Another significant step, which relied upon the functional properties of a specific component of an organic compound, was the isolation and purification of alkaloids ('alkali-like' substances) and an appreciation of their chemical and biological properties [13]. The isolation of the alkaloid, morphine, by Serturner in 1804 was a significant step and enabled an effective (but non-toxic) dose to be determined by

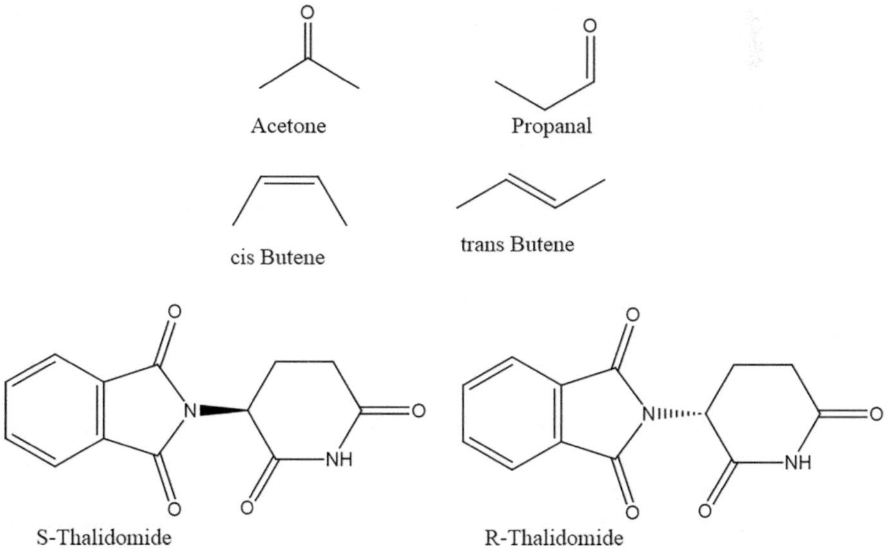

Fig. 3.1 Empirical formulae and isomers

weight. Not all bioactive alkaloids were beneficial (like morphine and, later, quinine) and some gave rise to poisoning. For example, the effects of ergot alkaloids were known in medieval times and ergotism (also known as St Anthony's Fire) was recognised [14]. Biological activity (sometimes by self-testing) was a useful discriminatory characteristic in the isolation of molecules from complex mixtures of natural products. The extension of this methodology to other properties (colour, smell, taste) was also developed [15].

The development of a better and more significant understanding of the nature of living systems can be obtained from the application of reductionism to consider organisms as a series of chemical reactions [7]. To study the nature of chemical reactions *in vivo*, a number of factors have to be considered. The first of these is specificity. Reactions in cells take place in the midst of a range of potential reactants, substrates, inhibitors and products. The use of enzymes can add a significant degree of selectivity to a complex reaction mix however this may not enable complete chemoselectivity and this has been recognised [16]. Another important consideration is reaction rate. In general, chemists favour reactions that proceed quickly from substrate(s) to product in the highest yield with the minimum of side reactions and undesired products. In the case of living systems with complex and interlinked pathways, a lower (optimal) rate of reaction may be desirable to prevent accumulation of some reaction products, for example and therefore a higher degree of control of reaction rates may be necessary. For example, in the case of the tricarboxylic acid (TCA) cycle, one of the important products from the cycle (NADH) acts as an inhibitor of many of the enzymes in the cycle and therefore provides feedback inhibition for the reaction system as a whole. In addition, some of the TCA cycle metabolites have a wider physiological function and may control other cellular functions [17]. Simply adding the cumulative impacts of optimised chemical reactions is, therefore, insufficient to fully describe a living system. The reverse of the paradigm of using a chemical reaction model to describe a living system is to use some aspects of a living system and to apply these to chemical systems. This type of biological copying has given rise to biomimetic systems in the chemistry laboratory. Aspects of this are used in both microorganism and plant mimicry—particularly in the synthesis of key industrial materials. In such systems, chemists seek to either copy reactions by using biological logic [18] or to use isolated (or manipulated) parts of a biological system for the production of specific chemical entities [19]. Such systems have been used to prepare biomimetic pigments [20] in contrast to the purely synthetic dyestuffs developed in the nineteenth century [21].

The concept of chemical synthesis is part of the analysis/synthesis cycle which developed as chemists studied natural products. The linking of analysis and synthesis was part of the alchemical tradition of chemistry [22]. The principle that matter was composed of three elements (the *tria prima* of salt, sulphur and mercury) gave the opportunity for a substance to be analysed, impurities removed and the three elements to be recombined into an improved version of the original substance. This concept was in essence the same as used by Lavoisier in his separation of water into its constituent parts (hydrogen and oxygen) and the subsequent recombining of these to give water again. This paradigm has a long tradition in chemistry [23] and the idea of breaking and recombining was carried forward into the nineteenth century. Chemists such as Perkin used empirical formulae (obtained by combustion analysis) and then sought to bring together arithmetically correct units from the combining of empirically known compounds to give the complex natural products whose empirical formula they discerned but whode full structure they did not know. In the case of Perkin, he synthesised an aniline dye (mauve) instead of the antimalarial drug quinine [24] from arithmetically correct starting materials (Fig. 3.2).

As the concept of functional groups became more established, methods were developed for the more detailed analysis at the functional group level in a given natural molecule [25]. As chemists began to understand the structures of molecules through analysis of fragments and reassembly (retrosynthesis), the assembly (synthesis) using biomimetic methods began to be investigated. The first generally accepted biomimetic synthesis that used retrosynthetic logic is considered to be that

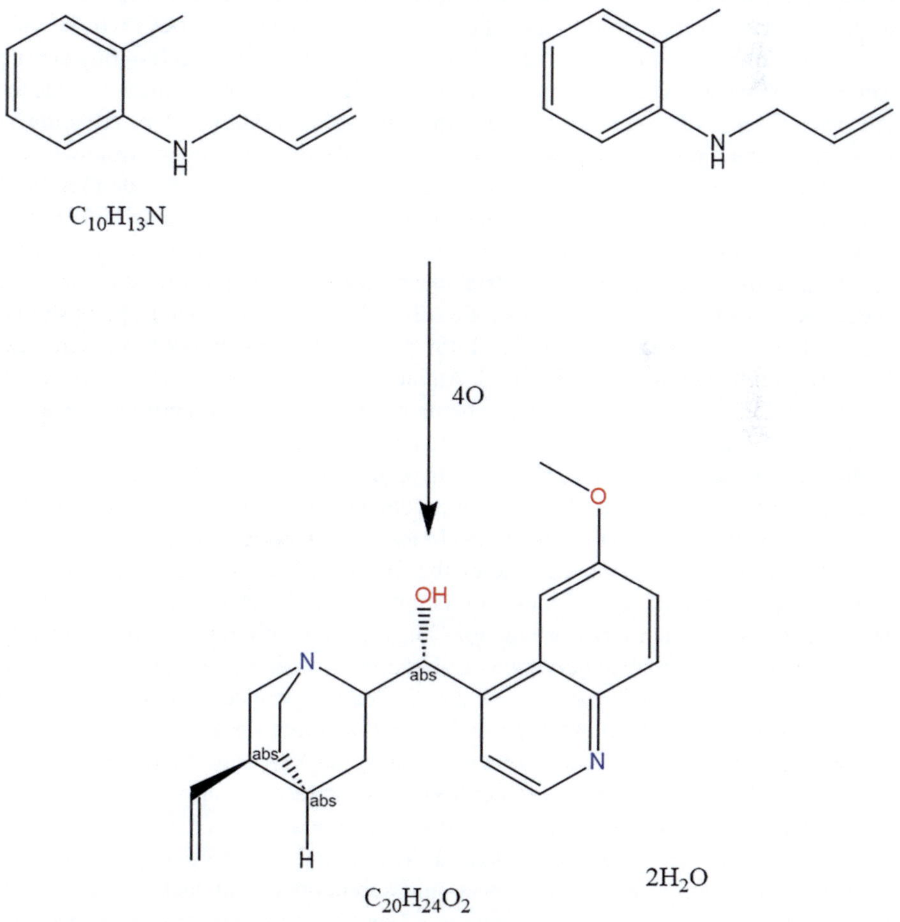

Fig. 3.2 The attempted synthesis of quinine by Perkin

of tropinone by Robinson in 1917 [26]. It is a one pot, multicomponent synthesis and as such was hugely influential [27] both in terms of the synthesis itself and how it impacted upon future synthetic strategies. The detailed enzymology of the biosynthesis of tropinone alkaloids has been investigated [28]. By understanding the details of the steps in the biosynthesis of valuable and interesting molecules such as tropinone through techniques such as transcriptome analysis, it is possible to envisage the planned modification of individual steps in the enzyme-catalysed sequence of reactions to give specific novel products [29].

While the use of plants for the isolation of valuable natural products has a significant historical pedigree, microbes are generally more favoured as sources of valuable products. They are generally easier to grow at scale, use fewer resources than plants and are able to produce the required product in a much shorter period of time. Fermentation remains the key technique in the use of microorganisms to derive useful products. The history of fermentation has been described alongside more modern modification processes such as omics technologies [30]. The use of fermentation to produce beer from a readily available resource in Norfolk (barley) will be outlined later however the key principles that underpin fermentative technology are the same for the production of foods and pharmaceuticals. Two of the most significant advances in fermentation were the use of industrial scale vessels and the adoption of these to produce pharmaceutical products (particularly antibiotics). Although fermentation has been practiced over many years, in general industrial scale fermentation is less cost effective (except for the production of high value medicinal entities and some isolated examples such as alcoholic beverages where there is not a viable non-fermentative option) than industrial chemistry or agricultural production [31]. Indeed some products such as quinine are still isolated primarily from a plant source [32] and the use of plants as sources of novel individual phytopharmaceuticals and bioactive mixtures remains viable [33]. An increasing emphasis on green processes has enabled industrial fermentation to be increasingly more competitive. There are a number of factors that can impact negatively upon industrial scale fermentations including contamination and genetic drift or instability [34]. The importance of scale in commercial industrial fermentation processes has been noted [35] and this remains a challenge for many processes. In terms of historical context, the production of acetone and butanol was one of the first industrial fermentation processes [36]. These are important feedstocks for a number of industrial processes including, in particular, synthetic rubber production. The organism of choice for the fermentation, *Clostridium acetobutylicum* was isolated by Weizmann and the process produced both acetone and butanol from starch. The importance of this fermentation declined with the increased use of petrochemicals as a feedstock in the middle of the twentieth century however it was reinvigorated by the drive towards greater sustainability [36]. Another important product from fermentation was citric acid which was used in the food industry in carbonated drinks and other uses [37]. The main producing organism was *Aspergillus niger* [38] although other microbes were also capable of production and both surface and submerged fermentations have been employed. Citric acid production provides a link to the production of pharmaceuticals by fermentation. Scheele was the first person to isolate citric acid from lemon juice in 1784 [39] however plant-derived citric acid was a challenging process and dependent upon the supply of fruit. The discovery of a microbial producer of citric acid (*A niger*) in 1917 [40] enabled Pfizer to become the biggest global producer of citric acid. Pfizer also developed deep tank fermentation for the production of citric acid and overcame the problem of growing an aerobic fungus in a deep tank fermenter. The producing species of penicillin was also aerobic and the process of production initially made use of surface cultures however these produced very small amounts and contamination was a major problem. Eventually Pfizer were able to mass produce penicillin using their deep tank technology and this was one of the first examples of the use of microorganisms in fermenters to produce a medicinally important drug and of the use of consortia of industry, academics and government to overcome an important health issue [41]. A range of drug molecules are now routinely produced by fermentation. In addition, the manipulation of microbes to produce medicinal agents is both accepted and convenient [42]. These include both conventional drugs [43] and biologics [44].

The development of plant-based systems for the synthesis and isolation of useful products presents a number of challenges. Traditional plant breeding is time consuming and a range of environmental factors can influence the yield of useful compounds (see Sect. 3.1 for a more detailed and specific example of traditional plant breeding). In particular, factors such as different types of stress can have a profound influence on the types and levels of secondary metabolites that are produced [45]. In addition, plants that produce larger amounts of useful bioactive substances are most frequently no different in morphology and external appearance to those which produce the compounds of interest at lower concentrations. Indeed, the successful production of bioactive compounds may be reflected in the effect these may have on surrounding organisms such as microbes [46] and hence on disease resistance or ecological effects rather than on the host plants themselves. Having to compete with other species for nutrients and space is a form of stress and the production of allelochemicals (as noted previously) is an important factor in the control of their immediate environment and the furnishing of a growth advantage for some plants [47]. One approach is the use of plant cells, tissues and organs in culture [48]. In principle this approach allows for the more controlled and rapid manipulation of plants in culture to optimise the yield of specific useful compounds. Manipulation of plants in culture followed by regeneration of the ‘improved’ variety as a plant lies at the core of genetic modification of commercially useful plants and there has been an increasing focus on the use of genetic manipulation to optimise both agronomic (e.g. yield and stress tolerance) and nutritional properties of crop plants [49]. This will be considered in more detail in Sect. 3.3. Useful plant secondary metabolites (unlike primary metabolites or nutrients) are often produced in small amounts and only in specific parts of the growing plant—for example, quinine production in the bark of the Cinchona tree [50] or taxol accumulation in the plant *Taxus media* [51]. This means that a large amount of ‘waste’ material must be produced (and subsequently disposed of after extraction of the bioactive secondary metabolite) in order to obtain commercially viable amounts of the target compounds. It is not only bioactive compound production from plants that generates waste products. Indigo is a major industrial product and valorisation of the waste products has been a significant challenge [52] which has stimulated the development of bacteria-based culture methods to produce the dye more efficiently and sustainably [53]. Many environment groups have an almost automatic and permanent opposition to genetic modification [54]—particularly if it involves the insertion of 'foreign' genes into a host (heterologous genetic modification). This then raises the question—is environmental protection (even if it involves the use of genetically modified organisms) the most important and overriding principle and, in the specific case of denim dyed with indigo, is sustainable production even possible? [55]. In a similar way the production of Tyrian Purple from sea snails (*Muricidae* and *Thaisidae* families) in which only a small part of the creature was used and the remainder discarded is highly inefficient and unsustainable [56]. This very inefficient means of production has been superceded by the production of synthetic dyes with a similar colour profile to Tyrian Purple (for example, the mauve aniline dye first prepared by Perkin) [57], by synthesis of the compound responsible for the colour of Tyrian Purple (6,6′-dibromoindole) (Fig. 3.3) and related compounds [58].

or by the production by heterologous gene expression of the biosynthetic pathway in a bacterium [59]. Sustainability is improved by the application of scientific ingenuity rather than relying on a natural source.

As noted above, there are significant challenges to the use of plants grown as vegetative material as a source of useful bioactive principles and this has resulted in the development of processes to make plants as responsive and useful as microbes for the optimised production of bioactive principles [60]. The use of 'plant molecular factories' to produce valuable bioactive principles has been reviewed [61] and a number of challenges were identified including, perhaps most significantly, the difficulty in controlling the level of expression of transgene expression in plant tissue, the correct folding of transgenic proteins and post-biosynthetic modification patterns. The use of culture techniques has enabled the production of molecules of interest at large scale and in a more efficient manner that relying upon the extraction of whole plants and the purification of specific components (which are often only present as minor components in a mixture of compounds). It has also facilitated the

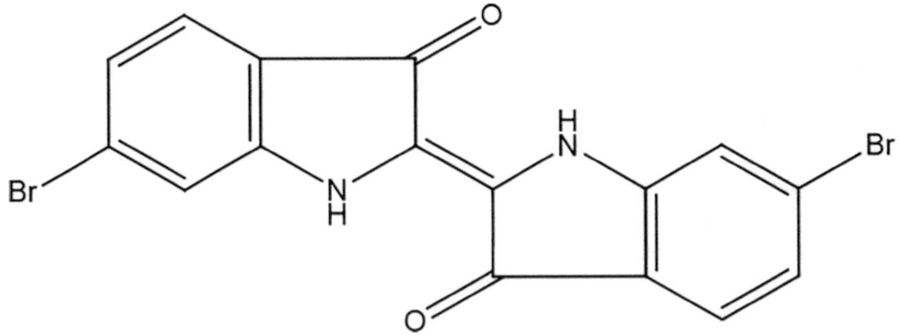

Fig. 3.3 The structures of 6,6′-dibromoindigo—responsible for the colour, Tyrian purple

rapid manipulation of plants in culture that can then be progressed to regenerate modified plants with defined characteristics. A range of techniques are available for the production of useful molecules from plant cultures and these have been the subject of a number of reviews [62]. In addition to the use of growing organisms in cell culture to produce valuable small molecules for a range of uses, biological species in cell culture can also be engineered to produce biological macromolecules such as proteins [63]. Culture based preservation of plants (and the regeneration of whole plants from culture samples) is also an important tool for plant conservation—in particular as a means of maintaining plant biodiversity [64].

The basis of plant breeding and the optimisation of crops (for whatever use) relies upon an understanding of the genetic architecture that underpins biosynthetic pathways, the control of production of specific metabolites and the impact of different genes and their manipulation on the physical appearance of a species and on the phylogenetic relationships between both closely related and distant members of biological families. A key aspect of investigating the chemical relationships within and between systems relies upon translating molecular interactions (which are unable to be observed directly and can only be examined on the basis of evidenced hypothesised postulates) into measurable endpoints. In the case of the link between different genes and their expression and the physical characteristics of plants, much useful work was carried out by hypothesis generation based on observation of traits even before genes were recognised as discoverable and manipulatable entities. The study of sexual combinations of different varieties of plants and their resulting phylogenetically stable biological progeny was the basis of the observations and recordings carried out by Mendel in the nineteenth century. This also enabled the development of 'black box' plant breeding which was used in Norfolk to improve the agronomic and processing characteristics of peas. These advances and the subsequent elucidation of the molecular events that underpins them is the subject of Sect. 3.1. Developing an understanding of what happens by careful observation is an essential aspect of crop manipulation (even if knowledge of why certain outcomes prevail is less clear). The idea that varieties of plants could be manipulated and the adoption of techniques that enabled this was, in some ways, an extension of the thinking that underpinned crop rotation—it was an extension from affecting the macro properties of crops to the manipulation of individual plants.

3.1 Legumes: Breeding and Crop Improvements

The beginning of inheritance as an area of predictive scientific study began, it can be argued, with the investigations of Gregor Mendel. While the studies that Mendel carried out were highly observational, it would be wrong to simply see them as those of an isolated individual observing the complexity of God's world. Mendel was an experimentalist and carefully designed a succession of studies that gradually provided the basis for the further understanding of how traits were passed on, in an empirical fashion, even if a mechanistic understanding remained a remote aspiration. Mendel chose the garden pea (*Pisum sativum*) as his model organism (he grew 28,000 individual plants over several years as he made his observations and drew conclusions about the nature of heredity). The pea had a number of advantages. Perhaps the most significant was that it had phenotypic variations that could easily be determined by observation and it grew relatively quickly so the consequences of crosses between particular plants could be quickly assessed. In addition, peas have remained an agriculturally important crop since they could be stored for long periods and were high in useable energy. A range of cultivars and varieties were available in the nineteenth century [65] and the use of legumes as a rotation crop continues [66]. The molecular (genetic) basis for some of the characteristics that Mendel studied have continued to be of interest to molecular biologists, geneticists and plant physiologists. The work of Mendel can be seen in the context of the corpuscular theory of matter as later outlined by Thompson [67] and earlier Boyle's atomic theory [68]. These theories suggested that matter existed in discrete entities (atoms) rather than being continuous. Molecules did not have the properties of a blend of their constituent elements but were distinctly different entities. In a similar fashion, before the work that Mendel carried out, the prevalent view was that the properties of progeny were a blend of those of their progenitors. Mendel's work on the characteristics of peas suggested that the blend hypothesis was unable to explain how characteristics were passed through generations and the concept of discrete packages of genetic material which was conceptually similar to atomic and molecular theory, was developed. Prior to the understanding precipitated by Mendel's studies about the corpuscular nature of genetic material and its effect on the acquisition of desired characteristics, animal breeding had been carried out for many years. Initially animal breeding was based around the identification of attributes that were associated with particular desirable traits (hardiness, higher meat or milk yield) and maintaining these traits through retaining and breeding animals that were exemplars for these characteristics and discarding those whose characteristics did not 'breed true'. One of the key figures in pre-Mendelian animal breeding was Robert Bakewell [69]. Bakewell was a celebrated animal breeder and his vigorous campaigns of inbreeding produced livestock with specific desirable characteristics [70]. The objects of Bakewell's breeding programmes was the production of early maturing animals (particularly sheep with high quality wool) and by working with other farmers who shared his interest, in producing more valuable animals [71]. His focus on specific characteristics and the recording of his breeding programmes was, in many ways, similar to the approach that Mendel took in designing and noting the results of experiments although Bakewell was primarily concerned with improving characteristics that would bring economic benefits while Mendel was more interested in understanding the process by which characteristics were passed to successive generations. It is important to recognise that there were three aspects to the breeding work carried out by Bakewell—selection, inbreeding and culling. All three aspects were essential and, in many cases, Bakewell started by selecting animals that had already been partially improved by others so there was a pre-selection of 'breedable' traits. The impact of selective breeding was seen among the more advanced agriculturalists including Coke at Holkham in Norfolk. Holkham became famous for its sheep shearings as well as for the quality of its livestock. The sheep shearings ran from 1790 until 1821and provided an opportunity for new agricultural ideas and implements to be showcased as well as for improvements in wool yield and quality to be advertised. The popularity of these shows spread among the farming community (nationally and internationally) and they are the precursor to the Royal Show which was held from 1839 until 2009 [72]. They also ensured that Coke and his Holkham estate had access to the latest thinking on agriculture and farming. It is against this background of agricultural innovation that the practical breeding trials of Bakewell and the scientific underpinning provided by Mendel (and Darwin) were delivered. Society, particularly in places with a pedigree of the adoption of agricultural scientific innovation like Norfolk, was ready for and receptive to new ideas. When improvements to animal husbandry and other agricultural practices were combined with the agronomic advances that came about with the introduction of crop rotation and new fodder crops and the over-wintering of livestock the resultant increases in agricultural production formed the underpinning for the growth in populations and non-agricultural industries.

Mendel focused upon specific characteristics of peas and reported his findings [73]. This resulted in a significant conceptual step from the 'trait-intensive' and inbreeding approach of animal breed development to a more granular approach. By concentrating on the accurate measurement of a set of defined characteristics and by ensuring that he controlled the production of hybrids from his pea plants with defined and stable characteristics, Mendel was able to derive rules regarding the inheritance of characteristics even if the mechanism (genes) that underpinned these rules was not known. The first principle was that, if a plant had two options for characteristics, one of these would be dominant and would be reflected in its progeny while the other would be suppressed (recessive). The second principle has been referred to as the 'Law of Segregation'. If two plants each had two options for a specific characteristic then they could be combined in such a way that there would be a defined mixture of dominant and recessive characteristic progeny and the proportions of these could be calculated in advance. It was obviously crucial that the selected characteristics were linked to specific genes and were able to be seen in the complete plants and therefore characteristics such as height, flower colour and shape of peas (wrinkled or smooth) were important measurable features. It is also clear that characteristics are binary and the progeny either show one characteristic or another—not an amalgam or a combination of them (see above). The third principle is the Law of Independent Assortment. This principle stated that characteristics were independent and not linked. The characteristic of flower colour, for example, was independent of the characteristic of pea shape or height of the plant. While these observations can be expressed in terms of genes, a purely trait-defined relationship can also be created which will predict the properties of progeny arising from specific crosses. This is called a Punnett Square (Fig. 3.4). The essence of a Punnett Square is that, for a single trait, a 4x4 square is constructed and the dominant trait (u/c letter) from one parent and the recessive trait (l/c letter) from the same parent are placed along the top of the grid, with one allele above each column. The

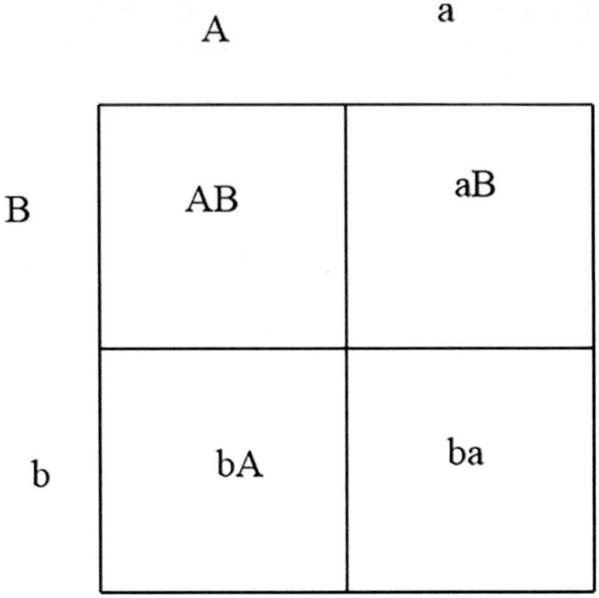

Fig. 3.4 The Punnett square

alleles from the second parent are placed along the left side, with one allele next to each row.

Each box represents a 25% probability for that specific genotype in the offspring. While the simple laws of Mendelian genetics have a significant historical importance and are an empirical precursor to the development of the idea of genes and their consequences, they are not invariably followed [74] and a number of examples of (particularly) phenotypic variance have been observed however some of these may remain within the boundaries of Mendelian thought [75] and there remains a degree of confusion. The fundamental principle of inheritance (that progeny is not a 'blend' of the characteristics of their precursors) is also a significant foundational principle of chemistry. Compounds cannot (usually) be described as having a mixture of the properties of their constituent elements. Sodium chloride is not characterised by being a blend of the properties of sodium and chlorine. In the organic sphere, ethyl acetate does not have properties that would enable an observer to determine that its component parts are ethanol and acetic acid. Where the similarity diverges is creates novelty and compounds with unanticipated properties. Combinations of genes for a specific trait, on the other hand, are dominant or recessive and do not create something that is a completely new entity with totally different properties (or functions) compared to its progenitors. The chemical combination of component parts (elements or molecules) can create something that is completely different from its precursors. Biology follows a principle of conservation of function while chemistry has access to functional creativity and novelty.

Despite the intellectual framework created by Mendel and others and the eventual identification of the mechanical framework necessary for the observed functional characteristics of hybridisation (chromosomes and genes), the development of new varieties of, for example, peas, remained an intensive process of trial and error. Complete gene: function relationships of even the seven basic characteristics identified by Mendel have not been completely delineated [76]. By using a preferred characteristic as a target, partial knowledge of the genetic underpinning of a trait can be sufficient for advances to be made in the application of specific breeding programmes. This is illustrated by work carried out on the development of pea varieties with improved agronomic characteristics in Norfolk culminating in the development of the 'leafless' and 'semi-leafless' pea.

The development of the 'leafless' pea was a significant breakthrough in applied breeding and much of the work (that built upon the agronomic improvement tradition carried out at Holkham) was carried out in Norfolk. Pea (*Pisum sativum*) has been found to be both a useful model organism (see above) and a plant of significant commercial importance [77]. In addition to studies of the genetic basis of the physiology of the traits of pea, structured hybridisation and breeding programmes have facilitated the development of better commercial varieties. One of the primary drivers for the development of new pea varieties in the 1960s was the industry requirement for a variety that was able to be harvested mechanically and contained less stalk and leaf material. One of the major advances was the development of the leafless (or semi-leafless) pea (see below) whose leaves were replaced by tendrils [78]. The leafless pea is the result of a recessive mutation in a particular gene (afila-Af) and this results in a plant with a dramatically reduced leaf morphology. Although spontaneous mutations had been reported as early as 1953 [79] (coincidentally the same year that Watson and Crick proposed the structure of DNA [80]), the exploitation of this characteristic remained undeveloped until there was a commercial impetus. Indeed, the initial observation of a modified pea variety with attenuated leaf morphology predated the discovery by Kujala by 30 years [81] and was reported in the context of reduced leaf stipules (St gene). The fully leafless pea was a double mutant (af,st) and was the first to be introduced commercially and following studies on crossing and selecting hybrids on the basis of phenotype [82] led to the development of the 'leafless' pea with improved agronomic and harvesting characteristics. A range of agronomic characteristics of 'leafless' peas compared to the non-mutant varieties have been assessed [83]. These include nitrogen balance. The facility of peas and other legumes to facilitate the fixing of atmospheric nitrogen is an important aspect of the agronomic attractiveness of these crops [84] including, as noted previously, as participants in crop rotations. This will be discussed further in Sect. 3.2. While the double mutant, 'leafless' variety became popular, it was found that its light conversion capacity was low resulting in poor yields. The development of a 'semi-leafless' variety—which also has tendrils instead of leaflets but retains the larger stipules—enabled growers to choose the most suitable of the variant lines for crop use. The semi-leafless pea has now become the most widely grown dry pea variety in the UK. In global terms, dry peas are an important crop and source of protein and their additional nutritional qualities (including through biofortification) have been recognised [85]. They have a dual function as a food and as an important aspect of agronomic health—particularly with respect to crop rotation and the growth of other crops. Some of these aspects, which are linked to the chemistry of plant and microbial metabolism, will be considered in Sect. 3.2.

3.2 Legumes: Nitrogen Fixing and the Chemistry of Crop Health

Plants, in common with other living organisms, contain a number of macro and micronutrients and deficiencies in these lead to observable symptoms that can be diagnostic and, importantly, reversed by addition of the deficient nutrient [86]. One of the most significant of the essential macronutrients for plant growth is nitrogen [87] and fertilisers containing nitrogen are heavily used in agriculture. Nitrogen status has a particular effect on the photosynthetic mechanisms that operate in plants [88] and is, therefore, an essential component of plant growth and development. Both organic and inorganic nitrogen can be taken up by plants [89] and the form of nitrogen that is preferentially absorbed depends on a number of factors including plant species and environmental conditions and, most importantly, the nitrogen availability. While much of the nitrogen status of the agricultural ecosystem is a consequence of the addition of nitrogen in fertilisers derived from the Haber-Bosch process [90], it is an energy intensive procedure and, in addition to biological solutions (see below), lower energy alternatives have been sought [91]. One alternative is the production of bio-ammonia [92], however in a biological system, ammonia is often a by-product of decomposition of organic matter and such a high entropy solution is an inefficient use of complex, nitrogenous metabolites. One alternative is the directed use of an enzymic (nitrogenase-based) system which reduce nitrogen (N_2) to ammonia (NH_3) thereby generating a useable nitrogen compound from a conveniently available gas and, while this has been attempted on a commercial scale [93], the reaction rate for nitrogenase is low and therefore production of significant amounts of ammonia using this route is problematic [94]. Nonetheless *Rhizobia* bacteria will colonise the roots of legumes inducing the formation of nodules [95] and they then use a nitrogenase-catalysed reaction to fix nitrogen [96]. The symbiosis between legumes and rhizobia is the most significant natural way of improving the bioavailable nitrogen status in the agricultural environment [97]. The overall reaction for nitrogen fixation, catalysed by nitrogenase, is illustrated in Fig. 3.5.

Studies have demonstrated that biological nitrogen fixation, through the use of the symbiotic properties of rhizobia, can increase the available nitrogen content of a depleted ecosystem [98] therefore the utilisation of legumes as part of a crop rotation system is clearly advantageous. Nonetheless, while many of the details of the formation (and regulation) of useable nitrogen in plants is known, the effect of biological nitrogen fixation on nitrogen and carbon balance remains an area of

$$N_2 + 16ATP + 16H_2O + 8e^- + 8H^+ \rightarrow 2NH_3 + H_2 + 16ADP + 16Pi$$

Fig. 3.5 Nitrogen fixation

investigation [99]. While there remains a reductionist tendency to prioritise some macro and micronutrients over others, the interlinked nature of biological cycles means that concentration on a single nutrient may cause imbalance both in the target plant and also in the environment as a whole. For example, phosphorous (alongside nitrogen) is a major macronutrient affecting optimal plant growth. The reliance of farming on the addition of (in particular) inorganic phosphate and its runoff from agricultural land has created significant environmental challenges [100]. The interlinked nature of the nitrogen and carbon cycles has fuelled studies into the use of techniques such as directed evolution, artificial intelligence (AI)-guided enzyme design and metabolic engineering to enhance the effectiveness of assimilation of both carbon and nitrogen and to improve crop health [101]. While the interlinked nature of nutrient cycles is an important factor, it remains possible to examine, to an extent, the component cycles in isolation and to determine some key factors that may influence their efficiency of operation [102]. Many of these studies are necessarily observational in nature and the impact of individual measures on overall plant health and crop yields may be challenging to disentangle. The collision of empiricism and complexity is an almost inevitable consequence of the application of chemistry and its manipulation to facilitate an impact on living systems.

3.3 Genetic Modification and Gene Editing

The progression of agricultural improvements can be seen as having a series of progressing foci which build upon successive levels of knowledge and understanding. In that sense it is similar (as will be explained later) to chemistry. There is no need for a complete shift in the tabula rasa as knowledge is acquired and applied. Crop rotation was highly dependent upon the correct recording of which sequence of crops gave the greatest benefit. This led to the 4-component rotation method that was developed in Europe and exported (and refined) by estates such as Holkham. Each set of rotations advised future experiments and the practices were gradually refined and optimised. Without the freedom to travel and the free exchange of ideas (that later facilitated the Enlightenment), the optimisation and the subsequent agricultural revolution that enabled the Industrial Revolution in England may have never occurred. The work that was carried out on the optimisation of specific crops was then able to be applied to multiple agricultural systems since differences between soil types and the impact of different agronomic practices was better understood and controlled and was no longer an uncontrollable confounding factor This work culminated in the development of methods for production of the hybrid 'semi-leafless' pea. The development of this variety was driven by the principles of crop manipulation for post harvest benefit and eventually advised some aspects of the breeding of animals for particular characteristics and welfare rather than solely for production traits [103]. While the mechanism of animal breeding remained little understood and hence optimisation of factors that could improve success rates (better selection criteria and less requirement to cull unsuccessful 'experiments') was not available, the consequences of selective inbreeding were measurable in the physiological parameters of the progeny from specific breeding and this trial-and-error approach was significantly successful. The intellectual underpinning for the planned creation of plant hybrids with particular characteristics was provided by the careful experiments carried out by Mendel in Austria. This eventually resulted in the development of the capability to produce a crop with significant commercial advantages in terms of harvesting and yield and that could also successfully be grown as part of an agricultural crop rotation system. It was only later that the molecular details of the nitrogen (and other) cycles emerged and therefore their manipulation in a structured way became increasingly feasible. In the case of peas (and other crops), the discovery of the role and importance of genes and DNA [104] enabled scientists to devise methods for the transfer of genes between species and, in particular, to create plants containing novel genes [105] Furthermore, the introduction of gene editing techniques [106] has further increased the capability of plant (and other) scientists to purposefully and precisely manipulate the genotype and the phenotype of target species. By using peas as an example, it is possible to examine the advances in crop manipulation and optimisation.

Genetic modification can conveniently be described as the transfer and manipulation of expression of a gene from one species into either another species or into the same species at different levels of expression. Traditional breeding operates within the confines of the existing genome of an organism—genetic modification can be the result either of manipulation of a host genome or the importation of new properties which are foreign to the host. In a sense, advances in genetic manipulation have been circular in nature and have followed the concept of increased focus. A particular trait is identified and the gene (or genes) responsible for the trait identified. The gene is then inserted into the host under a controlled expression system and a host with the new trait is produced. It is therefore a series in which a trait is reduced to a particular genetic characteristic which is modified to give a host with the targeted trait. The sequence of starting with a macro trait, carrying out reductionism to obtain the gene(s) then the expansion to the macro trait once again is fundamentally circular (and in the philosophical tradition of reductive chemistry) in nature. Traditional breeding enabled the transfer of genes between plants by a sexual mechanism without the need for the reductionist step [107]. While this was an effective method up to a point and screening of progeny for desired characteristics was possible, thereby improving the characteristics of the target plant, it was necessarily an imprecise mechanism with the possibility of undesirable traits also being transmitted. While some of these can be eliminated through discarding clones with undesirable properties, the focus on single traits can lead to a narrowing of the genetic base and the loss of responsive robustness in the bred species [108]. Genetic modification was a more precise technique which enabled genes that were responsible for specific traits to be transferred either in their entirety of after modification to alter some of their characteristics in a planned manner. The concept of cloning, modification and expression of plant genes is one that has increased the scope of methods available for manipulation of the properties of plants [109] and a range of techniques have been developed. Traditional plant breeding is a phenotypically driven transfer of large amounts of partially unspecified genetic material. Genetic modification (with or without gene modification by site-directed mutagenesis) is the precise transfer of genes responsible for a specific trait. The difference between the two techniques lies both in precision and in the level of interaction (multiple uncharacterised genes or molecular integration of a single gene). The first demonstrable (and patentable) technique for the practice of recombinant DNA technology was that by Boyer and Cohen and led to the commercialisation of said recombinant DNA technology [110]. Without a detailed exposition of the practical techniques of creation of transgenic organisms (which lies outwith the scope of this volume), the principles that underpin the production of organisms containing foreign (and/or modified) genes is conceptually relatively simple. The new piece of DNA is transferred into the plant cell in culture. The cells are then grown in tissue culture and they develop into whole plants the seeds of which contain the inserted DNA and plants grown from these seeds will have the functionality associated with the inserted DNA. Some bacteria (for example *Agrobacterium*) are able to transfer DNA directly into plants without human intervention however these are fairly rare events and will not be considered in any detail however it should be noted that *Agrobacterium*-mediated transfer of DNA can be used as a vehicle to insert desirable genes into plants.

Following the development of the proof in principle of genetic engineering by creating antibiotic resistant plants, the agricultural biotechnology industry began to develop specific solutions to known challenges facing crop plants [111]. The first significant, commercially grown genetically modified crop was a tomato with improved characteristics. One particular enzyme in tomato is associated with the degradation of the fruit. Polygalacturonase catalyses the degradation of pectin and hence the degradation of the cell wall and the softening of the tomato fruit [112]. Inhibition of the enzyme was initially carried out using RNA-interference [113]. More recently gene editing was adopted [114] and the technology that underpins this approach will be discussed later. Although the improved shelf life (and cost) of the genetically modified tomato helped it to gain market share, it eventually was withdrawn due, in part, to the hostility of environmental pressure groups and the growing impression that safety studies were not being fully disclosed. Following the creation, commercialisation and subsequent withdrawal of genetically modified tomatoes, agro-biotechnology moved away from quality and processing and towards more direct (and easier to demonstrate) benefits. The next generation of genetically modified crops contained single genes that offered the possibility of enhanced resistance to herbicides (particularly glyphosate—the target is 5-enolpyruvylshikimate-3-phosphate synthase and an engineered form of the enzyme confers resistance to glyphosate) viruses and pesticides (Bt toxin). The development of multiple traits (gene stacking) has enabled a series of benefits to be introduced into genetically modified plants. A number of technical challenges remain to be overcome if the use of genetic modification in plants is to be more widely adopted [115].

One approach that is both a reversion to previous paradigms and an application that requires a greater degree of understanding of the architecture that surrounds genetic modification is the use of gene editing. Traditional plant breeding and the selection of progeny with a particular desired characteristic can also introduce surplus genetic material whose function may be positive, negative or benign but will derive from the pre-cross parentage. Genetic modifications creates the opportunity to introduce genes which are not part of the genome of the original plant in a controlled fashion—even if the precise consequences of gene introduction cannot be fully anticipated. Gene editing is a technique that was developed as a consequence of the discovery of an anti-viral mechanism in bacteria. The CRISPR-Cas system operates by incorporating short DNA sequences from viruses (or plasmids) and integrating them into the CRISPR locus of the prokaryotic host. These spacer sequences are transcribed into short RNA guides that specify the cleavage site of Cas nucleases in the genome of the virus or plasmid [116]. Cleavage of the viral DNA renders it biologically inert. In a similar way, restriction endonucleases, which are an essential tool for the isolation and subsequent incorporation of whole genes using genetic modification, are produced by bacteria to cleave foreign DNA and thereby protect the bacterium from infection [117]. The development of restriction enzyme expression and the CRISPR-Cas system are linked as part of different aspects of the bacterial immune response to infection [118]. The discovery, development and application of the CRISPR-Cas gene editing tools has led to the validation of a significant tool for the manipulation of plants and other living systems [119]. Some of the main applications of gene editing have been focused upon human health and the possibility of using deletion or addition of specific functions into the genome in a precise and targeted way. By considering gene therapy, molecular methods may be brought to bear upon challenges in the diagnosis, prevention and treatment of disease [120]. In the plant biotechnology area, short deletions and inactivation of genes has been demonstrated however more advanced manipulations are also possible [121] with the aim of the complete and controlled reconstruction of plant genomes. In the case of peas, a number of possible and actual manipulations have been suggested. A key quality trait that limits the use of peas and other legumes is the production of volatile compounds (particularly hexanal) arising from the oxidation of polyunsaturated fatty acids (PUFAs) by the plant lipoxygenase [122]. By employing CRISPR/Cas9 gene editing, the lipoxygenase activity was substantially reduced and the flavour profile improved [123]. Although gene editing techniques can be developed and adopted [124], the successful development of new, stable varieties will still require challenges associated with other aspects of genetic manipulation of peas such as the regeneration of plants to be solved [125].

Some of the challenges that remain in plant biotechnology have analogues in the investigations of microbial systems—particularly the study of the production of polyketides by *Streptomyces*—work that was carried out in Norwich at the John Innes Institute [126] by David Hopwood and collaborators [127]. Some aspects of this groundbreaking work will be outlined in Sect. 3.4.

3.4 Soil Microorganisms: Polyketides and Antibiotics

While the discovery and subsequent commercialisation of penicillin has been reviewed and studied [128] the investigation of the antibiotics and other active principles from the soil organism *Streptomyces* has perhaps been of even greater significance and, while the beta lactam structure of penicillin is of fairly limited occurrence, the core structure of *Streptomyces* antibiotics (polyketides) has been at the core of investigations into the chemistry, enzymology and molecular biology of organisms in a wide arc.

Streptomyces are (unlike *Penicillium* and *Cephalosporium*) bacteria rather than fungi although their mode of growth (sporulation etc.) has a number of fungal characteristics. Indeed the production of specific classes of polyketides derived from arginine (arginoketides) (Fig. 3.6).

has been shown to be important for bacterial:fungal communication [129]. They are divided into two structural classes—linear and cyclic arginoketides and an example of a cyclic member of the class is shown in Fig. 3.6. *Streptomyces* reproduce by sporulation and, during this process, synthesise antibiotics. The concept that microorganisms synthesise secondary metabolites (and other substances) in order to compete directly with other microorganisms for resources and space (unlike most plant bioactive substances which have a range of effects based around the principle of deterrence of consumption) is substantiated in the case of *Streptomyces* which produce both anti-bacterial compounds (polyketide-derived macrolide antibiotics) and chitinases which degrade fungal and insect cell walls [130]. In addition *Streptomyces* species also produce small molecule anti-fungal compounds [131]. Many of the developments in the molecular biology and hence manipulation of *Streptomyces* arose from work carried out at the John Innes Institute in Norwich using the model organism *S. coelicolor* [132]. While the detailed description of the molecular biology and growth characteristics of *Streptomyces is* outwith the scope of this book, one particular aspect of the chemistry of the organism is of particular

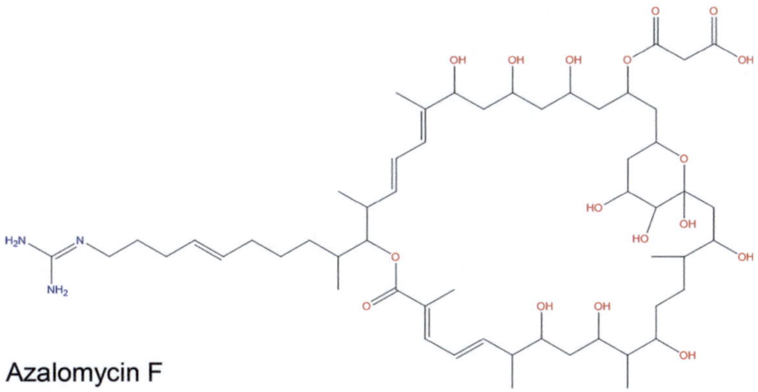

Fig. 3.6 Arginine-derived polyketides

interest. *Streptomyces* species lack any motile functionality (pili etc.) and seem to rely on sporulation for transmission to new sites of growth potential. One particular compound is linked to the sporulation process—geosmin (Fig. 3.7).

Geosmin is an attractant for arthropods who in turn aid in spore dispersal [133]. Geosmin is a sesquiterpenoid and is produced by a large proportion of *Streptomyces*. The other class of compounds produced by *Streptomyces* and which have been the subject of extensive study are polyketides. Indeed, it was the discovery of the important polyketide antibiotics such as streptomycin [134] that first aroused interest in the study of the cellular machinery of antibiotic production in the species and led to many of the fundamental molecular biology insights including cloning of pathways [135]. One of the factors in the use of *S. coelicolor* as a model organism is the different colours produced by the bacterium. Manipulation of the steps in the biosynthetic pathway of the ultimate colour (actinorhodin—Fig. 3.8).

at different stages in the biosynthetic pathway led to the production of intermediates of different colours so mutants could quickly and easily be identified. Work on the biosynthesis of actinorhodin led to the development of the concept that the genes for production of secondary metabolites are arranged in biosynthetic gene clusters (BGCs). In addition, actinorhodin arises from the action of a particular polyketide

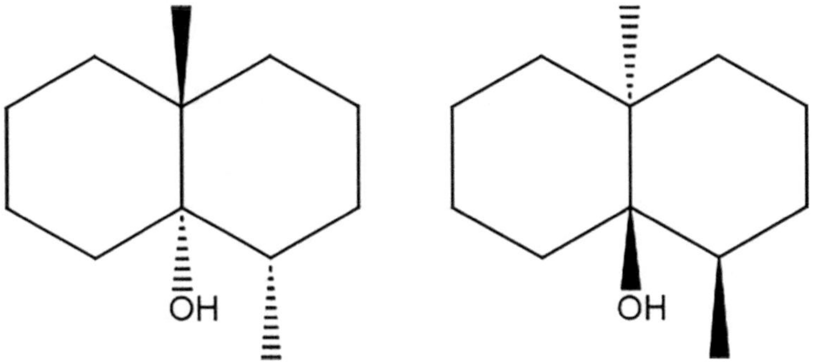

Fig. 3.7 Structure of geosmin

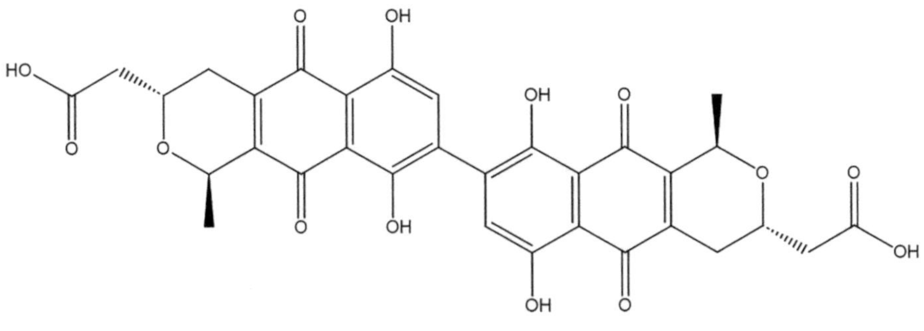

Fig. 3.8 Structure of actinorhodin

synthase, a member of a family of polyfunctional, modular enzymes which could be modified to give rise to novel antibiotics [136] by modulation of the key enzymes in the cluster [137]. The knowledge gained through the studies of *Streptomyces* provided much of the foundational basis for the manipulation of the genetic material of a range of organisms and enabled the development of strategies to produce compounds with a diverse range of properties [138]. In addition to the creation of novel products, genetic manipulation has also facilitated the discovery of potential secondary metabolites that are present in very small amounts but for which the genetic machinery exists [139]. The application of metabolomic and synthetic biology techniques to the field of natural products [140] has the potential to substantially increase the numbers and biological activities of molecules.

Another significant area is that of metagenomics. Many microbial species are difficult to isolate and to grow in culture. In the case of many of these species, their potential as a source of novel compounds remains untapped. Metagenomic analysis can be used to identify potential leads. In this technique, the total DNA from a sample is isolated and cloned to create a total DNA library. Individual clones are then screened either phenotypically or genetically (using sequence similarity) to identify potential candidate genes that may be involved in the biosynthesis of novel metabolites [141]. These techniques have been used to identify and isolate a range of potentially biologically useful structures the genes for which can be cloned into suitable bacterial or yeast hosts to facilitate the production of commercially useful amounts of product.

3.5 Soil Organisms: Characteristics and Control

As noted previously, the agronomic characteristics of a crop are heavily dependent upon the optimisation of the environment in which the crop grows. Crop rotation has already been discussed previously however the control of the growing environment requires an understanding of both the macro and the microenvironments. In particular, while macro nutrients (nitrogen, phosphorous and carbon) and their accompanying cycles are of significant importance to the growth of plants, microorganisms can be both symbiotic and deleterious and a few examples of these will serve to illustrate the general principles.

The presence of *Streptomyces* species in soil has been previously noted (Sect. 3.4). Like many other components of the soil microbiome, microorganisms are an important factor in the cycling of organic materials [142]. Some features that are specific to the ecology of the Norfolk environment (peat, fresh and seawater) will be considered later (Chap. 4) however some aspects of the nature of soil organisms will be mentioned.

While the soil provides an environment for organisms that cause a multitude of diseases in crops such as clubroot (caused by *Plasmodiophora brassicae*) and potato blight and potato scab. Interestingly the agent responsible for powdery scab in potatoes (*Spongospora subterranea* f. sp. *Subterranean*) is a single celled eukaryote—not to be confused with the agent responsible for common scab, which is a bacterial disease transmitted by *Streptomyces scabies*. *Spongospora subterranea* f. sp. *Subterranean* is a Plasmodiophorid [143], as is the agent responsible for clubroot in *Brassica* crops (*Plasmodiophora brassicae*). The resting spores of Plasmodiphorids are rich in chitin and the discovery of chitinase produced by *Streptomyces* species [144] suggests that, as a component of the soil microbiome, *Streptomyces* bacteria may have a role to play in limiting the spread of clubroot and potato powdery scab [145]. This in turn may impact upon the most beneficial crop rotation systems and the use of plants that encourage the proliferation of chitinase-producing bacteria being a potential phytosanitary measure [146]. The genes for chitinase have been identified from a phytopathogen-suppressive soil sample and studies designed to optimise the bacterial producers may have an impact on pathogen suppression [147]. The role of specific compounds in the protection of plants from microbial attack has been noted and reviewed [148]. Indeed microbial damage can induce the activation of the biosynthetic pathways (BGCs) that can result in the production of elevated levels of plant protective phytochemicals (phytoalexins). The structures of some representative phytoalexins are shown (Fig. 3.9). These compounds do not have particular structural similarities but are classified together on the basis of their functional properties.

Both plants themselves through the production of phytoalexins and associated microbes may have a role to play in limiting the consequences of pathogens in plant growth [149]. In addition, some of the phytochemicals whose production is elicited by pathogenic attack may also have a dual role in protecting consuming species (including humans) from the consequences of non-communicable diseases [33].

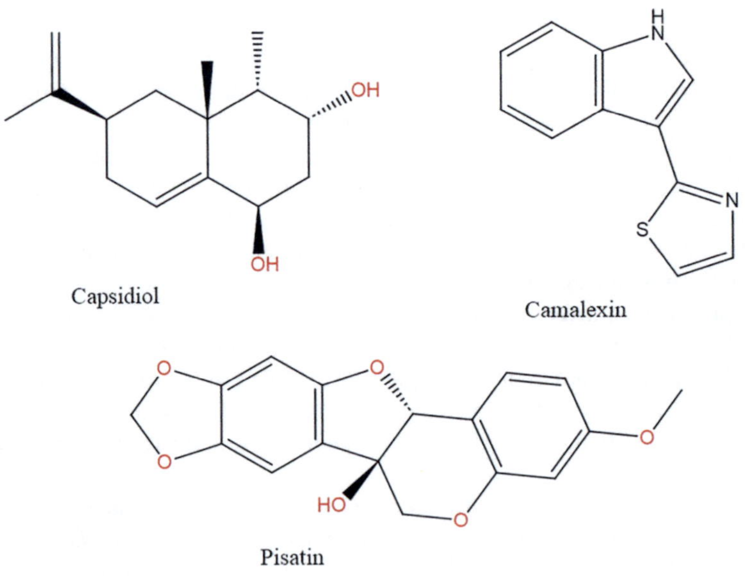

Fig. 3.9 Structures of some phytoalexins

The presence of disease-specific organisms (pathogens) and a chemical and biological response combined with the microbial competing requirements for nutrients and space suggests that the whole soil microbiome (both microbial and chemical aspects) may be an important aspect of plant and crop health.

The importance of various aspects of the soil microbiome in agriculture has been recognised [150]. Rapid sequencing techniques and probing of the soil metagenome has not only resulted in the discovery and isolation of novel chemical entities [151] but also an increasing awareness of the importance of a balanced microbiome to the soil ecosystem as a whole [152]. While the development of biological techniques such as DNA sequencing and the creation of biosynthetic gene cluster libraries to probe the biological properties of soil microbiome systems has made significant progress, particularly in the identification of potential biosynthetic products [151], the coupling of meta proteomics to microbial data through the use of mass spectrometry [153] and other techniques will increase the correlation between the chemical and biological properties upon which agriculture depends. Another important area which is relevant to all microbiomes, is the identification of signalling molecules and, in particular, quorum sensing initiators and inhibitors [154]. Bacterial quorum sensing has been linked to nitrogen cycling in the rhizosphere [155] and the study of such interrelationships will become increasingly possible as molecular techniques develop further [156].

3.6 Fertilisers: Natural and Synthetic Manipulations

As noted previously, the application of synthetic fertilisers (particularly nitrogenous agents produced by the Haber-Bosch process or even by biofertilisation) has had a significant impact on the yields of crops and on agronomic practices. It has been suggested, however that the use of synthetic fertilisers is unsustainable [157] and the development of a range of more environmentally friendly approaches is seen as a priority. The degradation of soil quality and the loss of microbial diversity, which are both regarded as consequences of the use of chemical fertilisers [158] are widely seen as being of concern however, chemical fertilisers are an important and convenient means to grow food and other industrial crops. The combination of fertilisers with biological methods is growing in importance [159] and is likely to be increasingly adopted if it can be demonstrated to be both effective [160] and cost neutral compared to the use of methodologies that rely on the use of synthetic chemicals. There is a perceived problem with eutrophication (loss of oxygen from waterways) caused by excessive nutrient accumulation as a consequence of agricultural practices [161] and this affects several agricultural areas including, in particular, Norfolk. The development of 'biofertilisers' is one promising approach [162] which may provide a solution however it remains questionable if such approaches are cost-effective and can be scaled up to the required level. In addition, the addition of organic material as a nutrient source may increase the microbial load of undesirable species. For example, the causative agent of potato scab mentioned previously

(*Streptomyces scabies*), can persist in soil for a considerable time due to the presence of an organic nutrient supply (not necessarily potatoes). While biocontrol methods have been advocated [163] with some degree of reported success, organic material content has also been found to correlate with disease severity [164] and this suggests that determining causality rather than correlation will be of increasing importance. The complexity of agronomic systems suggests that a reductionist approach to determine the most significant contributions to beneficial agronomy may not be possible and the broader approach, as typified by crop rotation, may be the most effective means of optimising yield, disease resistance and the spread of best practice.

References

1. Upcott EV, Henrys PA, Redhead JW, Jarvis SG, Pywell RF. A new approach to characterising and predicting crop rotations using national-scale annual crop maps. Sci Total Environ. 2023;860:160471. https://doi.org/10.1016/j.scitotenv.2022.160471.
2. Sterling AJ, Levine DS, Aldossary A, Head-Gordon M. Chemical bonding and the role of node-induced electron confinement. J Am Chem Soc. 2024;146(14):9532–43. https://doi.org/10.1021/jacs.3c10633.
3. Constable EC, Housecroft CE. Chemical bonding: the journey from miniature hooks to density functional theory. Molecules. 2020;25(11):2623. https://doi.org/10.3390/molecules25112623.
4. Paparazzo E. Vacuum: a void full of questions. Surf Interface Anal. 2008;40:450–3. https://doi.org/10.1002/sia.2608.
5. Nicolaides D. Atoms of matter and energy. In: Search of a theory of everything: the philosophy behind physics. New York: Oxford Academic; 2020. https://doi.org/10.1093/oso/9780190098353.003.0012.
6. Duerr PM, Wolf WJ. Methodological reflections on the MOND/dark matter debate. Stud Hist Phil Sci. 2023;101:1–23. https://doi.org/10.1016/j.shpsa.2023.07.001.
7. Schauenburg D, Weil T. Chemical reactions in living systems. Adv Sci. 2024;11(8):e2303396. https://doi.org/10.1002/advs.202303396.
8. Duranton F, Jankowski J, Więcek A, Argilés À. On the discovery of urea. Identification, synthesis and observations that let to establishing the first uraemic retention solute. G Ital Nefrol. 2016;33(66):S66.
9. Wöhler F. Ueber künstliche Bildung des Harnstoffs. Ann Phys Chem. 1828;88(2):253–6.
10. Innocenti B. Biomechanics: a fundamental tool with a long history (and even longer future!). Muscles Ligaments Tendons J. 2018;7(4):491–2. https://doi.org/10.11138/mltj/2017.7.4.491.
11. Bartley BA, Kim K, Medley JK, Sauro HM. Synthetic biology: engineering living systems from biophysical principles. Biophys J. 2017;112(6):1050–8. https://doi.org/10.1016/j.bpj.2017.02.013.
12. Lennartson A. Scheele's contribution to organic chemistry. In: Scheele CW, Bergman T, editors. Perspectives on the history of chemistry. Cham: Springer; 2020. https://doi.org/10.1007/978-3-030-49194-9_24.
13. Hosztafi S. The discovery of alkaloids. Pharmazie. 1997;52(7):546–50.
14. Hofmann A. Historical view on ergot alkaloids. Pharmacology. 1978;16(1):1–11. https://doi.org/10.1159/000136803.
15. Zhang HX, Wang ZZ, Du ZZ. Sensory-guided isolation and identification of new sweet-tasting dammarane-type saponins from Jiaogulan (Gynostemma pentaphyllum) herbal tea. Food Chem. 2022;15(388):132981. https://doi.org/10.1016/j.foodchem.2022.132981.

16. Peracchi A. The limits of enzyme specificity and the evolution of metabolism. Trends Biochem Sci. 2018;43(12):984–96. https://doi.org/10.1016/j.tibs.2018.09.015.
17. Martínez-Reyes I, Chandel NS. Mitochondrial TCA cycle metabolites control physiology and disease. Nat Commun. 2020;11(1):102. https://doi.org/10.1038/s41467-019-13668-3.
18. Mion D, Bunel L, Ramakrishnan S, Heo P, Pincet F. Biomimetic membrane in a microfluidic chip for the electrical and optical monitoring of biological reactions. Nat Protoc. 2025; https://doi.org/10.1038/s41596-025-01171-7.
19. Liang H, Ren P, Wang L, et al. Asymmetric biomimetic aldol reaction of glycinate enables highly efficient synthesis of chiral β-hydroxy-α-amino acid derivatives. Nat Catal. 2025;8:668–77. https://doi.org/10.1038/s41929-025-01364-z.
20. Shen Y, Su R, Altug H, Liu Z, Zhang X, Xu X, Liang Y, Kong J, Li Q, Wang Y, Qi W. Bioinspired three-component system to prepare full-color functional biomimetic pigments. ACS Appl Mater Interfaces. 2024;16(31):40531–42. https://doi.org/10.1021/acsami.4c06070.
21. Hagan E, Poulin J. Statistics of the early synthetic dye industry. Herit Sci. 2021;9:33. https://doi.org/10.1186/s40494-021-00493-5.
22. Klein JA, Ragland ER. Introduction analysis and synthesis in medieval and early modern Europe. Ambix. 2014;61(4):319–26.
23. Newman WR. Introduction: traditions of analysis and synthesis. In: Newman WR, Schickore J, editors. Traditions of analysis and synthesis. Archimedes, vol. 73. Cham: Springer; 2025.
24. Sousa MM, Melo MJ, Parola AJ, Morris PJ, Rzepa HS, de Melo JS. A study in mauve: unveiling Perkin's dye in historic samples. Chemistry. 2008;14(28):8507–13. https://doi.org/10.1002/chem.200800718.
25. Criddle WJ, Ellis GP. Functional group analysis. In: Qualitative organic chemical analysis. Boston: Springer; 1967. https://doi.org/10.1007/978-1-4899-6383-3_2.
26. Medley JW, Movassaghi M. Robinson's landmark synthesis of tropinone. Chem Commun. 2013;49(92):10775–7. https://doi.org/10.1039/c3cc44461a.
27. Birch AJ. Investigating a scientific legend: the tropinone synthesis of sir Robert Robinson, FRS. Notes Rec R Soc Lond. 1993;47:277–96.
28. Bedewitz MA, Jones AD, D'Auria JC, et al. Tropinone synthesis via an atypical polyketide synthase and P450-mediated cyclization. Nat Commun. 2018;9:5281. https://doi.org/10.1038/s41467-018-07671-3.
29. Paul A, Longchar B, Dkhar J. Tropane alkaloid biosynthesis in plants: insights from transcriptome analysis. In: Srivastava V, Mehrotra S, Mishra S, editors. Tropane alkaloids. Singapore: Springer; 2021. https://doi.org/10.1007/978-981-33-4535-5_8.
30. Mannaa M, Han G, Seo YS, Park I. Evolution of food fermentation processes and the use of multi-omics in deciphering the roles of the microbiota. Foods. 2021;10(11):2861. https://doi.org/10.3390/foods10112861.
31. Chen GQ, Liu X. On the future fermentation. Microb Biotechnol. 2021;14(1):18–21. https://doi.org/10.1111/1751-7915.13674.
32. Ochoa G, Armijos L, Figueroa JG, Jaramillo-Fierro X, Solano-Cueva N. Optimization of two methods for the rapid and effective extraction of quinine from Cinchona officinalis. Plants. 2025;14(3):364. https://doi.org/10.3390/plants14030364.
33. Nasim N, Sandeep IS, Mohanty S. Plant-derived natural products for drug discovery: current approaches and prospects. Nucleus. 2022;65(3):399–411. https://doi.org/10.1007/s13237-022-00405-3.
34. Duperray M, Delvenne M, François JM, Delvigne F, Capp JP. Genomic and metabolic instability during long-term fermentation of an industrial Saccharomyces cerevisiae strain engineered for C5 sugar utilization. Front Bioeng Biotechnol. 2024;26(12):1357671. https://doi.org/10.3389/fbioe.2024.1357671.
35. Schmidt FR. Optimization and scale up of industrial fermentation processes. Appl Microbiol Biotechnol. 2005;68:425–35. https://doi.org/10.1007/s00253-005-0003-0.

36. Sauer M. Industrial production of acetone and butanol by fermentation-100 years later. FEMS Microbiol Lett. 2016;363(13):fnw134. https://doi.org/10.1093/femsle/fnw134.
37. Książek E. Citric acid: properties, microbial production, and applications in industries. Molecules. 2023;29(1):22. https://doi.org/10.3390/molecules29010022.
38. Max B, Salgado JM, Rodríguez N, Cortés S, Converti A, Domínguez JM. Biotechnological production of citric acid. Braz J Microbiol. 2010;41(4):862–75. https://doi.org/10.1590/S1517-83822010000400005.
39. Scheele CW. Note about lemon juice, as well as ways to crystallize it. N Proc R Acad Sci. 1784;5:105–9.
40. Cairns TC, Nai C, Meyer V. How a fungus shapes biotechnology: 100 years of aspergillus Niger research. Fungal Biol Biotechnol. 2018;24(5):13. https://doi.org/10.1186/s40694-018-0054-5.
41. Quinn R. Rethinking antibiotic research and development: world war II and the penicillin collaborative. Am J Public Health. 2013;103(3):426–34. https://doi.org/10.2105/AJPH.2012.300693.
42. Harms H, König GM, Schäberle TF. Production of antimicrobial compounds by fermentation. Methods Mol Biol. 2017;1520:49–61. https://doi.org/10.1007/978-1-4939-6634-9_3.
43. Cummings M, Breitling R, Takano E. Steps towards the synthetic biology of polyketide biosynthesis. FEMS Microbiol Lett. 2014;351(2):116–25. https://doi.org/10.1111/1574-6968.12365.
44. Cao J, Perez-Pinera P, Lowenhaupt K, Wu MR, Purcell O, de la Fuente-Nunez C, Lu TK. Versatile and on-demand biologics co-production in yeast. Nat Commun. 2018;9(1):77. https://doi.org/10.1038/s41467-017-02587-w.
45. Yang L, Wen KS, Ruan X, Zhao YX, Wei F, Wang Q. Response of plant secondary metabolites to environmental factors. Molecules. 2018;23(4):762. https://doi.org/10.3390/molecules23040762.
46. Ketehouli T, Sossah FL, Panwala R, Suazo Tejada AK, Goss EM, Garcia FHS, Vallad GE, Martins SJ. Secondary metabolites in plant-microbe interactions. J Appl Microbiol. 2025;136(6):lxaf124. https://doi.org/10.1093/jambio/lxaf124.
47. Kong CH, Xuan TD, Khanh TD, Tran HD, Trung NT. Allelochemicals and signaling chemicals in plants. Molecules. 2019;24(15):2737. https://doi.org/10.3390/molecules24152737.
48. Loyola-Vargas VM, Vázquez-Flota F. An introduction to plant cell culture: back to the future. Methods Mol Biol. 2006;318:3–8. https://doi.org/10.1385/1-59259-959-1:003.
49. Hirschi KD. Genetically modified plants: nutritious, sustainable, yet underrated. J Nutr. 2020;150(10):2628–34. https://doi.org/10.1093/jn/nxaa220.
50. Eyal S. The fever tree: from malaria to neurological diseases. Toxins. 2018;10(12):491. https://doi.org/10.3390/toxins10120491.
51. Soliman SSM, Raizada MN. Sites of biosynthesis and storage of Taxol in Taxus media (Rehder) plants: mechanism of accumulation. Phytochemistry. 2020;175:112369. https://doi.org/10.1016/j.phytochem.2020.112369.
52. Pattanaik L, Naik SN, Hariprasad P. Valorization of waste Indigofera tinctoria L. biomass generated from indigo dye extraction process—potential towards biofuels and compost. Biomass Conv Bioref. 2019;9:445–57. https://doi.org/10.1007/s13399-018-0354-2.
53. Linke JA, Rayat A, Ward JM. Production of indigo by recombinant bacteria. Bioresour Bioprocess. 2023;10:20. https://doi.org/10.1186/s40643-023-00626-7.
54. Trewavas A, Leaver C. Is opposition to GM crops science or politics? An investigation into the arguments that GM crops pose a particular threat to the environment. EMBO Rep. 2001;2(6):455–9. https://doi.org/10.1093/embo-reports/kve123.
55. Periyasamy AP, Periyasami S. Critical review on sustainability in denim: a step toward sustainable production and consumption of denim. ACS Omega. 2023;8(5):4472–90. https://doi.org/10.1021/acsomega.2c06374.
56. Cooksey CJ. Tyrian purple: 6,6′-dibromoindigo and related compounds. Molecules. 2001;6(9):736–69. https://doi.org/10.3390/60900736.

57. Cova TFGG, Pais AACC, Seixas de Melo JS. Reconstructing the historical synthesis of mauveine from Perkin and Caro: procedure and details. Sci Rep. 2017;7:6806. https://doi.org/10.1038/s41598-017-07239-z.
58. Tanoue Y, Terada A, Sakata K, Hashimoto M, Morishita S-I, Hamada M, Kai N. A facile synthesis of Tyrian purple based on a biosynthetic pathway. Fish Sci. 2001;67:726–9. https://doi.org/10.1046/j.1444-2906.2001.00312.x.
59. Li F, Chen Q, Deng H, Ye S, Chen R, Keasling JD, Luo X. One-pot selective biosynthesis of Tyrian purple in Escherichia coli. Metab Eng. 2024;81:100–9. https://doi.org/10.1016/j.ymben.2023.11.003.
60. Ochoa-Villarreal M, Howat S, Hong S, Jang MO, Jin YW, Lee EK, Loake GJ. Plant cell culture strategies for the production of natural products. BMB Rep. 2016;49(3):149–58. https://doi.org/10.5483/bmbrep.2016.49.3.264.
61. Yao J, Weng Y, Dickey A, Wang KY. Plants as factories for human pharmaceuticals: applications and challenges. Int J Mol Sci. 2015;16(12):28549–65. https://doi.org/10.3390/ijms161226122.
62. Espinosa-Leal CA, Puente-Garza CA, García-Lara S. In vitro plant tissue culture: means for production of biological active compounds. Planta. 2018;248(1):1–18. https://doi.org/10.1007/s00425-018-2910-1.
63. Schillberg S, Raven N, Fischer R, Twyman RM, Schiermeyer A. Molecular farming of pharmaceutical proteins using plant suspension cell and tissue cultures. Curr Pharm Des. 2013;19(31):5531–42. https://doi.org/10.2174/1381612811319310008.
64. Cruz-Cruz CA, González-Arnao MT, Engelmann F. Biotechnology and conservation of plant biodiversity. Resources. 2013;2:73–95. https://doi.org/10.3390/resources2020073.
65. Lara SW, Ryan P. The current state of peas in the United Kingdom; diversity, heritage and food systems. Plants People Planet. 2025;7(5):1235–44. https://doi.org/10.1002/ppp3.70001.
66. Zhao J, Chen J, Beillouin D, Lambers H, Yang Y, Smith P, Zeng Z, Olesen JE, Zang H. Global systematic review with meta-analysis reveals yield advantage of legume-based rotations and its drivers. Nat Commun. 2022;13(1):4926. https://doi.org/10.1038/s41467-022-32464-0.
67. Author. The corpuscular theory of matter. Nature. 1908;77:505–6. https://doi.org/10.1038/077505a0.
68. Banchetti-Robino MP. The ontological complexity of Boyle's corpuscularian theory: microstructure, natural kinds, and essential form. In: The chemical philosophy of Robert Boyle: mechanicism, chymical atoms, and emergence. New York: Oxford Academic; 2020. https://doi.org/10.1093/oso/9780197502501.003.0004.
69. Wood RJ. Robert Bakewell (1725-1795), pioneer animal breeder, and his influence on Charles Darwin. Folia Mendeliana. 1973;58:231–42.
70. Milne A. Sentient genetics: breeding the animal breeder as fundamental other. J Eighteenth-Century Stud. 2010;33:583–97. https://doi.org/10.1111/j.1754-0208.2010.00324.x.
71. Spangler ML. Animal breeding and genetics: introduction. In: Spangler ML, editor. Animal breeding and genetics. Encyclopedia of sustainability science and technology series. New York: Springer; 2023. https://doi.org/10.1007/978-1-0716-2460-9_1112.
72. Miskell L. Putting on a show: the Royal Agricultural Society of England and the Victorian town, c.1840—1876. Agric Hist Rev. 2012;60(1):37–59.
73. Mendel G. Versuche über Plflanzenhybriden. In: Verhandlungen des naturforschenden Vereines in Brünn, Bd. IV für das Jahr 1865. Abhandlungen. 1866;4:3–47.
74. Ross L. From Mendel's laws to non-Mendelian inheritance. Nat Rev Genet. 2024;25:677. https://doi.org/10.1038/s41576-024-00753-7.
75. Strome S, Bhalla N, Kamakaka R, Sharma U, Sullivan W. Clarifying Mendelian vs non-Mendelian inheritance. Genetics. 2024;227(3):iyae078. https://doi.org/10.1093/genetics/iyae078.
76. Feng C, Chen B, Hofer J, Shi Y, Jiang M, Song B, Cheng H, Lu L, Wang L, Howard A, Bendahmane A, Fouchal A, Moreau C, Sawada C, LeSignor C, Zhang C, Vikeli E, Tsanakas G, Zhao H, Cheema J, Barclay JE, Hou J, Sayers L, Wingen L, Vigouroux M, Vickers M,

Ambrose M, Dalmais M, Higuera-Poveda P, Li P, Yuan Q, Spanner R, Horler R, Wouters R, Chundakkad S, Wu T, Zhao X, Li X, Sun Y, Huang Z, Wu Z, Deng XW, Steuernagel B, Domoney C, Ellis N, Chayut N, Cheng S. Genomic and genetic insights into Mendel's pea genes. Nature. 2025;642(8069):980–9. https://doi.org/10.1038/s41586-025-08891-6.
77. Sussmilch FC, Ross JJ, Reid JB. Mendel: from genes to genome. Plant Physiol. 2022;190(4):2103–14. https://doi.org/10.1093/plphys/kiac424.
78. Tayeh N, Hofer JMI, Aubert G, Jacquin F, Turner L, Kreplak J, Paajanen P, Le Signor C, Dalmais M, Pflieger S, Geffroy V, Ellis N, Burstin J. Afila, the origin and nature of a major innovation in the history of pea breeding. New Phytol. 2024;243:1247–61. https://doi.org/10.1111/nph.19800.
79. Kujala V. Felderbse bei welcher die ganze Blattspreite in Ranken umgewandelt ist. Arch Soc Zool Bot Fennicae. 1953;8:44–5.
80. Watson JD, Crick FH. Molecular structure of nucleic acids; a structure for deoxyribose nucleic acid. Nature. 1953;171(4356):737–8. https://doi.org/10.1038/171737a0.
81. Pellew C, Sverdrup A. New observations on the genetics of peas (pisum sativum). J Gen. 1923;13:125–31. https://doi.org/10.1007/BF02983043.
82. Snoad B. A preliminary assessment of 'leafless peas'. Euphytica. 1974;23:257–65. https://doi.org/10.1007/BF00035866.
83. Tran CT, Becker HC, Horneburg B. Agronomic performance of normal-leafed and semi-leafless pea (Pisum sativum L.) genotypes. Crop Sci. 2022;62:1430–42. https://doi.org/10.1002/csc2.20746.
84. Xiao Y, Yang D, Zhang SB, Mo YX, Dong YY, Wang KF, He LY, Dong B, Dossa GGO, Zhang JL. Nitrogen-fixing and non-nitrogen-fixing legume plants differ in leaf nutrient concentrations and relationships between photosynthetic and hydraulic traits. Tree Physiol. 2024;44(5):tpae048. https://doi.org/10.1093/treephys/tpae048.
85. Windsor N, Boatwright L, Boyles R, Bridges W, Rubiales D, Thavarajah D. Characterizing dry pea (Pisum sativum L.) for improved nutritional traits and the potential for biofortification. Legume Sci. 2024;6:e250. https://doi.org/10.1002/leg3.250.
86. de Bang TC, Husted S, Laursen KH, Persson DP, Schjoerring JK. The molecular-physiological functions of mineral macronutrients and their consequences for deficiency symptoms in plants. New Phytol. 2021;229(5):2446–69. https://doi.org/10.1111/nph.17074.
87. Weber K, Burow M. Nitrogen—essential macronutrient and signal controlling flowering time. Physiol Plant. 2018;162:251–60. https://doi.org/10.1111/ppl.12664.
88. Fañanás-Pueyo I, Carrera-Castaño G, Pernas M, Oñate-Sánchez L. Signalling and regulation of plant development by carbon/nitrogen balance. Physiol Plant. 2025;177(2):e70228. https://doi.org/10.1111/ppl.70228.
89. Liu M, Xu X, Wanek W, Sun J, Bardgett RD, Tian Y, Cui X, Jiang L, Ma Z, Kuzyakov Y, Ouyang H, Wang Y. Nitrogen availability in soil controls uptake of different nitrogen forms by plants. New Phytol. 2025;245:1450–67. https://doi.org/10.1111/nph.20335.
90. Smil V. Enriching the earth: Fritz Haber, Carl Bosch, and the transformation of world food production. Cambridge: MIT Press; 2004.
91. Wang T, Abild-Pedersen F. Achieving industrial ammonia synthesis rates at near-ambient conditions through modified scaling relations on a confined dual site. Proc Natl Acad Sci USA. 2021;118(30):e2106527118. https://doi.org/10.1073/pnas.2106527118.
92. Adeniyi A, Bello I, Mukaila T, Sarker NC, Hammed A. Trends in biological ammonia production. BioTech. 2023;12(2):41. https://doi.org/10.3390/biotech12020041.
93. Khosro M, Yousef S. Bacterial biofertilizers for sustainable crop production: a review. J Agric Biol Sci. 2012;7:307–16.
94. Rapson TD, Wood CC. Analysis of the ammonia production rates by Nitrogenase. Catalysts. 2022;12:844. https://doi.org/10.3390/catal12080844.
95. Ruiz B, Frostegård Å, Bruand C, Meilhoc E. Rhizobia: highways to NO. Biochem Soc Trans. 2021;49(1):495–505. https://doi.org/10.1042/BST20200989.

96. Rutten PJ, Poole PS. Oxygen regulatory mechanisms of nitrogen fixation in rhizobia. Adv Microb Physiol. 2019;75:325–89. https://doi.org/10.1016/bs.ampbs.2019.08.001.
97. Mathesius U. Are legumes different? Origins and consequences of evolving nitrogen fixing symbioses. J Plant Physiol. 2022;276:153765. https://doi.org/10.1016/j.jplph.2022.153765.
98. Liu L, Zhu Q, Wan Y, Yang R, Yang L, Meng L, Zheng M, Xu X, Zhu T, Elrys AS, Müller C, Dörsch P. Towards a mechanistic understanding of legume functioning in natural restoration of degraded ecosystem: legume-specific impacts on nitrogen transformation processes. Plant Cell Environ. 2025; https://doi.org/10.1111/pce.15550.
99. Chiurazzi M, Frugis G, Navazio L. Symbiotic nitrogen fixation: a launchpad for investigating old and new challenges. J Exp Bot. 2025;76(6):1473–7. https://doi.org/10.1093/jxb/erae510.
100. Hart MR, Quin BF, Nguyen ML. Phosphorus runoff from agricultural land and direct fertilizer effects: a review. J Environ Qual. 2004;33(6):1954–72. https://doi.org/10.2134/jeq2004.1954.
101. Zhao Z, Fernie AR, Zhang Y. Engineering nitrogen and carbon fixation for next-generation plants. Curr Opin Plant Biol. 2025;85:102699. https://doi.org/10.1016/j.pbi.2025.102699.
102. Ghafoor A, Javed H, Karim H, Studnicki M, Ali I, Yue H, Xiao P, Asghar M, Brock C, Wu Y. Biological nitrogen fixation for sustainable agriculture development under climate change–new insights from a meta-analysis. J Agro Crop Sci. 2024;210:e12754. https://doi.org/10.1111/jac.12754.
103. Rodenburg TB, Turner SP. The role of breeding and genetics in the welfare of farm animals. Anim Front. 2012;2(3):16–21. https://doi.org/10.2527/af.2012-0044.
104. Ciurea AV, Glavan LA, Costin HP, Covache-Busuioc RA, Brehar FM. Celebrating 70 years of DNA discovery: exploring the blueprint of life. J Med Life. 2024;17(4):387–91. https://doi.org/10.25122/jml-2024-1007.
105. Vyacheslavova AO, Berdichevets IN, Tyurin AA, et al. Expression of heterologous genes in plant systems: new possibilities. Russ J Genet. 2012;48:1067–79. https://doi.org/10.1134/S1022795412110130.
106. Gostimskaya I. CRISPR-Cas9: a history of its discovery and ethical considerations of its use in genome editing. Biochemistry. 2022;87(8):777–88. https://doi.org/10.1134/S0006297922080090.
107. Lorence A, Verpoorte R. Gene transfer and expression in plants. Methods Mol Biol. 2004;267:329–50. https://doi.org/10.1385/1-59259-774-2:329.
108. Singer SD, Laurie JD, Bilichak A, Kumar S, Singh J. Genetic variation and unintended risk in the context of old and new breeding techniques. Crit Rev Plant Sci. 2021;40(1):68–108. https://doi.org/10.1080/07352689.2021.1883826.
109. Atanassov II, Atanassov II, Etchells J, et al. A simple, flexible and efficient PCR-fusion/gateway cloning procedure for gene fusion, site-directed mutagenesis, short sequence insertion and domain deletions and swaps. Plant Methods. 2009;5:14. https://doi.org/10.1186/1746-4811-5-14.
110. Hughes SS. Making dollars out of DNA. The first major patent in biotechnology and the commercialization of molecular biology, 1974-1980. Isis. 2001;92(3):541–75. https://doi.org/10.1086/385281.
111. Raman R. The impact of genetically modified (GM) crops in modern agriculture: a review. GM Crops Food. 2017;8(4):195–208. https://doi.org/10.1080/21645698.2017.1413522.
112. Bird CR, Smith CJ, Ray JA, Moureau P, Bevan MW, Bird AS, Hughes S, Morris PC, Grierson D, Schuch W. The tomato polygalacturonase gene and ripening-specific expression in transgenic plants. Plant Mol Biol. 1988;11(5):651–62. https://doi.org/10.1007/BF00017465.
113. Sheehy RE, Kramer M, Hiatt WR. Reduction of polygalacturonase activity in tomato fruit by antisense RNA. Proc Natl Acad Sci USA. 1988;85(23):8805–9. https://doi.org/10.1073/pnas.85.23.8805.
114. Nie H, Shi Y, Geng X, Xing G. CRISRP/Cas9-mediated targeted mutagenesis of tomato polygalacturonase gene (SlPG) delays fruit softening. Front Plant Sci. 2022;19(13):729128. https://doi.org/10.3389/fpls.2022.729128.

115. Li XX, Gao YH, Ma HW, et al. Non-tissue culture genetic modifications for plant improvement. Plant Mol Biol. 2025;115:67. https://doi.org/10.1007/s11103-025-01594-6.
116. Modell JW, Jiang W, Marraffini LA. CRISPR-Cas systems exploit viral DNA injection to establish and maintain adaptive immunity. Nature. 2017;544(7648):101–4. https://doi.org/10.1038/nature21719.
117. Arias CF, Acosta FJ, Bertocchini F, Herrero MA, Fernández-Arias C. The coordination of anti-phage immunity mechanisms in bacterial cells. Nat Commun. 2022;13(1):7412. https://doi.org/10.1038/s41467-022-35203-7.
118. Maguin P, Varble A, Modell JW, Marraffini LA. Cleavage of viral DNA by restriction endonucleases stimulates the type II CRISPR-Cas immune response. Mol Cell. 2022;82(5):907–919.e7. https://doi.org/10.1016/j.molcel.2022.01.012.
119. Carroll D. Genome editing: past, present, and future. Yale J Biol Med. 2017;90(4):653–9.
120. Li G, Li X, Zhuang S, Wang L, Zhu Y, Chen Y, Sun W, Wu Z, Zhou Z, Chen J, Huang X, Wang J, Li D, Li W, Wang H, Wei W. Gene editing and its applications in biomedicine. Sci China Life Sci. 2022;65(4):660–700. https://doi.org/10.1007/s11427-021-2057-0.
121. Gilbertson L, Puchta H, Slotkin RK. The future of genome editing in plants. Nat Plants. 2025;11:680–5. https://doi.org/10.1038/s41477-025-01956-4.
122. Bi S, Lao F, Pan X, Shen Q, Liu Y, Wu J. Flavor formation and regulation of peas (Pisum sativum L.) seed milk via enzyme activity inhibition and off-flavor compounds control release. Food Chem. 2022;30(380):132203. https://doi.org/10.1016/j.foodchem.2022.132203.
123. Bhowmik P, Yan W, Hodgins C, Polley B, Warkentin T, Nickerson M, Ro DK, Marsolais F, Domoney C, Shariati-Ievari S, Aliani M. CRISPR/Cas9-mediated lipoxygenase gene-editing in yellow pea leads to major changes in fatty acid and flavor profiles. Front Plant Sci. 2023;21(14):1246905. https://doi.org/10.3389/fpls.2023.1246905.
124. Soulard C, Monfort M, Pillot J-P, Gery C, Chauvin L, Morlière S, Lesné A, Sugio A, Pilet-Nayel M-L, Rameau C, Nogué F, de Saint Germain A. Efficient and heritable gene editing through CRISPR-Cas9 in Pisum sativum. Plant Biotechnol J. 2025; https://doi.org/10.1111/pbi.70091.
125. Kaur R, Donoso T, Scheske C, Lefsrud M, Singh J. Highly efficient and reproducible genetic transformation in pea for targeted trait improvement. ACS Agric Sci Technol. 2022;2(4):780–7. https://doi.org/10.1021/acsagscitech.2c00084.
126. Chater K. David Hopwood and the emergence of streptomyces genetics. Int Microbiol. 1999;2(2):61–8.
127. Hopwood DA. Highlights of streptomyces genetics. Heredity. 2019;123:23–32. https://doi.org/10.1038/s41437-019-0196-0.
128. Ligon BL. Penicillin: its discovery and early development. Semin Pediatr Infect Dis. 2004;15(1):52–7. https://doi.org/10.1053/j.spid.2004.02.001.
129. Krespach MKC, Stroe MC, Netzker T, et al. Streptomyces polyketides mediate bacteria–fungi interactions across soil environments. Nat Microbiol. 2023;8:1348–61. https://doi.org/10.1038/s41564-023-01382-2.
130. Gomaa EZ. Microbial chitinases: properties, enhancement and potential applications. Protoplasma. 2021;258(4):695–710. https://doi.org/10.1007/s00709-021-01612-6.
131. Cowen A, Yiu B, Fallah S, Meyer KJ, Puumala E, Lee Y, Zubyk HL, Robbins N, Nodwell JR, MacAlpine J, Cowen LE. Characterizing antimicrobial activity of environmental Streptomyces spp. and oral bacterial and fungal isolates from Canis familiaris and Felis catus. mSphere. 2025;10:e00098. https://doi.org/10.1128/msphere.00098-25.
132. Schlimpert S, Elliot MA. The best of both worlds—Streptomyces coelicolor and Streptomyces venezuelae as model species for studying antibiotic production and bacterial multicellular development. J Bacteriol. 2023;205:e00153. https://doi.org/10.1128/jb.00153-23.
133. Becher PG, Verschut V, Bibb MJ, Bush MJ, Molnár BP, Barane E, Al-Bassam MM, Chandra G, Song L, Challis GL, Buttner MJ, Flärdh K. Developmentally regulated volatiles geosmin and 2-methylisoborneol attract a soil arthropod to Streptomyces bacteria promoting spore dispersal. Nat Microbiol. 2020;5:821–9.

134. Wainwright M. Streptomycin: discovery and resultant controversy. Hist Philos Life Sci. 1991;13(1):97–124.
135. Thompson CJ, Ward JM, Hopwood DA. DNA cloning in Streptomyces: resistance genes from antibiotic-producing species. Nature. 1980;286:525–7.
136. Hopwood DA, Malpartida F, Kieser HM, Ikeda H, Duncan J, Fujii I, Rudd BA, Floss HG, Omura S. Production of "hybrid" antibiotics by genetic engineering. Nature. 1985;314:642–4.
137. McDaniel R, Ebert-Khosla S, Hopwood DA, Khosla C. Engineered biosynthesis of novel polyketides. Science. 1993;262:1546–50.
138. Chen J, Wu Q, Hawas UW, Wang H. Genetic regulation and manipulation for natural product discovery. Appl Microbiol Biotechnol. 2016;100(7):2953–65. https://doi.org/10.1007/s00253-016-7357-3.
139. Chen J, Wu Q, Hawas UW, et al. Genetic regulation and manipulation for natural product discovery. Appl Microbiol Biotechnol. 2016;100:2953–65. https://doi.org/10.1007/s00253-016-7357-30.
140. Nath S, Dias LG. Metabolomics in natural product discovery and their applications. In: Soni V, Hartman TE, editors. Metabolomics. Cham: Springer; 2023. https://doi.org/10.1007/978-3-031-39094-4_9.
141. Brady SF, Simmons L, Kim JH, Schmidt EW. Metagenomic approaches to natural products from free-living and symbiotic organisms. Nat Prod Rep. 2009;26(11):1488–503. https://doi.org/10.1039/b817078a.
142. Sokol NW, Slessarev E, Marschmann GL, et al. Life and death in the soil microbiome: how ecological processes influence biogeochemistry. Nat Rev Microbiol. 2022;20:415–30. https://doi.org/10.1038/s41579-022-00695-z.
143. Falloon RE, Merz U, Butler RC, Curtin D, Lister RA, Thomas SM. Root infection of potato by Spongospora subterranea: knowledge review and evidence for decreased plant productivity. Plant Pathol. 2016;65:422–34.
144. Ekundayo FO, Folorunsho AE, Ibisanmi TA, et al. Antifungal activity of chitinase produced by Streptomyces species isolated from grassland soils in Futa area. Akure Bull Natl Res Cent. 2022;46:95. https://doi.org/10.1186/s42269-022-00782-4.
145. Khodashenas Rudsari M, Zouhar M, Manasova M, et al. Biocontrol potential of cell-free supernatant of Paenibacillus chitinolyticus against Plasmodiophora brassicae in two important brassica species. Eur J Plant Pathol. 2024;170:119–30. https://doi.org/10.1007/s10658-024-02885-2.
146. Chen S, Zheng Y, Wang Q, Mu R, Sun X, Ma G, Liu L, Ren J, Huang K, Chen G. Harnessing streptomyces for the management of clubroot disease of Chinese cabbage (Brassica rapa subsp pekinensis). Plants. 2025;14:2195. https://doi.org/10.3390/plants14142195.
147. Hjort K, Bergström M, Adesina MF, Jansson JK, Smalla K, Sjöling S. Chitinase genes revealed and compared in bacterial isolates, DNA extracts and a metagenomic library from a phytopathogen-suppressive soil. FEMS Microbiol Ecol. 2009;71(2):197–207. https://doi.org/10.1111/j.1574-6941.2009.00801.x.
148. Dixon RA. Natural products and plant disease resistance. Nature. 2001;411(6839):843–7. https://doi.org/10.1038/35081178.
149. Chen S-J, Yu M, Qiu Y-K, Shi J-Y, Gong X, Huang G-L, Zheng C-J. Natural products from plant-derived fungi against agricultural pathogenic fungal. Chem Biodivers. 2025:e00979. https://doi.org/10.1002/cbdv.202500979.
150. Frąc M, Hannula ES, Bełka M, Salles JF, Jedryczka M. Soil mycobiome in sustainable agriculture. Front Microbiol. 2022;28(13):1033824. https://doi.org/10.3389/fmicb.2022.1033824.
151. Crits-Christoph A, Diamond S, Butterfield CN, et al. Novel soil bacteria possess diverse genes for secondary metabolite biosynthesis. Nature. 2018;558:440–4. https://doi.org/10.1038/s41586-018-0207-y.
152. Hermans SM, Lear G, Case BS, Buckley HL. The soil microbiome: an essential, but neglected, component of regenerative agroecosystems. iScience. 2023;26(2):106028. https://doi.org/10.1016/j.isci.2023.106028.

153. Lee JY, Mitchell HD, Burnet MC, Wu R, Jenson SC, Merkley ED, Nakayasu ES, Nicora CD, Jansson JK, Burnum-Johnson KE, Payne SH. Uncovering hidden members and functions of the soil microbiome using de novo metaproteomics. J Proteome Res. 2022;21(8):2023–35. https://doi.org/10.1021/acs.jproteome.2c00334.
154. Alum EU, Gulumbe BH, Izah SC, Uti DE, Aja PM, Igwenyi IO, Offor CE. Natural product-based inhibitors of quorum sensing: a novel approach to combat antibiotic resistance. Biochem Biophys Rep. 2025;24(43):102111. https://doi.org/10.1016/j.bbrep.2025.102111.
155. DeAngelis KM, Lindow SE, Firestone MK. Bacterial quorum sensing and nitrogen cycling in rhizosphere soil. FEMS Microbiol Ecol. 2008;66(2):197–207. https://doi.org/10.1111/j.1574-6941.2008.00550.x.
156. Zeng X, Zou Y, Zheng J, Qiu S, Liu L, Wei C. Quorum sensing-mediated microbial interactions: mechanisms, applications, challenges and perspectives. Microbiol Res. 2023;273:127414. https://doi.org/10.1016/j.micres.2023.127414.
157. Nicolopoulou-Stamati P, Maipas S, Kotampasi C, Stamatis P, Hens L. Chemical pesticides and human health: the urgent need for a new concept in agriculture. Front Public Health. 2016;18(4):148. https://doi.org/10.3389/fpubh.2016.00148.
158. Pahalvi HN, Rafiya L, Rashid S, Nisar B, Kamili AN. Chemical fertilizers and their impact on soil health. In: Dar GH, Bhat RA, Mehmood MA, Hakeem KR, editors. Microbiota and biofertilizers, vol. 2. Cham: Springer; 2021. https://doi.org/10.1007/978-3-030-61010-4_1.
159. Hu HW, Chen QL, He JZ. The end of hunger: fertilizers, microbes and plant productivity. Microb Biotechnol. 2022;15(4):1050–4. https://doi.org/10.1111/1751-7915.13973.
160. Hei Z, Peng Y, Hao S, Li Y, Yang X, Zhu T, Müller C, Zhang H, Hu H, Chen Y. Full substitution of chemical fertilizer by organic manure decreases soil N2 O emissions driven by ammonia oxidizers and gross nitrogen transformations. Glob Chang Biol. 2023;29(24):7117–30. https://doi.org/10.1111/gcb.16957.
161. Harper D. A case study in restoration: shallow eutrophic lakes in the Norfolk broads. In: Eutrophication of freshwaters. Dordrecht: Springer; 1992. https://doi.org/10.1007/978-94-011-3082-0_8.
162. Schütz L, Gattinger A, Meier M, Müller A, Boller T, Mäder P, Mathimaran N. Improving crop yield and nutrient use efficiency via biofertilization-a global meta-analysis. Front Plant Sci. 2018;12(8):2204. https://doi.org/10.3389/fpls.2017.02204.
163. Haq IU, Munir M, Qadir S, et al. Suppression of common scab disease in potato via application of antagonistic Trichoderma harzianum and Bacillus subtilis. J Plant Dis Prot. 2025;132:126. https://doi.org/10.1007/s41348-025-01123-z.
164. Lazarovits G, Hill J, Patterson G, Conn KL, Crump NS. Edaphic soil levels of mineral nutrients, pH, organic matter, and cationic exchange capacity in the geocaulosphere associated with potato common scab. Phytopathology. 2007;97(9):1071–82. https://doi.org/10.1094/PHYTO-97-9-1071.

Part II
Chemistry and Commercialisation

Chapter 4
Chemistry and Commercialisation. Natural Products and their Exploitation. From Woad to Marine Harvests

Abstract Developments in the exploitation of knowledge of chemistry and the spread of technology in Europe through trade routes (linking societies), links between institutions (universities) and individuals. The development of the Hanseatic League as a way of acquiring wealth but also spreading expertise and technology. Redox chemistry and woad. The landscape of Norfolk—irrigation, peat and ports. The preservation and deterioration of material in peat bogs. The influence of pH, aerobic and anaerobic conditions. Land-based and marine harvests. Secondary metabolites from plants and microbes and biological activity. Medicinal chemistry and quorum sensing (and quenching). Deleterious effects of marine toxins. The production and benefits of long-chain polyunsaturated fatty acids (LC-PUFAs).

Keywords Woad (*Isatis tinctoria*) · Isatans · Indoxyl · Indigo · Redox in living systems · Quorum sensing

While the philosophical basis of chemistry lies with the consequences of the collision between alchemy and the atomic theories of Democritus and others and the replacement of Aristolean thinking by the corpuscular and atomic hypotheses which were developed throughout the Renaissance and the Enlightenment, the underlying motivation has frequently been to create added value. In the case of the alchemists, this was achieved by the transmutation of base metal into gold or the quest for eternal life. Later, as more precise methods for measuring the proportions of the constituent atoms in a molecule were developed in conjunction with the theory of atoms and organic chemistry, in particular, benefitted from the calculation of empirical formulae (the proportions of the different components that were present in a substance) and the philosophical tradition of degradation of compounds into their constituent parts to determine their structure followed by recombination to form the original authentic material [1]. In addition to the transformative tradition of alchemy and early chemistry another important enabling focus was the alleviation of disease and this is reflected in the isolation and purification of morphine [2] and quinine [3]. These objective of the treatment of disease and ways of measuring the proportions of different atoms in a substance enabled chemists to consider the possibility of

B. Hanley, *The Chemistry of a County*,
https://doi.org/10.1007/978-3-032-21491-1_4

reacting together substances whose additive formulae were the same as that of a target compound in order to create the compound of interest. Empirical formulae were, however an imprecise guide to the composition of organic compounds since they did not provide information about the structures of compounds or the relationships between their component parts. Indeed it was not until the beginning of the nineteenth century that Dalton began to expound his atomic theory [4] and the aromatic structure was not proposed until 1865 [5]. The possibility of combining empirical formulae to create compounds led W H Perkin to attempt a synthesis of quinine from N-allyl-toluidine (Fig. 4.1) using the empirical formulae.

Perkin did not synthesise his target molecule but, with a presence of mind that has subsequently eluded many synthetic organic chemists faced with the almost ubiquitous 'black intractable tar', he isolated the first, stable aniline dye, mauve in 1856 [6]. While this product was not an effective treatment for malaria (unlike quinine), it led to the creation of the dyestuff industry [7]—particularly in England [8] and Germany—and, because he was the sole inventor named on the patent Perkin became very wealthy such that he was able to retire very early in 1874 and to work on chemistry research. In addition to mauve, Perkin also synthesised a number of other organic molecules including alizarin and the perfume ingredient, coumarin. His son (W H Perkin Jr) was a notable academic and was Professor of Chemistry at Manchester and Oxford [9]. The development of organisations such as the Society of Chemical Industry in the UK (1881) and the American Chemical Society (1876) with the aim of promulgating the application of chemistry were a sign of the importance of chemistry to industry in the nineteenth century and this continued into the twentieth century with the establishment of organisations such as the European Chemical Industry Council (Cefic), founded in 1972 and recent innovations such as European Federation of Pharmaceutical Industries and Associations (EFPIA)

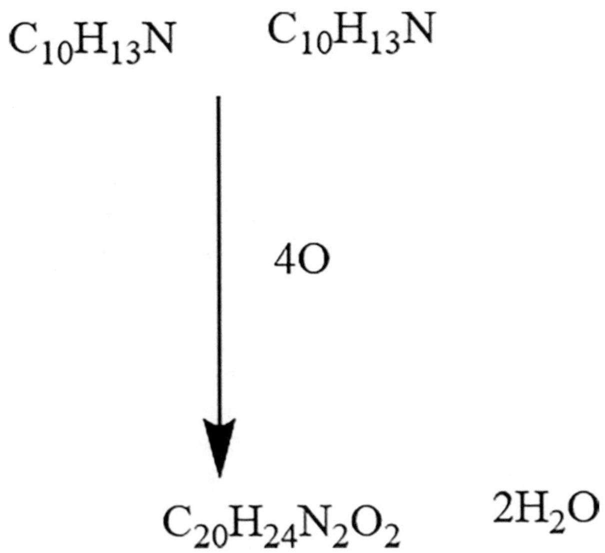

Fig. 4.1 The proposed synthesis of quinine by W H Perkin

founded in 1978. While there is a significant history of the use of chemistry in preserving health and combatting disease—for example in the development of the 'magic bullet' concept enunciated by Ehrlich [10] whereby compounds were developed to specifically act against disease vectors but leave the host unharmed, the application of chemistry in other areas has, as indicated above, created significant and impactful developments. Some of these will be illustrated in the following chapters. The development and exploitation of chemistry requires involvement at individual (chemist), organisational (commercial or academic setting) and societal (appreciation of benefits) levels. By harnessing these, chemistry can be the basis for successive developments in all three. This will become increasingly clear as successive innovations underpin advances and wealth creation in Norfolk.

As observed previously in the consideration of crop rotation, an empirical approach to the development of new technology can yield favourable results even if the mechanism of action of the technology remains obscure. In the case of the use of woad (*Isatis tinctoria*) as the source of a colouring agent, the method of preparation far outstripped any understanding of the chemistry involved. The use of dyes was common in pre-medieval Europe and both woad and indigo (*Indigofera tinctoria*) were common sources of blue colour. Recipes for the extraction and use of colours in Medieval times are available in manuscripts and the languages used (often German, English or Italian) gives some indication of the regional spread of dye making. A number of manuscripts included instructions for the preparation of medicinal products as well as extractions of dyes, suggesting a crossover between methodologies [11]. Merchants and manufacturers who dealt in dyestuffs were able to amass considerable wealth. Indeed in Germany in the Middle Ages, one group of very wealthy merchants were called waid herren (gentlemen of woad), and the town of Erfurt (Thuringia, Germany) had gained enough wealth through the woad trade to establish its own university in 1379 (before even Heidelberg) and it had an important medical school [12] however it was closed in 1816 when Erfurt became part of Prussia before being re-established in the twentieth century (in 1994). The importance of the colour blue increased in the Middle Ages for economic and religious reasons and, since woad could be grown in the region around it, Erfurt became an important market town with strong links to the Hanseatic League however the importation of indigo (*Indigofera tinctoria*) from Asia in the sixteenth and seventeenth centuries resulted in the gradual diminishing of the wealth of Erfurt and other centres of the woad industry however protectionism in parts of Europe limited the spread of indigo and allowed the woad trade to continue to prosper for a period after it was no longer economically competitive.

4.1 Woad Blue: Plants and Processing

In England woad-derived blue colour was used in pre-Roman times by iron age tribes and has traditionally been common in East Anglia. In part, the later production of a blue pigment from woad may have increased due to the influx of Flemish

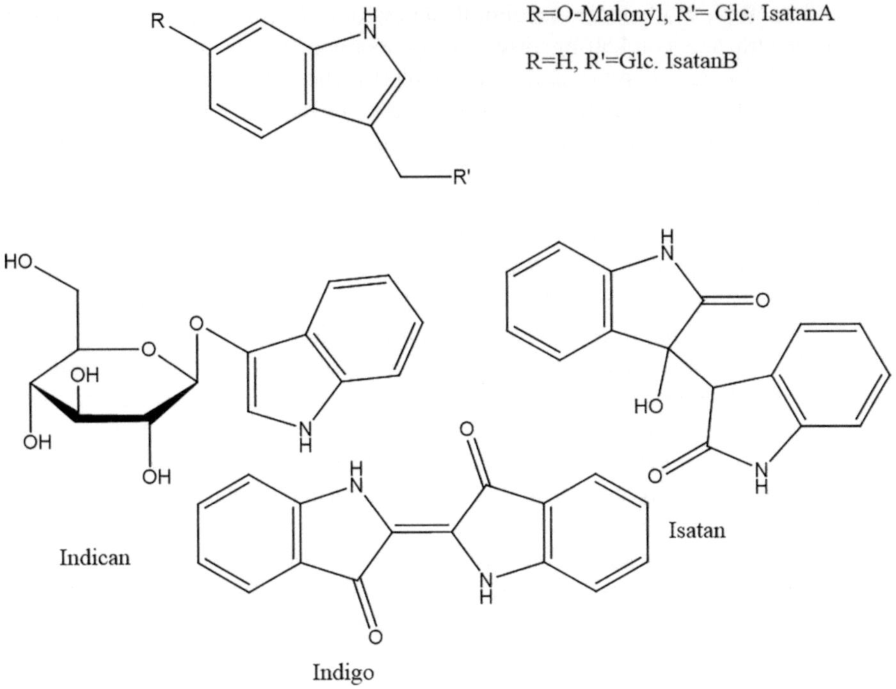

Fig. 4.2 Structures of Isatans and other Indigo precursors in Woad

weavers in the sixteenth century (see Theme Two, Chap. 6) however woad processing was already established and this, together with the needs of the textile industry, led to the need to import woad from Europe. A significant number of myths and legends have grown up around the development and use of woad. Some of these relate to the skin-dyeing properties of extracts of the plant [13] while others deal with the use of woad-derived colours in the production of dyed fabrics and illuminated manuscripts. The indigo precursors present in woad (isatan A and isatan B) have been isolated and characterised [14]. Both are glycosides and undergo hydrolysis to give blue indigo (Fig. 4.2).

In addition to the indigo precursors, woad also contains a number of bioactive ingredients [15] and this may account for its reported ethnobotanical use as an antiseptic and for other medicinal uses [16]. The use of agricultural land to grow woad was limited and in England woad cultivation became strictly regulated in the late 1500s in a period of food shortage prior to the famine of 1586 and concerns that too much land was being devoted to woad rather than to cereals. Queen Elizabeth I issued a "Proclamation against the sowing of woade" on 14 October 1585. This led to an increase in the importation of woad from Europe. Interestingly one of the major trade routes in the 15th and 16th centuries connected Erfurt with the Baltic Sea ports of the Hanseatic League including Lubeck and hence with Norfolk.

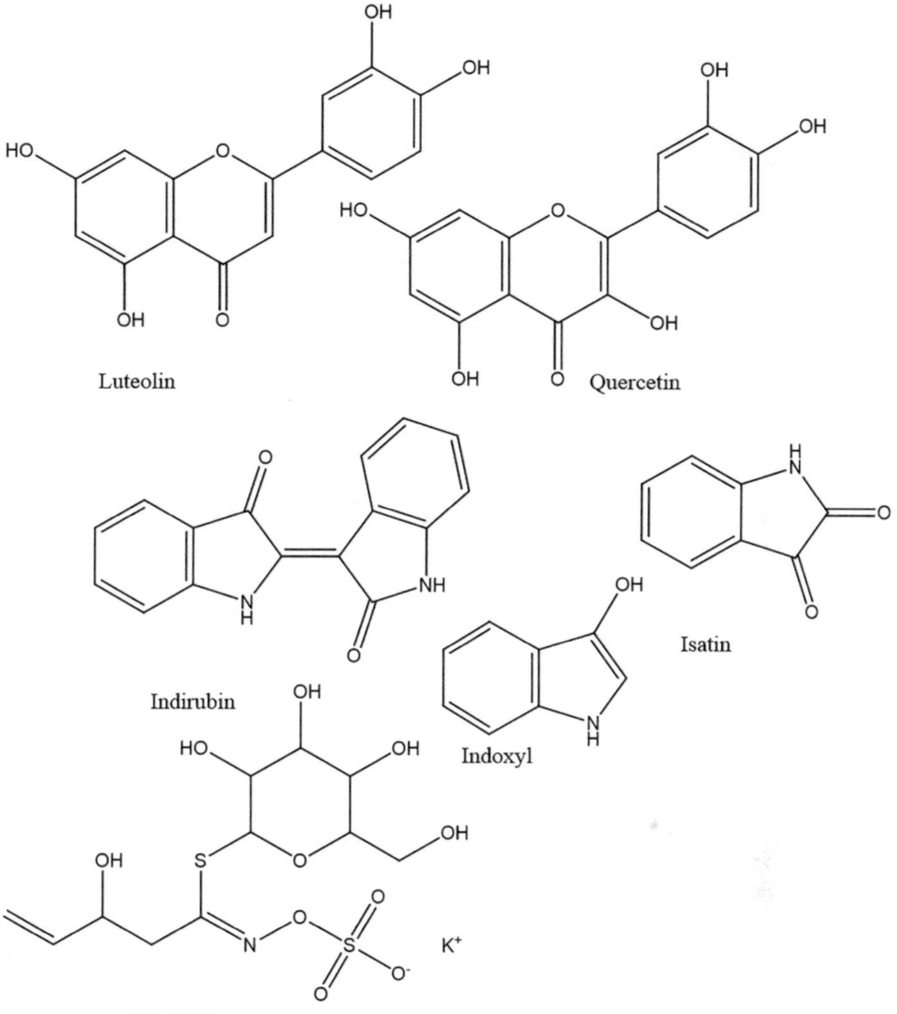

Fig. 4.3 Structures of phytochemicals from Woad

The presence of indole compounds (some as precursors of indigo) has been mentioned. Woad contains a number of indole compounds of various classes including alkaloids, phytoalexins and glucosinolates (Fig. 4.3).

and a number of these have been characterised [17, 18]. The principal commercially interesting feature of woad remains the production of a blue dye and the chemistry of this process has been studied. The production of indigo occurs from indoxyl [19]. Briefly the two indigo (or indoxyl) precursors, indican (~25%) and isatan B (~75%) are present and hydrolysis of isatan B and indican releases the intermediate compound indoxyl from which indigo or indigotin is subsequently formed (Fig. 4.4).

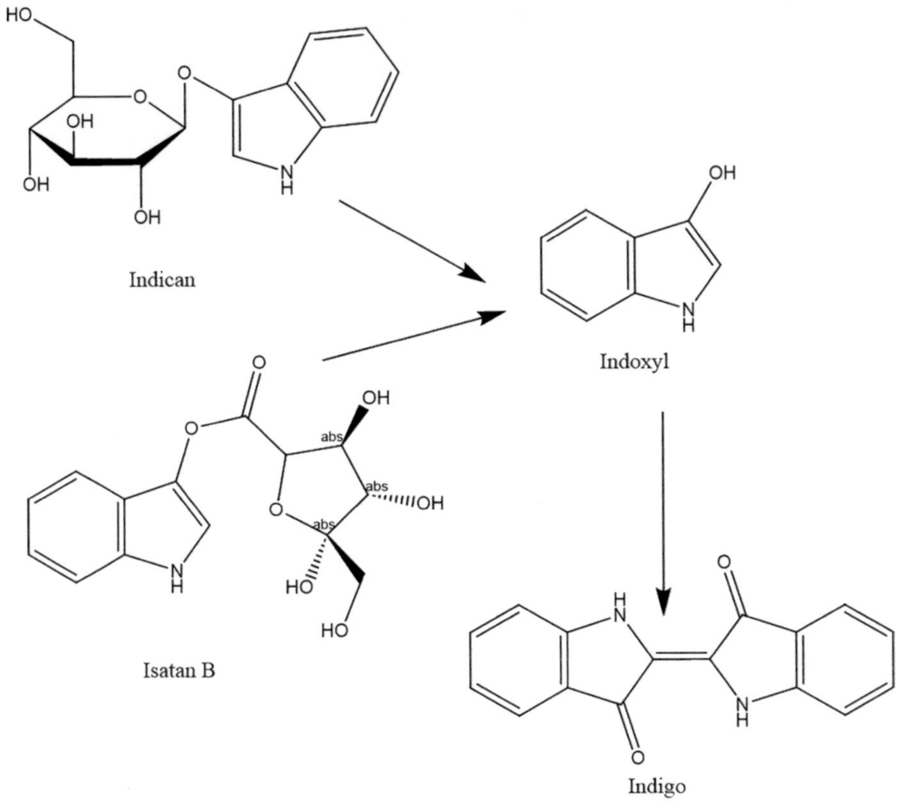

Fig. 4.4 The production of Indigo from precursors from Woad

although the precise details of the transformation remain unclear [20] although a spontaneous dimerization of indoxyl has been suggested [21]. In addition to providing the source of colour, the indole compounds have been suggested to be useful metabolomic markers to identify different members of the *Isatis* family [22]. The potential biological properties of extracts of woad have also been noted (see above) and this may also help to explain its use beyond being a source of dye. Part of the medicinal mechanism of action may be a consequence of anti-inflammatory activity [23] however anti-ageing effects have also been noted [24] and anti-viral (COVID 19) activity suggested based upon molecular modelling [25]. Woad was the primary source of indigo in Europe before the opening of trade routes to Asia and the importing of *Indigofera tinctoria*. The processing of the leaves of woad to give blue indigo is not a trivial process and, like crop rotation, it was likely developed as a result of trial and error. The process involves grinding the plant leaves to a smooth paste. This will allow hydrolytic enzymes in the plant to come into contact with substrate molecules and lead to the release of the aglycones. Since woad also contains indole glucosinolates, a range of hydrolysis products are also likely to be formed [26] including indole-3-acetonitrile and indolyl dimers, trimers and other compounds. In addition woad also contains other members of the glucosinolate family [27] and the paste will therefore produce a range of pungent principles including isothiocyanates. The extract is made alkaline (traditionally by addition of urine or the use of water derived from chalk streams which occur throughout Norfolk) and transformed (reduced) into a water-soluble form (leuco-indicotin) which is converted into indigo by air oxidation (see above). A range of factors can influence the nature of the compounds that are formed (and hence the colour) and some of these have been studied [28]. Clearly the traditional production of indigo from woad will be accompanied by the release of volatile odour compounds produced by rotting brassicaceous vegetation [29]. While the more recent uses of woad are medicinal rather than for the production of colouring material, the levels of bioactive secondary metabolites present is low and therefore a considerable amount of extracted vegetative material must be discarded and disposed of. This is even the case for extraction from Indigofera species and therefore the use of recombinant [30] or wild type [31] microbial strains has been investigated. This requires the use of fermentation rather than the growth of plants. The development of fermentative technology for the production of useful secondary metabolites was achieved, amongst others, by Pfizer [32]. In an important breakthrough, Pfizer developed deep tank fermentation for the production of citric acid using a strain of *Aspergillus niger* and developed conditions which were later used for the production of penicillin [33]. The extension of fermentative techniques in more recent times has been greatly assisted by the application of technology to identify active and silent biosynthetic gene clusters (BGCs) from a range of species [34] and their insertion into appropriate host organisms whose fermentative properties are known and can be manipulated to maximise the yield of the desired metabolite [35]. This is a significant advance over the use of plants as a host for metabolite production or also for the production of food bioactive constituents [36].

The processing of woad to give a range of colours and for its medicinal properties is, in some ways, a microcosm of the development of natural product chemistry. The importance of redox chemistry in the generation of a stable blue colour was appreciated in a pragmatic and empirical sense before the terms oxidation and reduction came into usage by chemists. Other similar processes were used in, for example, the production of Tyrian Purple (another indolic blue dye) [37] and the limited number of processes that were available relied upon material that was able to be accessed (chalk, urine, air oxidation etc.) and microbial digestion. The medicinal properties of woad extracts was appreciated before methods were developed for the precise molecular study of its active principles had been developed [15] and later, for example, the molecular basis of its anti-inflammatory activity was investigated [23]. The macroenvironment in Norfolk including the long coastline and the inland waterways that developed as a consequence of extensive peat extraction [38] also had an impact upon the chemistry that was developed and, as with the adoption of crop rotation, advances and expertise imported from Europe rather than from within England was significant.

4.2 Peat Bogs and Irrigation

Much of the environment of Norfolk was created as a consequence of the digging out of peat for fuel and the subsequent flooding of low-lying land. This not only affected the development of agriculture but also impacted upon transport and the processing of crops. There is an analogy with the low-lying parts of near-Europe and the solutions developed there came to be applied in Eastern England. In pre-Roman times, a large part of Norfolk was submerged by alkaline, mineral rich water. Subsequent excavations suggested that much of the land was subjected to successive farming activity and floods [39]. Local tribes (particularly the Iceni, pre-dominated and the low lying nature of the land discouraged the Romans from building roads or other communication networks. More significantly, rather than looking to the rest of England, the Romans built ports although a fall in water levels led to the construction of the Great Fen Causeway which ran from Peterborough into Norfolk [40]. Later, the building of the cathedral in Norwich by the Normans was aided by the importation of stone from Caen (France) rather than from elsewhere in England. The stone to build Ely cathedral, however was quarried from just north of Peterborough and transported *via* internal waterways. Further east of the Causeway, the land was less well drained and the open sea was in evidence as far inland as Acle (currently about 8 miles from the sea) and many of the rivers were navigable which meant that supplies and even building materials could be sourced from Europe (see above). In the times before the Norman Conquest and the establishment of Norwich as a key administrative and trading centre (and a navigable port), the east part of Norfolk came to be increasingly settled and farmed. Later, during the Middle Ages, arable farming and the wool trade increased the wealth of Norfolk and this is evidenced by the presence of records of the existence of over a thousand medieval churches. The importance of the region (despite the relatively poor land transport network) was shown by the growth in the population and the establishment (later) of significant estates. It is against this background that the evolution of agricultural improvements such as crop rotation and the import of weaving technology should be noted. Both were established in an area with a long tradition of agriculture and of looking both to Europe and to the rest of England for innovation. While rich agricultural land contributed greatly to wealth generation, the less manicured harvest from the sea also provided significant benefits. The difference between the chemistry of fresh and sea water helps to explain the economic benefits of each. While the sea conditions were suitable for inshore fishing and the harvesting of crabs and other shellfish, a major source of income for Norfolk was the annual migration of the herring shoals down the east coast of Scotland and England [41]. Great Yarmouth was recorded as a settlement in the Domesday Book and by the fourteenth century, it was one of the wealthiest towns in England. It fell into decline due to competition from Baltic ports and the silting up of its harbour but remained a significant presence on the East coast [42]. In addition to local processing and consumption, herring, being an oily fish, required to be salted or pickled before the fat in it was oxidised and the fish became rancid and techniques for preservation were imported from the Netherlands and Scotland. The export market to northern Europe was an important resource and both the east coast of Scotland and England and northern European ports benefitted [43] both from the herring itself and the migratory labour that followed it [44]. In addition to their role as a marine crop, herring also attract other fish such as tuna and higher predators and so have an important impact upon the whole sea ecosystem [45]. The presence of genomic adaptation in Baltic and Atlantic herring has been noted [46] and this is, at least in part, due to adaptation between salt (Atlantic) and brackish (Baltic) water. Herring are a foraging species and are more prone to extensive population shifts (collapse and recovery) than other species [47]. The genetic stability of herring over time has also been examined using single nucleotide polymorphism (SNP) analysis [48] of fish recovered from ancient sites and the impact of human activity on the different distinct populations of herring and the relative proportions of, for example, Spring and Autumn-spawning populations [49]. In addition to the evolution of fishing techniques, the development of processing had a significant impact on medieval communities [50] and the fishing habits provided an impetus for processing of fish at scale (primarily, in the case of England, for export) and this fuelled the expansion of fishing fleets [51]. A significant transient population, therefore, followed the herring shoals along the east coasts of Scotland and England.

As was the case with the natural resources of the sea, the trade in agricultural crops in Norfolk and elsewhere in East Anglia was dominated by Europe. By the end of the eighteenth century, a significant proportion of the land in Norfolk was used for arable farming [52]. This situation was encouraged by the adoption of the practice of enclosure (as outlined in the Inclosure Act of 1773) whereby small holders lost their land and it was amalgamated into larger, more efficient units. This facilitated the development of a mature export market and by the end of the eighteenth century, significant amounts of grain were being exported from Norfolk (and elsewhere in England, although Norfolk because of its advanced agricultural practice and nearness to Northern Europe was at the centre of trade) [53]. The previous practice of peat digging (discontinued by the fourteenth century due to rising sea levels and flooding) enabled the creation of navigable channels and this also facilitated the export of agricultural goods to Europe. In addition to grain and related crops, Norfolk also had a thriving weaving and textile industry—in part predating the influx of skills from Flanders during the reign of Elizabeth I (see Theme Three, Chap. 6). The worsted wool industry of the fourteenth century produced a lighter grade of cloth than the material (broadcloth) that was traditionally exported to the colder countries of northern Europe and so was exported to more southern countries (Portugal, Spain, Italy, France) [54]. While this export market began to be lost to locally produced cloth in Europe, it was revitalised by the importing of Flemish technology (and craftspeople) in the sixteenth and seventeenth centuries. Technology for food production, preservation and manufacture of goods came about as a consequence of the links between Norfolk and Europe. These advances (and the wealth and expertise they generated) were an important pre-requisite of the growth of industry and the subsequent increase in industrial output in England as a whole [55]. While the coastal waters provided a beneficial environment for harvesting fish and the agricultural land in Norfolk was fertile, there were other aspects of the local environment that facilitated the adoption and growth of technology. As noted previously, peat digging combined with rising water levels and flooding created the vast areas of navigable rivers and lakes called the Norfolk Broads. However, in addition, other rivers in Norfolk have particular characteristics. Among the most significant are those which derive from chalk aquifers. Only slightly over 200 such streams occur globally, and a high proportion exist in Northern Europe [56] and in Norfolk. Freshwater streams—and those derived from chalk beds in particular—are under threat both from climate change [57] and from the impact of an increasing human population and abstraction of water. The chalk aquifers of northern Europe have been studied more generally and the impact of their reduction assessed [56]. In the case of Norfolk, chalk streams were used to power mills and for other uses. Chalk streams in Norfolk include the Rivers Bure, Glaven, Stiffkey, Burn, Heacham, Ingol, Hun, Babingley and Gaywood, but the longest, biggest and most significant is the River Wensum which flows through Norwich. The river provided the basis for the industrial growth of Norwich including, in particular, brewing (Theme 4 Chap. 9), light engineering and manufacturing (Theme 3 Chap. 7) and dyeing (Theme 2 Chap. 5). The high levels of calcium and other minerals (see below) in the Wensum and other chalk streams may also have health benefits for consumers [58]. A number of mineral water manufacturers were in business throughout Norfolk in the nineteenth century including Caleys which later diversified into chocolate and confectionery manufacture before being taken over, successively by Mackintosh, Rowntree and, eventually the Swiss conglomerate Nestle (which also has a significant mineral water business and has investigated the effects of high mineral content water on human health [59]). The importance of mineral nutrition in the growth of plants has been noted historically [60] and the mineral content (particularly calcium) of the freshwater streams, aquifers and groundwater of Norfolk will likely contribute to the success in the growing of arable crops [61]. The realisation of the importance of calcium for plant growth is a relatively recent development and, while observations about which conditions best support plant growth will have been recorded during the development (for example) crop rotation, this will not have been derived from a mechanistic understanding [62]. The water supply of Norfolk from underground sources was reviewed in 1921 [63]. This survey defined Norfolk as a maritime county with a significant sea coast and with other county boundaries made up primarily of river courses, with the Ouse and the Yare as the most predominant river basins. Chalk springs were found to come to the surface at various places in the county and underground rivers were also found to be present. In a few places, canals were dug to improve communication and the transit of goods. In terms of transport *via* water, Norfolk rivers (and canals) either flow to the west to enter the sea at the Wash by way of the Ouse or, if they originate further east, they are part of the Yare basin and reach the sea at Yarmouth. In either case, the principal sea routes were traditionally towards Europe. Near the coast, it has been suggested in the survey of 1921 [63], that "It is evident that the fresh water is floating on the salt, and it is remarkable that it should not undergo appreciable contamination." The same analysis suggested that the waters found in Norfolk are "extremely hard, from the presence of lime and magnesia salts, oxide of iron, alumina, and silica". These significant levels of minerals, together with the addition of organic matter from decomposing vegetation (implicated in the formation of peat) and from domesticated animals, may help to combat the nutrient deficiency of peat-containing soils found in Norfolk but also change the carbon accumulation [64], may impact upon the soil microbiome (which is responsive to nutrient content) and balance the acidic pH. The interaction of peat with soluble nutrients is complex and has been the subject of considerable study and different types of peat have different characteristics [65]. The combination of an adequate, mineral-rich water supply and good drainage may help to explain the fertility of the soil in Norfolk and its growth as a centre for arable farming during the medieval period [66]. Other aspects of the industrial growth of Norwich (as noted previously and later) also made use of the supply of water with particular characteristics and rivers both for use and for navigation. While transport and access to new ideas and technology from Europe and the facilitation of agriculture are key aspects of the value of a water-dependent economy, other industries make direct use of a source of water, and these will be examined in more detail in later segments.

4.3 Peat and Preservation

The importance of peat as a medium for preservation has been noted. Part of this attribute is a consequence of the particular microbiome of peat [67], and this has also been increasingly appreciated as a way of optimising carbon capture by peat. In addition the anaerobic and acidic properties of peat limit the growth of microbial communities that may decompose organic material. The exception to this is bone which, under acidic conditions in peat, is more rapidly dissolved and loses structural integrity. One of the major challenges to determining the functional properties of peat as an environment is the accurate determination of the microbial communities that are present [68]. Once the microbial communities (and their functional properties) have been determined then the preservative characteristics can be examined in more detail. Peat itself has been found to be one of the few types of environmental conditions that can preserve historical artefacts [69]. In addition to those articles that are composed of reactive metals (for example, iron goods) which are often lost to oxidation, many artefacts are, as noted above, organic. Not only can peat act as a preservation environment, because peat bogs are laid down over many years, but they can also provide an extended time capsule of preservation. Radiocarbon dating can determine when an object was no longer living (i.e. the radioactive carbon was no longer being renewed) and this, combined with non-invasive methods such as x-ray fluorescence, can provide significant information [70].

Peat comprises a number of components derived from decomposing organic matter and it is those properties that make a significant contribution to the value of peat as an adjunct to the growth of agricultural crops. The most significant components in peat are humic and fulvic acids [71]. Humic acids comprise a number of structural types depending on source material, treatment and microbial activity. In the context of the last of these, humic acids, themselves have a number of effects on microbial populations including bacteria [72] and viruses [73] and these properties suggest a potential pharmacological use as anti-microbial agents. Humic substances more generally occur extensively as a product of decomposed organic material and are classified according both to physicochemical parameters and derivation [74]. Both biotic and abiotic degradation processes can take place and these give rise to the building blocks of humic substances including oxidised lignins, aromatic components, carboxylic acids, peptide complexes, quinones and others. The microbiome of peat has been studied, as noted previously and is affected by vegetation type [75] and also by the period of time of fermentation and, obviously by the population of microbes present [76]. The microbiome in turn affects the properties of the peat [77]. While peat is acidic and this limits the release of nutrients, the addition of lime (and presumably alkaline chalk irrigation) increases the pH [78] but only to a value of about 7.5. Because of its constituents, peat has an intrinsic buffering capacity however its capacity to bind to and store nutrients is impeded by a raised pH and some nutrients (and minerals) will become increasingly bioavailable, however the simple correlation between soil pH and nutrient bioavailability is not necessarily justified [79]. This combination of peat and chalk springs may in part explain the fertile nature of agricultural soil in Norfolk. In addition a positive effect of peat on the rhizobium-dependent nitrogen fixation in legumes has been found with an increased microbial tolerance to stress and an increase in abundance in proteins involved in amino acid and carbohydrate transport and metabolism [80]. In addition to nutrient release, peat (as noted above) can also assist in the preservation of artefacts and other material. The preservation of artefacts comprising different materials in soil has been studied [81]. The low pH of peat means that material containing calcium (for example, bone) is dissolved more readily than other material. The replacement of calcium by other elements present in soil is also possible. In the case of glass, medieval material made from wood ash (containing potassium) is more susceptible to acid attack that earlier glass made with soda ash [82] which is resistant to acid attack. Decorative features on glass made with potash-lime-silica and hence more alkaline are also fairly rapidly lost [83] compared to earlier (soda-lime-silica) glass. Organic material is rapidly decomposed by damp, aerobic conditions since these encourage microbial growth and biological oxidation. Fewer bacteria (and more fungi) are found in more acidic soil [84] and this also decreases the rate of decomposition. If a soil solution is high in dissolved organic matter (greater than 2%), then this can react with buried organic material and this process may confer resistance to biological degradation, as occurs when skin and soft tissues are deposited in waterlogged peat and become biologically resilient. One exception to this occurs when soil is prepared for agricultural exploitation and oxygen is introduced. The concomitant release of nutrients can lead to an increase in microbial activity and the subsequent accelerated degradation of organic material. Clearly the preservation (and hence rate of degradation) of material in soil is dependent upon the nature of the material and the stability of the environment as well as the features of the soil itself. Cycles of wetness and drying is likely to accelerate degradation. In the case of peat, as long as it is wet and undisturbed, its low pH and high organic content (and the exclusion of oxygen) means that organic material may be relatively well preserved [85]. The increasingly widespread adoption of agricultural practices will result in the accelerated degradation of any buried organic material. While this suggests that leaving ancient sites undisturbed has generally been considered to be good archaeological practice, this has been questioned and the deterioration of organic material preserved undisturbed in peat has been found to occur [86]. In the case of Norfolk (and other sites in East Anglia), the extensive spread of drainage practices (with, in many cases, technology imported from Europe) to create rich agricultural land, has diminished the preservation of historical artefacts [87]. The study of peatlands in Norfolk where extensive agriculture has been carried out [88] suggests that drainage causes more mineralisation of organic matter resulting in changes in microbial colonisation (particularly by sulphate reducing bacteria) results in more acid sulphate soils and increased degradation of peat and organic material. While this may have detrimental effects on the preservation qualities of peat, it may also increase soil fertility and the bioavailability of essential nutrients. The multivariate and complex effects of a range of factors on soil and its fertility makes extrapolation between different sets of conditions problematic and partial explanations are more likely. By determining the effects of different conditions (pH, level of aeration, water content etc.) on surrogate measures (the level of preservation of organic material and nutrient release from peat), it is possible to arrive at some generalisations concerning some aspects of the characteristics of some types of soil and an explanation of its agricultural consequences.

4.4 Fresh and Sea Water

As noted above, the economic and industrial relevance of Norfolk is intrinsically linked to the marine and inland aquiferous environments. The nature of the sea around Norfolk is a consequence of the geology of the shore. The coastline is underlaid by chalk bedrock composed of the remains of marine organisms. While chalk is a relatively soft material, it is more resistant to erosion than many of the other deposits found on the North Norfolk coast [89]. The cliffs are relatively soft and heterogenous and are, therefore, prone to erosion caused both by waves and by undercutting caused by leaching of groundwater. Different types of erosion can occur depending on the nature of the cliffs and the beach below them. A number of methods have been adopted to prevent erosion (and loss of agricultural and settled land) including physical sea defences (which can be effective against wave-induced erosion but less so as a protection against groundwater-induced erosion) and investment in salt marshes [90]. The interaction between the various components of the shore is essential to developing an understanding of factors that influence erosion. The use of Light Detection And Ranging (LiDAR) together with photographic evidence and information about wave activity provides a more comprehensive picture of soft-cliff erosion and mass movement processes that occur in these (and similar) coastal areas [91]. The tidal area around the Wash is of national and international significance and, while significant areas have been periodically drained and converted for agricultural use, the remaining intertidal area is a significant ecosystem [92]. There is a long standing and significant tradition of inshore fishing in Norfolk which is in part a consequence of the sandbanks and natural features of the coast and this, together with processing and the export of products, helped the region accumulate wealth in the past. The decline in this harvest has, more recently, had a negative effect on the social activities of a number of coastal towns [93]. In addition to fishing and agricultural land reclamation, there is some evidence, in parts of Norfolk, of the harvesting of sea salt in Saxon and medieval times [94]. While this was a largely transient activity, it occurred at various coastal sites throughout England [95] and Scotland [96]. The methods used in some places have been documented and the activity continued at coastal sites until the discovery and mining of inland salt deposits in Cheshire in the eighteenth century [97]. In addition to the production of salt and the exploitation of near to shore shellfish resources (including crabs) there is some evidence of fishing for cod and herring in medieval times [98] and, while this was noted previously (see above) it would also have required the building and launching of seagoing ships thereby impacting upon ancillary industries. In addition, fish was an important resource in areas such as Norfolk with its large number of religious communities [99] where there were religious constraints on the consumption of meat [100]. The nature of the water (particularly its chemical content) and the geology of the seabed near to the Norfolk coast is an important factor in the establishment, growth and development of edible crustaceans. The Cromer shoal chalk bed which can be found just offshore, and stretches along the coast for 20 miles from Weybourne, west of Cromer, to Happisburgh in the east is an important feature that facilitates the growth of a particular, economically important species of brown crab (*Cancer pagurus)*, for example. The wealth generated by the fishing communities and the adoption of European innovations [101] were significant factors in the development of wealth in Norfolk. This, combined with the growing sea trade, provided significant income and wealth based on fishing and on the associated skills and trades (ship building, sailing, food preservation etc.).

The growth in agriculture as a consequence of the availability of arable land and the adoption of certain agronomic practices has already been mentioned. Arable land in Norfolk was productive and land reclamation was an important factor in ongoing wealth creation. The genesis of fresh water in Norfolk was a consequence of geology and the temperate climate. Chalk springs and rivers were important sources of fresh water and this, combined with the surface collection of rainwater, facilitated the growth of crops and animals. While both plants and animals were primarily a source of macronutrients and selective breeding was employed to maximise these characteristics, the later importance of crops such as mustard and mint was dependent upon secondary metabolites (particularly isothiocyanates and terpenoids) while factors that affected processing characteristics (ease of harvesting) led to the development of peas, in particular, with specific agronomic characteristics (semi-leafless varieties). These have already been noted and will be considered in more detail in Sects. 4.5 and 4.6 and in Theme 4, Chap. 8, alongside the small molecule communication strategies (quorum sensing) that are important aspects of the health of ecosystems, which provide some mechanistic basis for many of the observational studies that led to the development of crop rotation and are also a key function of plant and microbial secondary natural products.

4.5 Natural Products and Quorum Sensing

The importance of natural products has been apparent for a considerable time. The simplest interpretation of their use for a producing organism is that they provide a growth or survival advantage over other species either by directly acting as competitive growth inhibitors or deterring predation by grazing species. In addition to the isolation of individual (or groups of) compounds, natural product biosynthetic gene clusters (BGCs) can be studied and provide information at a gene expression level. This information may be used, among other things, to predict the potential for natural product formation and has enabled ecosystem advantages for specific organisms to be delineated [102]. While the potential for the production of compounds and even in some cases their likely structures and perhaps some indication of likely biological activity can be predicted through the investigation of BGCs, the specific advantageous mechanism of action of secondary natural products requires the isolation of the products themselves and their testing in a relevant activity screen—this has traditionally been the role of the natural product chemist. The application of reductionist principles in the isolation and characterisation of individual compounds has led to the isolation of specific molecular entities. The advantage that is gained by an organism that invests in the machinery to produce secondary metabolites can be through the development of specific toxic entities that inhibit competing organisms directly (for example the antimicrobial compounds produced by some fungi), the use of some compounds to prevent pathogen attack [103] and the production of antifeedants—particularly by plants—to discourage predation [104]. These can all be considered to be defensive strategies since their objective is to prevent or inhibit the proliferation of competitor species. Such secondary functions provide organisms with a distinct evolutionary advantage over other species that compete for resources or space.

Some secondary natural products also had other serendipitous biological activities which were found to be of particular (often medicinal) value [105]. The discovery (and purification) of compounds as diverse as morphine, quinine and statins led to the discovery of a use for them which was not part of their primary function for the producing organism. Optimisation of this exceptional biological activity and of the production of natural products and structural variants with enhanced properties, provided an impetus for synthetic chemistry and, subsequently, for the genomic manipulation that led to the development of synthetic biology [106].

In addition to these 'accidental' advantages, some natural products are of apparent benefit both to the producing and the consuming organism. In the case of mustard (noted above), one of the principal classes of secondary natural products are

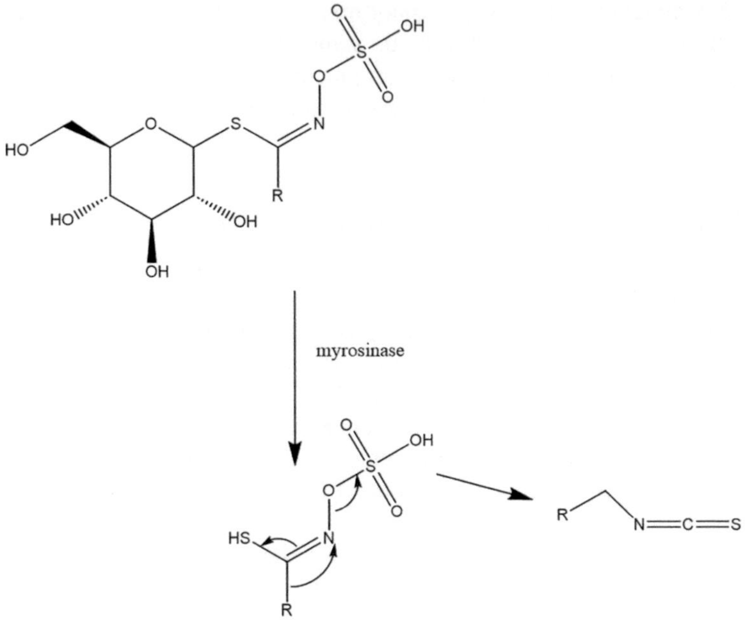

Fig. 4.5 Hydrolysis of glucosinolates by myrosinase and subsequent rearrangement

glucosinolates (previously referred to as mustard oil glycosides). These involatile thioglucosides which are largely biologically inert are stored in particular types of plant cells [107] and are physically separated from a co-occurring thioglucosidase enzyme [108]. When tissue damage occurs (by, for example, insect or herbivore attack) the interior cell wall structures are disrupted and the enzyme and substrate come into contact and the thioglucose moiety is cleaved giving rise (after a spontaneous rearrangement) to a range of products including isothiocyanates (Fig. 4.5).

Most of the isothiocyanates are pungent and have insecticidal activity [109] and this can limit the damage done to the producing plant and can deter consumption by herbivores. In addition to their protective activity for the producing organism, some isothiocyanates appear to be of some benefit to some consuming organisms [110]. In particular, one such compound, sulforaphane, has been implicated as an anti-cancer agent [111] and may have other properties [112]. In addition (and perhaps more significantly from its potential use as a therapy for chronic disease) sulforaphane has documented activity as an anti-inflammatory agent [113] and is proposed to act as a negative regulator of the NF-κB pathway. While general anti-inflammatory activity has been postulated as a mode of action for a range of secondary metabolites, an effect on specific disease-related clinical endpoints is more challenging to establish. In this context the molecular mechanism that underpins the potential of sulforaphane as a preventative treatment for Alzheimer's disease may be more interesting [114] and may provide testable hypotheses. While chemistry is primarily concerned with molecular mechanisms, traditional plant breeding has been, to a greater extent, focused upon the microbiological endpoints of combining the properties of different varieties of the same (or related) plants. The granular nature of the principle that underpin plant breeding (rather than the production of varieties with blended properties) was demonstrated by Mendel and this formed the basis of the production of plants with more desirable properties (for example, as has been shown previously, the more easily harvested semi-leafless pea). While the molecular interactions involved in gene regulation are an important factor for plant breeding, it is not necessary to understand genetic molecular interactions in order to empirically engineer the transfer of desirable traits. The use of advanced methods such as molecular trait loci mapping can greatly help in the analysis of the functional consequences of genetic variations [115] and this can both increase understanding and the development of desirable traits. In the case of a major *Brassica* crop, rapeseed, varietal breeding without understanding molecular mechanisms enabled varieties with lower levels of undesirable secondary metabolites to be developed. The impetus for the development of 'double zero' rapeseed varieties was the concerns over relatively high levels of erucic acid (which may have long term health effects) [116] and glucosinolates which can both produce goitrogens and limit the palatability of rapeseed [117]. While detailed genomic mapping and methods to monitor and engineer the reduced function of specific genes are now available [118], rapeseed with low erucic acid and low glucosinolate levels was first developed using traditional breeding techniques [119]. While challenges were noted due to the narrow genetic variability of the 'double zero' variety, molecular genomic trait analysis will likely enable the production of varieties with greater genetic variability. As molecular techniques are developed, the production of more sophisticated and focussed species will become possible.

As noted in the case of sulforaphane, compounds can have different (or multiple) modes of action on different biological systems within a single organism and pathways including modulation of carcinogenesis or inflammation can be affected, for example. An alternative paradigm is the production of compounds that are designed to have an effect on the biological systems in multiple organisms. An example of this is the case of quorum sensing. Quorum sensing is a communication system, traditionally observed in different species of bacteria, which enables organisms to coordinate their gene expression based upon population density and mediated by small molecules (autoinducers) [120]. The term 'autoinducers' is to an extent historical since the process of quorum sensing was originally referred to as autoinduction and was first observed in *Vibrio fischeri.* Quorum sensing has been implicated in a number of bacterial behaviours including swarming and biofilm formation [121]. While a detailed description of the relevance and molecular basis for quorum sensing lies outwith this chapter, for a historical perspective, a number of review articles have been produced and can be consulted [122, 123]. Quorum sensing has primarily been studied as a bacterial phenomenon, it has become clear that other organisms are able to manipulate microbial quorum sensing systems and cross-species effects are observed [124]. One important area of research is the inhibition of quorum sensing (quorum quenching) in which non-microbial species can interfere with quorum sensing and affect microbial characteristics such as biofilm formation [125]. In particular some plants can produce molecules that can act as quorum quenchers [126] thereby controlling the growth and spread of pathogens. Furthermore, the use of exogenous agents (nematodes, fungi) as a source of quorum sensing modulators by plants [124] may have significant effects on plant/microbial interactions and on the ecology of crop growth. A range of phytochemicals have been found to influence microbial quorum sensing and this could be significant in the development of strategies to control food-borne microbial disease [127]. Indeed, some of the reported anti-microbial effects of plant extracts may be a consequence of their quorum quenching capability. For example, it has been suggested that sulforaphane has significand quorum quenching capability and can prevent the formation of biofilms by pathogenic bacteria [128]. It has been suggested that biofilm-associated beta amyloid proteins could be involved in neurodegenerative disease [129] and that biofilm-forming pathogens may be implicated [130]. The quorum quenching and biofilm inhibitory activities of sulforaphane may prevent inflammation, inhibit beta amyloid plaque over-expression in response to pathogen infiltration and inhibit the protection of pathogens from removal by the creation of biofilms [131] and thereby slow the progression of conditions such as Alzheimer's disease that have a pathology involving beta amyloid plaque formation. Such an infective hypothesis together with modulation using quorum quenching may provide opportunities for new therapies for neurodegenerative diseases [132].

4.6 Marine Natural Products: Toxins, Polymers and Fatty Acids

As noted above, natural products have a number of functions including, in particular, the creation of an enhanced capability for a host to create a growth or survival advantage. In addition, some natural products also have biological activities that are distinct from those of relevance to the host and are of value to non-host species including anti-inflammatory and target receptor (including enzyme) binding. While marine natural toxins have been extensively studied, the range of activities identified has been relatively limited (not least because of challenges in accessing sufficient material) however there has been some particular activities identified including anti-infective agents [133] and the treatment of diseases with a prominent inflammatory component [134]. There is also a substantial and long-standing interest in discovering pharmaceuticals with anti-cancer properties [135] and others [136]. While these types of activity have not led to the development of significant pharmaceutical industry development in Norfolk, this is more a consequence of the traditional concentration of such industry in the area near to London in the UK for logistical reasons than a consideration of the potential source of marine bioactives. There is however an increased level of interest in the risks posed by toxins and in the production of polymers and nutritional fatty acids.

The spectrum of activity of marine toxins can include both toxic and beneficial effects and, while the emphasis in the next section will be on compounds that cause adverse effects, some of the potential beneficial effects will also be noted. In some cases (for example cyanobacteria) the causative organism can exist in either a marine or an inland environment and this should be noted. In the context of the chemistry of marine toxins, considerable efforts have been made to identify and develop analytical procedures for causative agents and these will, where appropriate, be referenced. Toxins represent a considerable threat to industry that relies upon quality and safety (as is the case for the food industry) and the identification of hazards and the quantification of risk is an important requirement. The drivers for the discovery of marine toxins are their toxicity and hence hazard to organisms that consume them but also their potential as medicinal agents [137, 138].

The structural divergence between marine and terrestrial natural toxins has been noted [139] although some exceptions such as tetrodotoxin (Fig. 4.6) have been noted. Tetrodoxin is found in some fish of the order *Tetraodontiformes*, but it is found in other species and in some bacteria.

Some of the most notable marine toxins include domoic acid, kainic acid, saxitoxin, okadaic acid tetrodotoxin, and polyether polyketides such as brevetoxin (Fig. 4.7) and while they form distinct structural types, they have some features in common with terrestrial secondary metabolites.

Fig. 4.6 The structure of tetrodotoxin

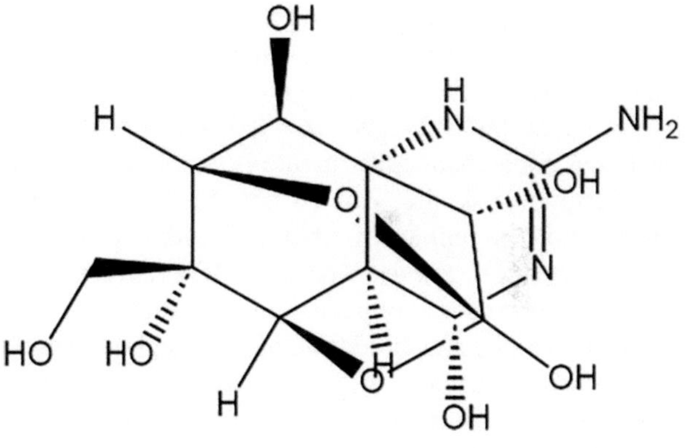

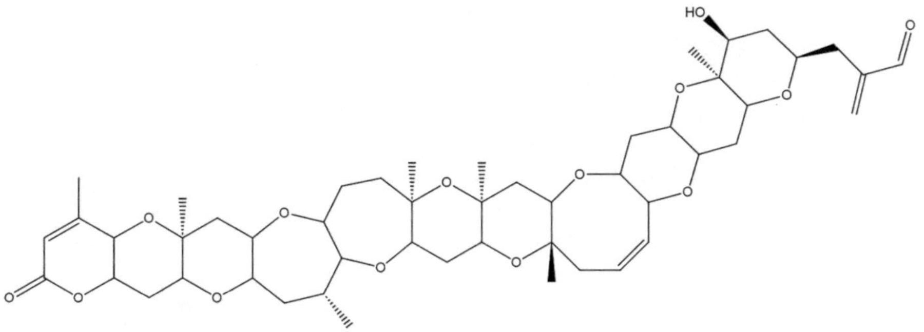

Fig. 4.7 The structure of brevetoxin

Another aspect of marine toxins is that they are commonly produced by algae and then accumulate in shellfish and other organisms that are consumed [140]. They are commonly referred to by their physiological effects—therefore domoic acid, is commonly referred to as Amnesic Shellfish Poisoning (ASP), saxitoxin causes Paralytic Shellfish Poisoning (PSP), okadaic acid causes Diarrheic Shellfish Poisoning (DSP) and azaspiracids cause Azaspiracid Shellfish Poisoning (AZP). Brevetoxins are responsible for Neurologic Shellfish Poisoning (NSP). While their production is widespread and their effect on consuming organisms (particularly humans) can be severe, it is thought that their production (like that of penicillin) provides some nutrient or resource advantage over other phytoplanktonic species. It appears that the toxins accumulate in the digestive organs of shellfish which consume them, and their effects are considerably less severe than in the case of humans. Shellfish toxins can be hydrophilic or lipophilic and this affects both their analysis and their bioavailability. The analysis of marine toxins has made use of both analytical and bioanalytical assays. One of the challenges to analysis is the need for authentic standards. Because of the complex nature of the molecular structures of the toxins, the most efficacious method was the isolation of authentic samples from natural sources rather than synthesis. In addition, the growing number of characterised toxins has revealed gaps in analytical methodology [141]. A number of techniques have been developed to remove marine toxins including depuration (washing repeatedly with clean water to remove toxins) and other methods [142]. Analytical methodology has relied heavily on animal-based bioassays which were the standard analytical control method for a number of years, however other analytical [143] and bioanalytical methods have been developed [144]. One of the problems with analysis based upon chemical structure is that it is only an indirect measure of the potential harm or risk that can be caused by known entities and does not provide information on unknown toxins or mixtures of toxic entities. Nonetheless the replacement of the mouse bioassay (MBA) as reference method in the EU Legislation by liquid chromatography coupled to tandem mass spectrometry (LC–MS/MS) was a major advance and validation of new methods remains a priority [145].

While maintenance of confidence in the safety of marine organisms that are to be consumed by humans is a major priority in marine counties such as Norfolk, other products from the sea are also an important resource. This includes, in particular biopolymers. The impetus for the use of marine biopolymers [146] is, in part, a consequence of the problems (perceived and actual) associated with synthetic polymers. One of the important factors in the development of synthetic polymers was their robust non-degradative properties. In addition they, were able to be manipulated and had multiple applications. Indeed, without the use of plastics and other polymers in a medical setting, control of infection would be far more challenging and storage of biological samples for later use (for example, blood and plasma) would be much more difficult. Synthetic polymers can degrade under certain conditions—either as a planned strategy for their removal [147] or as a desired/designed functional property of the polymer [148]. In addition, some polymers are prone to photodegradation [149] and this can lead to the uncontrolled production of chemical artefacts which may also be of concern. The degradation of synthetic polymers has increased awareness of, and concern about, the environmental impact of microplastics. Analytical methods for microplastics have outstripped work on the biological consequences of their ingestion and a number of correlative relationships between microplastic ingestion and clinical endpoints have been suggested [150, 151]. It is important to realise that many of the studies are correlative rather than causal however the evidence base is building [152]. The definition of microplastics is generally considered to include particles derived from the purposeful production (for example for cosmetic use) or accidental degradation of plastics (or other synthetic polymers) to give rise to particles that are less than 5 mm in size. Those which are produced from larger polymer pieces (i.e. excluding those which are produced to have a specific function in paints and cosmetics) arise from abrasion in tyres, textiles and other sources. The detection and characterisation of microplastics depends on the material being analysed (biological materials are more challenging) [153] and the degree of precision required—methods exist with varying levels of sensitivity and degrees of quantitation [154] however until a dose/response relationship has been established and the extent of bioaccumulation determined, analytical chemists will simply try to develop increasingly sensitive techniques.

Natural and marine-derived polymers are considered to be an alternative to synthetic plastics. One of the main perceived advantages is that they are biodegradable and hence do not accumulate in the environment or cause long-term adverse biological effects. Nonetheless, even biodegradable microplastics are not completely risk free and their uncontrolled degradation may have environmental consequences [155]. Further work on the microstructural aspects of synthetic polymers and in particular the stability and survival of microcrystalline phases [156] may, together with the development of new materials reduce the formation of micro and nano particles.

Marine-derived biopolymers in particular have been suggested to be a biodegradable and less environmentally challenging prospect than even biodegradable microplastics [157]. In addition, unlike terrestrial-based bioplastics which often make use of food crop plants such as soya and maize, marine biopolymers are not inherently non-sustainable. There are a number of sectors which may make use of marine-derived biopolymers including, in particular, food and agriculture and biomedicine. In the biomedical area, marine biopolymers such as alginate, carrageenan, collagen, chitosan and silica have properties that make them suitable for drug delivery [158]. In addition, the development of analytical techniques for specific classes of marine biopolymers (such as polysaccharides, for example) has facilitated their isolation and their characterisation as bioactive agents [159]. In addition to being a useful source of biopolymers for specific biological uses, the marine environment also provides biopolymers that are able to be used (once characterised) for a variety of other purposes [160]. The use of marine biopolymers for food use has attracted considerable attention. Since they are generally biodegradable, they do not have the sustainability and disposal issues associated with many synthetic polymers. One area of particular interest has been in food packaging materials [161]. A number of marine biopolymers are derived from waste products (including those from which collagen and chitosan can be derived) [162] and this is another argument in favour of their use (sustainability). The use of marine alginates specifically as a sustainable packaging resource in food systems has been investigated [163] and their use as a carrier for other molecules provides certain targeted advantages. One of the most long-standing marine polymers is isinglass which is isolated traditionally from the swim bladder of the sturgeon and used to clarify wine [164]. Isinglass consists predominantly of fish collagen [165] and is denatured at a relatively low temperature (29 °C) and this has encouraged the development of alternative fining solutions.

While chemists have sought to use complicated chemistry to construct polymers with combinations of desirable properties while being able to be biodegraded and hence to have a smaller environmental footprint, these often seem to focus on the demonstration of technical prowess than providing useful (and scalable) solutions.

Carrageenans are a class of sulphated, linear polysaccharides derived from seaweed. They resemble mammalian glycosaminoglycans and have found applications in medicine. In addition, different members of the family of molecules have different gelling properties and they are used as thickening and gelling agents in a number of foods. Due to their history of use carrageenans are generally regarded as safe to use in food products however it has been suggested that this designation is itself unsafe due to the lack of relevant toxicity and other studies [166]. More generally, algae related hydrocolloids are found in various foods and in different combinations and further studies will likely be found to be necessary in future [167].

Fatty acids are an important class of marine compounds. From a human nutritional perspective, the most significant are the long chain-polyunsaturated fatty acids (LC-PUFAs) such as arachidonic acid (AA), α-linolenic acid (ALA), docosahexaenoic acid (DHA) and eicosatetraenoic acid (EPA) (Fig. 4.8).

The particular class of LC-PUFAs that are of greatest interest are those designated omega-3 (n-3) LC-PUFAs and these have been extensively studied [168]. The n-3 designation, which applies to DHA and refers to the first double bond in the molecule which occurs at the third carbon atom from the methyl end of the molecule. It has been generally considered that oily fish provide a good source of n-3 PUFAs but it has been demonstrated that microalgae are a more sustainable source of DHA and EPA [169]. While there have been a significant number of studies that have attempted to find a correlation between LC-PUFA consumption and a range of non-communicable diseases however, in most cases, fish consumption has been used as a surrogate for LC-PUFA consumption and there are a range of confounders both in terms of other dietary constituents and disease endpoints. In a meta-analysis, an inverse correlation between fish consumption and death from cardiovascular disease was noted [170] however a mechanistic basis for the effect was not explored. A correlation between resistance to obesity and LC-PUFA consumption was suggested to be due to an increase in GLP-1 [171] however this was a rodent (mouse) study and its relevance to humans is a matter for debate with differences in diet-induced gene expression between different taxa being observed [172].

Theme 1 has examined the correlation between chemistry and agriculture in Norfolk using correlation and observation (crop rotation), specific breeding and species manipulation (peas) followed by molecular manipulation of genes and natural products. Theme 2 explored the beginnings of chemical manipulation and

Fig. 4.8 Structures of the major LC-PUFAs

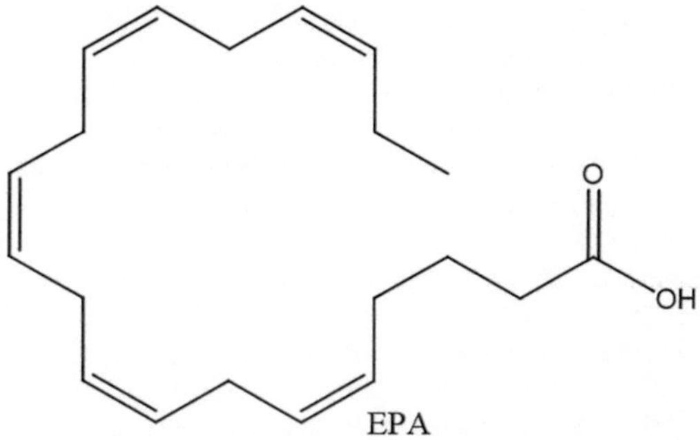

EPA

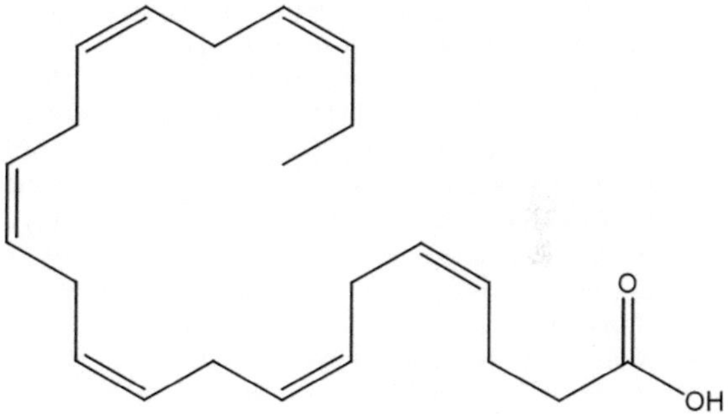

DHA

commercialisation and how the natural marine and terrestrial harvests were influenced by the chemistry of each. The next stage in Theme 2 (Chap. 5) will examine further the chemical manipulation of systems to give new and better products with a focus on colours and dyes, synthetic biology and sustainability.

References

1. Hattab H. Philosophical methods of analysis and synthesis from medieval scholasticism to Descartes and Hobbes. In: Newman WR, Schickore J, editors. Traditions of analysis and synthesis. Archimedes, vol. 73. Cham: Springer; 2025. https://doi.org/10.1007/978-3-031-76398-4_5.
2. Schmitz R. Friedrich Wilhelm Sertürner and the discovery of morphine. Pharm Hist. 1985;27(2):61–74.
3. Pelletier PJ, Caventou JB. Recherches chimiques sur les quinquinas. Ann Chim Phys. 1820;15:337–67.
4. Pratt HT. A letter signed: the very beginnings of Dalton's atomic theory. Ambix. 2010;57(3):301–10. https://doi.org/10.1179/174582310X12849808295742.
5. Rocke AJ. It began with a daydream: the 150th anniversary of the Kekulé benzene structure. Angew Chem Int Ed Engl. 2015;54(1):46–50. https://doi.org/10.1002/anie.201408034.
6. Perkin WH. The origin of the coal-tar colour industry, and the contributions of Hofmann and his pupils. J Chem Soc. 1896;69:596–637.

7. Mellor CM, Cardwell DSL. Dyes and dyeing 1775–1860. Br J Hist Sci. 1963;1(3):265–79. https://doi.org/10.1017/S0007087400001503.
8. Fox MR. Dye-makers of Great Britain, 1856–1976: a history of chemists, companies, products and changes. Slough: ICI; 1987. p. 94–101.
9. Seeman JI. The relationship of William Henry Perkin, Jr. and sir Robert Robinson: teacher and student, then student and teacher. Chemistry. 2021;27(5):1576–91. https://doi.org/10.1002/chem.202002924.
10. Strebhardt K, Ullrich A. Paul Ehrlich's magic bullet concept: 100 years of progress. Nat Rev Cancer. 2008;8(6):473–80. https://doi.org/10.1038/nrc2394.
11. Nurse J. Pigmented remedies: the pharmacy of colour in early modern Europe. Can Bull Med Hist. 2021;38(S1):S93–S117. https://doi.org/10.3138/cbmh.474-102020.
12. Abe HR. Die Universität Erfurt und ihre medizinhistorischen Traditionen. Dtsch Gesundheitsw. 1967;22(20):913–9.
13. Carr G. Woad, tattooing and identity in later iron age and early Roman Britain. Oxf J Archaeol. 2005;24:273–92. https://doi.org/10.1111/j.1468-0092.2005.00236.x.
14. Oberthür C, Schneider B, Graf H, Hamburger M. The elusive indigo precursors in woad (Isatis tinctoria L.)–identification of the major indigo precursor, isatan A, and a structure revision of isatan B. Chem Biodivers. 2004;1(1):174–82. https://doi.org/10.1002/cbdv.200490009.
15. Speranza J, Miceli N, Taviano MF, Ragusa S, Kwiecień I, Szopa A, Ekiert H, Isatis tinctoria L. (Woad): a review of its botany, ethnobotanical uses, phytochemistry, biological activities, and biotechnological studies. Plants. 2020;9(3):298. https://doi.org/10.3390/plants9030298.
16. Branca F. Isatis tinctoria L.: an ancient dye plant of interest as a multifunctional crop. Chron Horticult. 2015;55:20–4.
17. Nguyen TK, Marcelo P, Gontier E, Dauwe R. Metabolic markers for the yield of lipophilic indole alkaloids in dried woad leaves (Isatis tinctoria L.). Phytochemistry. 2019;163:89–98. https://doi.org/10.1016/j.phytochem.2019.04.006.
18. Elliott MC, Stowe BB. Indole compounds related to auxins and Goitrogens of Woad (Isatis tinctoria L.). Plant Physiol. 1971;47(3):366–72. https://doi.org/10.1104/pp.47.3.366.
19. Garcia-Macias P, John P. Formation of natural indigo derived from woad (Isatis tinctoria L.) in relation to product purity. J Agric Food Chem. 2004;52(26):7891–6. https://doi.org/10.1021/jf0486803.
20. Mocquard J, Le Lamer A-C, Fabre P-L, Mathieu C, Chastrette C, et al. Indigo dyeing from Isatis tinctoria L.: from medieval to modern use. Dyes Pigments. 2022;207:110675. https://doi.org/10.1016/j.dyepig.2022.110675.
21. Fabara AN, Fraaije MW. An overview of microbial indigo-forming enzymes. Appl Microbiol Biotechnol. 2020;104(3):925–33. https://doi.org/10.1007/s00253-019-10292-5.
22. Huang D, Zhang C, Chen J, Xiao Y, Li M, Sun L, Qiu S, Chen W. Computational workflow to study the diversity of secondary metabolites in fourteen different Isatis species. Cells. 2022;11(5):907. https://doi.org/10.3390/cells11050907.
23. Lotts T, Kabrodt K, Hummel J, Binder D, Schellenberg I, Ständer S, Agelopoulos K. Isatis tinctoria L.-derived petroleum ether extract mediates anti-inflammatory effects via inhibition of Interleukin-6, Interleukin-33 and mast cell degranulation. Acta Derm Venereol. 2020;100(10):adv00131. https://doi.org/10.2340/00015555-3476.
24. Kim EJ, Woo J, Shin S, Choi H, Kim Y, Kim J, Kang C. A focused natural compound screen reveals senolytic and senostatic effects of Isatis tinctoria. Animal Cells Syst. 2022;26(6):310–7. https://doi.org/10.1080/19768354.2022.2143895.
25. Kazemi F, Mojarrab M, Bahrami G, Miraghaei S, Shahram H, Saba M, Mohammad B. The alkaloids of Isatis indigotica as promising candidates against COVID-19: a molecular docking simulation for drug development. J Rep Pharm Sci. 2022;11(2):165–81. https://doi.org/10.4103/jrptps.JRPTPS_113_21.
26. Elliott MC, Stowe BB. Distribution and variation of indole Glucosinolates in Woad (Isatis tinctoria L.). Plant Physiol. 1971;48(4):498–503. https://doi.org/10.1104/pp.48.4.498.

27. Mohn T, Cutting B, Ernst B, Hamburger M. Extraction and analysis of intact glucosinolates–a validated pressurized liquid extraction/liquid chromatography-mass spectrometry protocol for Isatis tinctoria, and qualitative analysis of other cruciferous plants. J Chromatogr A. 2007;1166(1-2):142–51. https://doi.org/10.1016/j.chroma.2007.08.028.
28. Vauquelin R, Juillard-Condat L, Joly N, Jullian N, Choque E, Martin P. Study of Woad (Isatis tinctoria L.)-extracted Indoxyl precursors conversion into dyes: influence of the oxidative media on indigo recovery yields and Indigotin/Indirubin ratio measured by HPLC-DAD method. Molecules. 2024;29(20):4804. https://doi.org/10.3390/molecules29204804.
29. Bell L, Kitsopanou E, Oloyede OO, Lignou S. Important odorants of four Brassicaceae species, and discrepancies between glucosinolate profiles and observed hydrolysis products. Foods. 2021;10(5):1055. https://doi.org/10.3390/foods10051055.
30. Linke JA, Rayat A, Ward JM. Production of indigo by recombinant bacteria. Bioresour Bioprocess. 2023;10(1):20. https://doi.org/10.1186/s40643-023-00626-7.
31. Chandel N, Singh BB, Dureja C, Yang YH, Bhatia SK. Indigo production goes green: a review on opportunities and challenges of fermentative production. World J Microbiol Biotechnol. 2024;40(2):62. https://doi.org/10.1007/s11274-023-03871-2.
32. Książek E. Citric acid: properties, microbial production, and applications in industries. Molecules. 2023;29(1):22. https://doi.org/10.3390/molecules29010022.
33. Bud R. Innovators, deep fermentation and antibiotics: promoting applied science before and after the second world war. Dynamis. 2011;31(2):323–41. https://doi.org/10.4321/s0211-95362011000200004.
34. Mao D, Okada BK, Wu Y, Xu F, Seyedsayamdost MR. Recent advances in activating silent biosynthetic gene clusters in bacteria. Curr Opin Microbiol. 2018;45:156–63. https://doi.org/10.1016/j.mib.2018.05.001.
35. Walker KB, Bacon J. The role of fermentation in BCG manufacture: challenges and ways forward. In: Christodoulides M, editor. Vaccines for neglected pathogens: strategies, achievements and challenges. Cham: Springer; 2023. https://doi.org/10.1007/978-3-031-24355-4_9.
36. Gundogan R, Karaca AC, Jafari SM. An overview of microbial production of food bioactive compounds. In: Jafari SM, Harzevili FD, Karaca AC, editors. Microbial production of food bioactive compounds. Cham: Springer; 2025. https://doi.org/10.1007/978-3-030-81403-8_1-1.
37. Cooksey C. Tyrian purple: the first four thousand years. Sci Prog. 2013;96(Pt 2):171–86. https://doi.org/10.3184/003685013X13680345111425.
38. Smith CT. Dutch peat digging and the origin of the Norfolk broads. Geogr J. 1966;132(1):69–72.
39. Wells CE (1989) Historical and palaeoecological investigations of some Norfolk Broadland flood-plain mires and post medieval turf cutting
40. Roman routeways across the Fens, Heather Wallis, EAA Occasional Paper 10 (2002)
41. Wendy R. Childs, cod and herring: the archaeology and history of medieval sea fishing. In: Barrett JH, Orton DC, editors. The English historical review. Oxford: Oxford Academic; 2018. p. 668–70. https://doi.org/10.1093/ehr/cey071.
42. Saul AR, Lewis. English towns in the late middle ages: the case of Great Yarmouth. J Mediev Hist. 1982;8:75–88.
43. Poulsen B, Lyddon D, Vandaele K. Producers and production of salted herring. In: Venbrux E, editor. Dutch herring: an environmental history, c. 1600–1860. Berlin: Amsterdam University Press; 2009. p. 40–75.
44. Utterström G. Migratory labour and the herring fisheries of Western Sweden in the 18th century. Scand Econ Hist Rev. 1959;7(1):3–40.
45. Lindegren M, Ostman O, Gårdmark A. Interacting trophic forcing and the population dynamics of herring. Ecology. 2011;92(7):1407–13. https://doi.org/10.1890/10-2229.1.
46. Martinez Barrio A, Lamichhaney S, Fan G, Rafati N, Pettersson M, Zhang H, Dainat J, Ekman D, Höppner M, Jern P, Martin M, Nystedt B, Liu X, Chen W, Liang X, Shi C, Fu Y, Ma K, Zhan X, Feng C, Gustafson U, Rubin CJ, Sällman Almén M, Blass M, Casini M, Folkvord A, Laikre L, Ryman N, Ming-Yuen Lee S, Xu X, Andersson L. The genetic basis for ecological

adaptation of the Atlantic herring revealed by genome sequencing. elife. 2016;3(5):e12081. https://doi.org/10.7554/eLife.12081.
47. Trochta JT, Branch TA, Shelton AO, Hay DE. The highs and lows of herring: a meta-analysis of patterns and factors in herring collapse and recovery. Fish Fish. 2020;21:639–62. https://doi.org/10.1111/faf.12452.
48. Atmore LM, van der Jagt I, Boilard A, Häberle S, Blevis R, Dierickx K, Quinlan LM, Orton DC, Hufthammer AK, Barrett JH, Star B. The once and future fish: assessing a millennium of Atlantic herring exploitation through mixed-stock analysis and ancient DNA. Glob Chang Biol. 2024;30(12):e70010. https://doi.org/10.1111/gcb.70010.
49. Atmore LM, et al. Population dynamics of baltic herring since the viking age revealed by ancient DNA and genomics. Proc Natl Acad Sci U S A. 2022;119(45):1–11.
50. Hoffmann RC. A brief history of aquatic resource use in medieval Europe. Helgol Mar Res. 2005;59:22–30. https://doi.org/10.1007/s10152-004-0203-5.
51. Holm P, Nicholls J, Hayes PW, Ivinson J, Allaire B. Accelerated extractions of North Atlantic cod and herring, 1520–1790. Fish Fish. 2022;23:54–72. https://doi.org/10.1111/faf.12598.
52. Williamson T, Macnair A. William Faden and Norfolk's 18th-century landscape. Oxford: Oxbow Books; 2010.
53. Berlandi M. Market positioning over productivity: rethinking England's eighteenth-century grain export boom. English Hist Rev. 2025:ceaf139. https://doi.org/10.1093/ehr/ceaf139.
54. Allison KJ. The Norfolk worsted industry in the sixteenth and seventeenth centuries. Bull Econ Res. 1960;12:73–83. https://doi.org/10.1111/j.1467-8586.1960.tb00007.x.
55. Ang JB, Rajabrata B, Madsen JB. Innovation and productivity advances in British agriculture: 1620–1850. Southern Econ J. 2013;80(1):162–86.
56. Farrell RP, Massei N, Foley AE, Howlett PR, West LJ, editors. The chalk aquifers of Northern Europe. London: Geological Society of London; 2023.
57. Visser A, Beevers L, Patidar S. The impact of climate change on Hydroecological response in chalk streams. Water. 2019;11(3):596. https://doi.org/10.3390/w11030596.
58. Pop MS, Cheregi DC, Onose G, Munteanu C, Popescu C, Rotariu M, Turnea MA, Dogaru G, Ionescu EV, Oprea D, Iliescu MG, Minea M, Stanciu LE, Silişteanu SC, Oprea C. Exploring the potential benefits of natural calcium-rich mineral waters for health and wellness: a systematic review. Nutrients. 2023;15(14):3126. https://doi.org/10.3390/nu15143126.
59. Dupont C, Campagne A, Constant F. Efficacy and safety of a magnesium sulfate-rich natural mineral water for patients with functional constipation. Clin Gastroenterol Hepatol. 2014;12(8):1280–7. https://doi.org/10.1016/j.cgh.2013.12.005.
60. Pennazio S. Mineral nutrition of plants: a short history of plant physiology. Riv Biol. 2005;98(2):215–36.
61. Hepler PK. Calcium: a central regulator of plant growth and development. Plant Cell. 2005;17(8):2142–55. https://doi.org/10.1105/tpc.105.032508.
62. White PJ, Broadly MR. Calcium in plants. Ann Bot. 2003;92(4):487–511. https://doi.org/10.1093/aob/mcg164.
63. Whitaker W. The water supply of Norfolk from underground sources. London: Geological Survey of Great Britain; 1921.
64. Ehnvall B, Ratcliffe JL, Olid C, et al. Carbon accumulation in recently deposited peat is reduced by increased nutrient supply. Nat Commun. 2025;16:4271. https://doi.org/10.1038/s41467-025-59387-w.
65. Lourenco M, Fitchett JM, Woodborne S. Peat definitions: a critical review. Prog Phys Geograph Earth Environ. 2022;47(4):506–20. https://doi.org/10.1177/03091333221118353.
66. Campbell BMS. Arable productivity in medieval England: some evidence from Norfolk. J Econ Hist. 1983;43(2):379–404. https://doi.org/10.1017/S0022050700029697.
67. Zhang Y, Huang X, Zhao B, et al. Microbial responses to changing plant community protect peatland carbon stores during Holocene drying. Nat Commun. 2025;16:6912. https://doi.org/10.1038/s41467-025-62175-1.

68. Reynolds MC, Cadillo-Quiroz H. Microbial DNA sample preservation and possible artifacts for field-based research in remote tropical peatlands. J Microbiol Methods. 2024;224:106997. https://doi.org/10.1016/j.mimet.2024.106997.
69. Beck M. Site and artifact preservation: natural and cultural formation processes. In: Smith C, editor. Encyclopedia of global archaeology. Cham: Springer; 2020. https://doi.org/10.1007/978-3-030-30018-0_1491.
70. Analytical Methods Committee Amctb No. Hand-held X-ray fluorescence analysis of archaeological artefacts: challenges, advantages and limitations. Anal Methods. 2021; https://doi.org/10.1039/d1ay90094c.
71. de Melo BA, Motta FL, Santana MH. Humic acids: structural properties and multiple functionalities for novel technological developments. Mater Sci Eng C Mater Biol Appl. 2016;62:967–74. https://doi.org/10.1016/j.msec.2015.12.001.
72. Swidsinski A, Dörffel Y, Loening-Baucke V, Gille C, Reißhauer A, Göktas O, Krüger M, Neuhaus J, Schrödl W. Impact of humic acids on the colonic microbiome in healthy volunteers. World J Gastroenterol. 2017;23(5):885–90. https://doi.org/10.3748/wjg.v23.i5.885.
73. Socol DC. Clinical review of humic acid as an antiviral: leadup to translational applications in clinical humeomics. Front Pharmacol. 2023;4(13):1018904. https://doi.org/10.3389/fphar.2022.1018904.
74. Orsi M. Molecular dynamics simulation of humic substances. Chem Biol Technol Agric. 2014;1:10. https://doi.org/10.1186/s40538-014-0010-4.
75. Girkin NT, Lopes dos Santos RA, Vane CH, et al. Peat properties, dominant vegetation type and microbial community structure in a tropical peatland. Wetlands. 2020;40:1367–77. https://doi.org/10.1007/s13157-020-01287-4.
76. Zhou X, Zhang Z, Tian L, et al. Microbial communities in peatlands along a chronosequence on the Sanjiang plain. China Sci Rep. 2017;7:9567. https://doi.org/10.1038/s41598-017-10436-5.
77. Bai Y, Wang J, Zhan Z, Guan L, Liang J, Zheng G. Investigation of bacterial communities in peat land of the Gahai Lake natural conservation area. AIP Conf Proc. 2017;1890(1):040024. https://doi.org/10.1063/1.5005226.
78. Kanang KD, Mohidin H, Petrus AC, Liuan MCW, Joo GK, Melling L. Effects of liming on the chemical properties of tropical peat soils: an incubation-based analysis. Arch Agron Soil Sci. 2025;71(1):1–17. https://doi.org/10.1080/03650340.2025.2527094.
79. Barrow NJ, Hartemink AE. The effects of pH on nutrient availability depend on both soils and plants. Plant Soil. 2023;487:21–37. https://doi.org/10.1007/s11104-023-05960-5.
80. Atieno M, Wilson N, Casteriano A, Crossett B, Lesueur D, Deaker R. Aqueous peat extract exposes rhizobia to sub-lethal stress which may prime cells for improved desiccation tolerance. Appl Microbiol Biotechnol. 2018;102(17):7521–39. https://doi.org/10.1007/s00253-018-9086-2.
81. Kibblewhite M, Tóth G, Hermann T. Predicting the preservation of cultural artefacts and buried materials in soil. Sci Total Environ. 2015;1(529):249–63. https://doi.org/10.1016/j.scitotenv.2015.04.036.
82. Mukhamedbaeva Z, Yulchieva S, Alieva R, Naimov S. Acid-resistant liquid glass coatings using soda ash production waste. Int J Prog Sci Technol. 2023;38(1):415–21.
83. Melcher M, Wiesinger R, Schreiner M. Degradation of glass artifacts: application of modern surface analytical techniques. Acc Chem Res. 2010;43(6):916–26. https://doi.org/10.1021/ar9002009.
84. Rousk J, Brookes PC, Bååth E. Contrasting soil pH effects on fungal and bacterial growth suggest functional redundancy in carbon mineralization. Appl Environ Microbiol. 2009;75(6):1589–96. https://doi.org/10.1128/AEM.02775-08.
85. Spriggs JA. Anaerobic conditions (bogs, waterlogged, subaquatic): preservation and conservation. In: Encyclopedia of global archaeology. Cham: Springer; 2018. https://doi.org/10.1007/978-3-319-51726-1_465-2.

86. Boethius A, Kjällquist M, Magnell O, Apel J. Human encroachment, climate change and the loss of our archaeological organic cultural heritage: accelerated bone deterioration at Ageröd, a revisited Scandinavian Mesolithic key-site in despair. PLoS One. 2020;15(7):e0236105. https://doi.org/10.1371/journal.pone.0236105.
87. French C. Some thoughts on the monitoring and preservation of waterlogged archaeological sites in eastern England. Wiley Interdiscip Rev Water. 2017;4(3):e1204.
88. Dawson Q, et al. Subsidence and degradation of agricultural peatlands in the fenlands of Norfolk, UK. Geoderma. 2010;154:181–7.
89. Bird ECF, editor. Encyclopedia of the world's coastal landforms. Dordrecht: Springer; 2010. https://doi.org/10.1007/978-1-4020-8639-7_78.
90. Spencer T, French JR, Leggett DJ, Dixon M. The sea-defence value of salt marshes: field evidence from North Norfolk. Water Environ J. 2001;15:109–16. https://doi.org/10.1111/j.1747-6593.2001.tb00315.x.
91. Pennington C, Shaw M, Brooks T, et al. Nine months of daily LiDAR, orthophotos and MetOcean data from the eroding soft cliff coast at Happisburgh, UK. Sci Data. 2024;11:846. https://doi.org/10.1038/s41597-024-03499-3.
92. Davidson NC. The Wash estuary and North Norfolk coast (UK). In: Finlayson C, Milton G, Prentice R, Davidson N, editors. The wetland book. Dordrecht: Springer; 2018. https://doi.org/10.1007/978-94-007-4001-3_20).
93. White CS. Symbols of resilience and contested place identity in the coastal fishing towns of Cromer and Sheringham, Norfolk, UK: implications for Social wellbeing. In: Johnson D, Acott T, Stacey N, Urquhart J, editors. Social wellbeing and the values of small-scale fisheries. MARE publication series, vol. 17. Cham: Springer; 2018. https://doi.org/10.1007/978-3-319-60750-4_3.
94. Kwiatkowska M. Late Saxon to early medieval Saltern at Lynnsport 3: land south of front way, King's Lynn. Norfolk: PXA and UPD. Archaeology Data Service; 2020. https://doi.org/10.5284/1110105.
95. Keen L. Coastal salt production in Norman England. Anglo Norman Stud. 1988;11:133–79.
96. Oram R. The sea-salt industry in medieval Scotland. Stud Medieval Renaissance Hist. 2012;3(IX):1. http://acmrs.org/publications/journals/smrh
97. Shannon WD. Sleech salt making from the Solway to the Mersey in the medieval and early-modern period. North Hist. 2023;60(1):28–51. https://doi.org/10.1080/0078172X.2022.2146562.
98. Barrett JH, Locker AM, Roberts CM. The origins of intensive marine fishing in medieval Europe: the English evidence. Proc Biol Sci. 2004;271(1556):2417–21. https://doi.org/10.1098/rspb.2004.2885.
99. Rawcliffe C, Harper-Bill C. The religious houses. In: Medieval Norwich. New York: Hambledon Continuum; 2004. p. 301–24.
100. Social RM. Life and religious culture in twelfth-century Norwich and Norfolk. In: van Houts E, editor. Anglo-Norman studies XXXVIII: Proceedings of the battle conference 2015. Anglo-Norman Studies. Woodbridge: Boydell & Brewer; 2016. p. 155–68.
101. Smith PJ. Chapter 1 introduction: towards a cultural history of early modern ichthyology (1500–1880). In: Ichthyology in context (1500–1880). Leiden: Brill; 2023. https://doi.org/10.1163/9789004681187_002.
102. Salamzade R, Kalan LR. Context matters: assessing the impacts of genomic background and ecology on microbial biosynthetic gene cluster evolution. mSystems. 2025;10(3):e0153824. https://doi.org/10.1128/msystems.01538-24.
103. Kaur S, Samota MK, Choudhary M, Choudhary M, Pandey AK, Sharma A, Thakur J. How do plants defend themselves against pathogens-Biochemical mechanisms and genetic interventions. Physiol Mol Biol Plants. 2022;28(2):485–504. https://doi.org/10.1007/s12298-022-01146-y.
104. Barik A. Phyto-antifeedants. In: Omkar, editor. Molecular approaches for sustainable insect pest management. Singapore: Springer; 2021. https://doi.org/10.1007/978-981-16-3591-5_9.

105. Guo L, Yao H, Chen W, Wang X, Ye P, Xu Z, Zhang S, Wu H. Natural products of medicinal plants: biosynthesis and bioengineering in post-genomic era. Hortic Res. 2022;28(9):uhac223. https://doi.org/10.1093/hr/uhac223.
106. Wang J, Nielsen J, Liu Z. Synthetic biology advanced natural product discovery. Meta. 2021;11:785. https://doi.org/10.3390/metabo11110785.
107. Koroleva OA, Gibson TM, Cramer R, Stain C. Glucosinolate-accumulating S-cells in Arabidopsis leaves and flower stalks undergo programmed cell death at early stages of differentiation. Plant J. 2010;64:456–69. https://doi.org/10.1111/j.1365-313X.2010.04339.x.
108. Bhat R, Vyas D. Myrosinase: insights on structural, catalytic, regulatory, and environmental interactions. Crit Rev Biotechnol. 2019;39(4):508–23. https://doi.org/10.1080/07388551.2019.1576024.
109. Du Y, Grodowitz MJ, Chen J. Insecticidal and enzyme inhibitory activities of Isothiocyanates against red imported fire ants, Solenopsis invicta. Biomolecules. 2020;10(5):716. https://doi.org/10.3390/biom10050716.
110. Fimognari C, Turrini E, Ferruzzi L, Lenzi M, Hrelia P. Natural isothiocyanates: genotoxic potential versus chemoprevention. Mutat Res. 2012;750(2):107–31. https://doi.org/10.1016/j.mrrev.2011.12.001.
111. Liu P, Zhang B, Li Y, et al. Potential mechanisms of cancer prevention and treatment by sulforaphane, a natural small molecule compound of plant-derived. Mol Med. 2024;30:94. https://doi.org/10.1186/s10020-024-00842-7.
112. Mangla B, Javed S, Sultan MH, Kumar P, Kohli K, Najmi A, Alhazmi HA, Al Bratty M, Ahsan W. Sulforaphane: a review of its therapeutic potentials, advances in its nanodelivery, recent patents, and clinical trials. Phytother Res. 2021;35(10):5440–58. https://doi.org/10.1002/ptr.7176.
113. Treasure K, Harris J, Williamson G. Exploring the anti-inflammatory activity of sulforaphane. Immunol Cell Biol. 2023;101(9):805–28. https://doi.org/10.1111/imcb.12686.
114. Vu GH, Nguyen HD. Molecular mechanisms of sulforaphane in Alzheimer's disease: insights from an in-silico study. In Silico Pharmacol. 2024;12(2):96. https://doi.org/10.1007/s40203-024-00267-4.
115. Ko BS, Lee SB, Kim TK. A brief guide to analyzing expression quantitative trait loci. Mol Cells. 2024;47(11):100139. https://doi.org/10.1016/j.mocell.2024.100139.
116. Mohanty S, Mehrotra N, Khan MT, Sharma S, Tripathi P. Paradoxical effects of erucic acid—a fatty acid with two-faced implications. Nutr Rev. 2025:nuaf032. https://doi.org/10.1093/nutrit/nuaf032.
117. Miklavčič Višnjevec A, Tamayo Tenorio A, Steenkjær Hastrup AC, Hansen NML, Peeters K, Schwarzkopf M. Glucosinolates and isothiocyantes in processed rapeseed determined by HPLC-DAD-qTOF. Plants. 2021;10(11):2548. https://doi.org/10.3390/plants10112548.
118. Jhingan S, Harloff HJ, Abbadi A, Welsch C, Blümel M, Tasdemir D, Jung C. Reduced glucosinolate content in oilseed rape (Brassica napus L.) by random mutagenesis of BnMYB28 and BnCYP79F1 genes. Sci Rep. 2023;13(1):2344. https://doi.org/10.1038/s41598-023-28661-6.
119. Goyal A, Tanwar B, Sihag MK, Kumar V, Sharma V, Soni S. Rapeseed/canola (Brassica napus) seed. In: Tanwar B, Goyal A, editors. Oilseeds: health attributes and food applications. Singapore: Springer; 2021. https://doi.org/10.1007/978-981-15-4194-0_2.
120. Giannakara M, Koumandou VL. Evolution of two-component quorum sensing systems. Access Microbiol. 2022;4(1):000303. https://doi.org/10.1099/acmi.0.000303.
121. Nadell CD, Xavier JB, Levin SA, Foster KR. The evolution of quorum sensing in bacterial biofilms. PLoS Biol. 2008;6(1):e14. https://doi.org/10.1371/journal.pbio.0060014.
122. Whiteley M, Diggle SP, Greenberg EP. Progress in and promise of bacterial quorum sensing research. Nature. 2017;551(7680):313–20. https://doi.org/10.1038/nature24624.
123. Turovskiy Y, Kashtanov D, Paskhover B, Chikindas ML. Quorum sensing: fact, fiction, and everything in between. Adv Appl Microbiol. 2007;62:191–234. https://doi.org/10.1016/S0065-2164(07)62007-3.

124. Zheng X, Liu J, Wang X. Quorum signaling molecules: interactions between plants and associated pathogens. Int J Mol Sci. 2025;26(11):5235. https://doi.org/10.3390/ijms26115235.
125. Paluch E, Rewak-Soroczyńska J, Jędrusik I, Mazurkiewicz E, Jermakow K. Prevention of biofilm formation by quorum quenching. Appl Microbiol Biotechnol. 2020;104(5):1871–81. https://doi.org/10.1007/s00253-020-10349-w.
126. Rathnaprabha D, Ganesh A, Srividya A. Quorum quenching a chemical biological approach for microbial biofilm mitigation and drug development, vol. 22. London: Royal Society of Chemistry; 2023. p. 354–67.
127. Nazzaro F, Fratianni F, Coppola R. Quorum sensing and phytochemicals. Int J Mol Sci. 2013;14:12607–19.
128. Bendary MM, Ali MAM, Abdel Halim AS, Boufahja F, Chaudhary AA, Elkelish A, Soliman RHM, Hegazy WAH. Investigating sulforaphane's anti-virulence and anti-quorum sensing properties against Pseudomonas aeruginosa. Front Pharmacol. 2024;21(15):1406653. https://doi.org/10.3389/fphar.2024.1406653.
129. Singh AA, Khan F, Song M. Biofilm-associated amyloid proteins linked with the progression of neurodegenerative diseases. Int J Mol Sci. 2025;26(6):2695. https://doi.org/10.3390/ijms26062695.
130. Chakravarthi ST, Joshi SG. An association of pathogens and biofilms with Alzheimer's disease. Microorganisms. 2021;10(1):56. https://doi.org/10.3390/microorganisms10010056.
131. Vigasova D, Nemergut M, Liskova B, et al. Multi-pathogen infections and Alzheimer's disease. Microb Cell Factories. 2021;20:25. https://doi.org/10.1186/s12934-021-01520-7.
132. Vojtechova I, Machacek T, Kristofikova Z, Stuchlik A, Petrasek T. Infectious origin of Alzheimer's disease: amyloid beta as a component of brain antimicrobial immunity. PLoS Pathog. 2022;18(11):e1010929. https://doi.org/10.1371/journal.ppat.1010929.
133. Donia M, Hamann MT. Marine natural products and their potential applications as anti-infective agents. Lancet Infect Dis. 2003;3(6):338–48. https://doi.org/10.1016/s1473-3099(03)00655-8.
134. Zhang T, Ouyang Z, Zhang Y, Sun H, Kong L, Xu Q, Qu J, Sun Y. Marine natural products in inflammation-related diseases: opportunities and challenges. Med Res Rev. 2025;45:1375–406. https://doi.org/10.1002/med.22109.
135. El-Seedi HR, Refaey MS, Elias N, El-Mallah MF, Albaqami FMK, Dergaa I, Du M, Salem MF, Tahir HE, Dagliaa M, Yosri N, Zhang H, El-Seedi AH, Guo Z, Khalifa SAM. Marine natural products as a source of novel anticancer drugs: an updated review (2019-2023). Nat Prod Bioprospect. 2025;15(1):13. https://doi.org/10.1007/s13659-024-00493-5.
136. Haque N, Parveen S, Tang T, Wei J, Huang Z. Marine natural products in clinical use. Mar Drugs. 2022;20:528. https://doi.org/10.3390/md20080528.
137. McConnell OJ, Longley RE, Koehn FE. The discovery of marine natural products with therapeutic potential. Biotechnology. 1994;26:109–74.
138. Molinski T, Dalisay D, Lievens S, et al. Drug development from marine natural products. Nat Rev Drug Discov. 2009;8:69–85. https://doi.org/10.1038/nrd2487.
139. Daly JW. Marine toxins and nonmarine toxins: convergence or symbiotic organisms? J Nat Prod. 2004;67(8):1211–5. https://doi.org/10.1021/np040016t.
140. Daranas AH, Norte M, Fernandez JJ. Toxic marine microalgae. Toxicon. 2001;39:1101–32. https://doi.org/10.1016/s0041-0101(00)00255-5.
141. Otero P, Alfonso C. Analysis of marine toxins: gaps on food safety control of marine toxins. In: Phycotoxins: chemistry and biochemistry. Hoboken: Wiley; 2015. https://doi.org/10.1002/9781118500354.ch1.
142. Bian Y, Feng XS, Zhang Y, Du C, Wen YQ. Marine toxins in environment: recent updates on depuration techniques. Ecotoxicol Environ Saf. 2024;284:116990. https://doi.org/10.1016/j.ecoenv.2024.116990.
143. Goya AB, Tarnovius S, Hatfield RG, Coates L, Lewis AM, Turner AD. Paralytic shellfish toxins and associated toxin profiles in bivalve mollusc shellfish from Argentina. Harmful Algae. 2020;99:101910. https://doi.org/10.1016/j.hal.2020.101910.

144. Vilariño N, Louzao MC, Vieytes MR, Botana LM. Biological methods for marine toxin detection. Anal Bioanal Chem. 2010;397(5):1673–81. https://doi.org/10.1007/s00216-010-3782-9.
145. Martins JML, Martínez AG. Analytical alternatives for the analysis of marine biotoxins. In: Encyclopedia of analytical chemistry. Hoboken: Wiley; 2025. https://doi.org/10.1002/9780470027318.a9798.
146. Bose I, Nousheen RS, Yaduvanshi P, Sharma S, Chandel V, Biswas D. Unveiling the potential of marine biopolymers: sources, classification, and diverse food applications. Materials. 2023;16(13):4840. https://doi.org/10.3390/ma16134840.
147. Mohanan N, Montazer Z, Sharma PK, Levin DB. Microbial and enzymatic degradation of synthetic plastics. Front Microbiol. 2020;26(11):580709. https://doi.org/10.3389/fmicb.2020.580709.
148. Xu Y, Kim CS, Saylor DM, Koo D. Polymer degradation and drug delivery in PLGA-based drug-polymer applications: a review of experiments and theories. J Biomed Mater Res B Appl Biomater. 2017;105(6):1692–716. https://doi.org/10.1002/jbm.b.33648.
149. Yousif E, Haddad R. Photodegradation and photostabilization of polymers, especially polystyrene: review. Springerplus. 2013;2:398. https://doi.org/10.1186/2193-1801-2-398.
150. Kannan K, Vimalkumar K. A review of human exposure to microplastics and insights into microplastics as obesogens. Front Endocrinol. 2021;18(12):724989. https://doi.org/10.3389/fendo.2021.724989.
151. Blackburn K, Green D. The potential effects of microplastics on human health: what is known and what is unknown. Ambio. 2022;51(3):518–30. https://doi.org/10.1007/s13280-021-01589-9.
152. Ziani K, Ioniță-Mîndrican CB, Mititelu M, Neacşu SM, Negrei C, Moroşan E, Drăgănescu D, Preda OT. Microplastics: a real global threat for environment and food safety: a state of the art review. Nutrients. 2023;15(3):617. https://doi.org/10.3390/nu15030617.
153. Zhao J, Lan R, Tan H, et al. Detection and characterization of microplastics and nanoplastics in biological samples. Nat Rev Bioeng. 2025; https://doi.org/10.1038/s44222-025-00335-0.
154. Thompson RC, Courtene-Jones W, Boucher J, Pahl S, Raubenheimer K, Koelmans AA. Twenty years of microplastic pollution research-what have we learned? Science. 2024;386(6720):eadl2746. https://doi.org/10.1126/science.adl2746.
155. Piao Z, Agyei Boakye AA, Yao Y. Environmental impacts of biodegradable microplastics. Nat Chem Eng. 2024;1:661–9. https://doi.org/10.1038/s44286-024-00127-0.
156. Mendez NF, Sharma V, Valsecchi M, et al. Mechanism of quiescent nanoplastic formation from semicrystalline polymers. Nat Commun. 2025;16:3051. https://doi.org/10.1038/s41467-025-58233-3.
157. Tennakoon P, Chandika P, Yi M, Jung WK. Marine-derived biopolymers as potential bioplastics, an eco-friendly alternative. iScience. 2023;26(4):106404. https://doi.org/10.1016/j.isci.2023.106404.
158. Galasso C, Ruocco N, Mutalipassi M, Barra L, Costa V, Giommi C, Dinoi A, Genovese M, Pica D, Romano C, Greco S, Pennesi C. Marine polysaccharides, proteins, lipids, and silica for drug delivery systems: a review. Int J Biol Macromol. 2023;253(Pt 5):127145. https://doi.org/10.1016/j.ijbiomac.2023.127145.
159. Lesco KC, Williams SKR, Laurens LML. Marine algae polysaccharides: an overview of characterization techniques for structural and molecular elucidation. Mar Drugs. 2025;23:105. https://doi.org/10.3390/md23030105.
160. Carvalho DN, Gonçalves C, Sousa RO, Reis RL, Oliveira JM, Silva TH. Extraction and purification of biopolymers from marine origin sources envisaging their use for biotechnological applications. Mar Biotechnol. 2024;26(6):1079–119. https://doi.org/10.1007/s10126-024-10361-5.
161. Mahmud N, Islam J, Tahergorabi R. Marine biopolymers: applications in food packaging. PRO. 2021;9:2245. https://doi.org/10.3390/pr9122245.

162. Uranga J, Zarandona I, Andonegi M, Guerrero P, de la Caba K. Biopolymers derived from marine sources for food packaging applications. In: Sustainable food packaging technology. Hoboken: Wiley; 2021. https://doi.org/10.1002/9783527820078.ch2.
163. Kontominas MG. Use of alginates as food packaging materials. Foods. 2020;9(10):1440. https://doi.org/10.3390/foods9101440.
164. Ribéreau-Gayon P, Glories Y, Maujean A, Dubourdieu D. Clarification and stabilization treatments: fining wine. In: Ribéreau-Gayon P, Glories Y, Maujean A, Dubourdieu D, editors. Handbook of enology. Hoboken: Wiley; 2006. https://doi.org/10.1002/0470010398.ch10.
165. Hickman D, Sims TJ, Miles CA, Bailey AJ, de Mari M, Koopmans M. Isinglass/collagen: denaturation and functionality. J Biotechnol. 2000;79(3):245–57. https://doi.org/10.1016/s0168-1656(00)00241-8.
166. David S, Shani Levi C, Fahoum L, Ungar Y, Meyron-Holtz EG, Shpigelman A, Lesmes U. Revisiting the carrageenan controversy: do we really understand the digestive fate and safety of carrageenan in our foods? Food Funct. 2018;9(3):1344–52. https://doi.org/10.1039/c7fo01721a.
167. Liao YC, Chang CC, Nagarajan D, Chen CY, Chang JS. Algae-derived hydrocolloids in foods: applications and health-related issues. Bioengineered. 2021;12(1):3787–801. https://doi.org/10.1080/21655979.2021.1946359.
168. Khan I, Hussain M, Jiang B, Zheng L, Pan Y, Hu J, Khan A, Ashraf A, Zou X. Omega-3 long-chain polyunsaturated fatty acids: metabolism and health implications. Prog Lipid Res. 2023 Nov;92:101255. https://doi.org/10.1016/j.plipres.2023.101255.
169. Martins DA, Custódio L, Barreira L, Pereira H, Ben-Hamadou R, Varela J, Abu-Salah KM. Alternative sources of n-3 long-chain polyunsaturated fatty acids in marine microalgae. Mar Drugs. 2013;11(7):2259–81. https://doi.org/10.3390/md11072259.
170. Zhang B, Xiong K, Cai J, Ma A. Fish consumption and coronary heart disease: a meta-analysis. Nutrients. 2020;12(8):2278. https://doi.org/10.3390/nu12082278.
171. Haneishi Y, Furuya Y, Hasegawa M, et al. Polyunsaturated fatty acids-rich dietary lipid prevents high fat diet-induced obesity in mice. Sci Rep. 2023;13:5556. https://doi.org/10.1038/s41598-023-32851-7.
172. Weng K, Hu H, Xu AG, Khaitovich P, Somel M. Mechanisms of dietary response in mice and primates: a role for EGR1 in regulating the reaction to human-specific nutritional content. PLoS One. 2012;7(8):e43915. https://doi.org/10.1371/journal.pone.0043915.

Chapter 5
Chemistry and Commercialisation. Colours, Natural and Synthetic. Impressionism. Synthetic Biology. Sustainability in Colour and Dye Production

Abstract The use of Gram-positive and Gram-negative stains for the classification of bacteria. The link between colour molecules and bioactivity. The analysis and characterisation of Lake pigments. The perception of malaria in Europe and the history of the isolation and use of quinine. Attempted synthesis of quinine and the production of aniline dyes. The importance of, and communication between European chemists including Boerhaave, Stahl, Liebig and Lavoisier. The establishment of the Royal College of Chemistry. Synthetic colours and en plein air painting and Impressionism. The Red Wensum in Norwich and Rose Madder. Production of natural and synthetic dyes and sustainability. The use of synthetic biology and fermentation to produce dyestuffs.

Keywords Natural and synthetic colours · Tria prima · Gram stain · Quinine · Malaria · Synthetic biology

While the isolation and exploitation of medicinal products from a range of living systems (for example anti-microbial compounds, enzyme inhibitors, anti-inflammatory substances) has driven a substantial research effort [1], other properties of natural products have also been of considerable interest. In some cases, there is a clear synergy between chemistry and other industries and this will be examined in the following chapter. In addition, the importance of some types of chemistry can change due to other influences. For example, the extraction and use of natural colours was of particular importance. As the synthetic dyestuffs industry developed in the nineteenth and early twentieth centuries, natural colours were superseded. Concerns about the use of petrochemicals and sustainability led to a renewed interest in using biological systems (but more efficiently than in the past) for the production of chemicals with colouring properties. The importance of marine-derived polymers has already been noted (Sect. 4.6) and identified as an area where natural products are of particular commercial value. Another area of significance is the production and use of natural colouring materials and the history of the extraction and use of chemicals as colouring materials is illustrative both of the progression of

B. Hanley, *The Chemistry of a County*,
https://doi.org/10.1007/978-3-032-21491-1_5

natural product chemistry and the ingenuity of chemists and the development of techniques and skills that had consequences in other areas.

There are a number of challenges to the effective development and use of colouring agents and dyes. First of all, they must be able to adhere to the material that is to be coloured (paper, wood, stone, textiles etc.). Secondly, they have to be sufficiently robust to be stable under the conditions of use. This requires pH, chemical and light stability, in particular. Finally they must be able to be spread throughout the medium that is used to apply them in order that the colour is evenly distributed throughout the target material. In addition to visual appeal, colours have also been associated with particular societal status in the hierarchy and the importance of some roles. In medieval society there were seven primary colours—white, yellow, red, green, blue, purple, and black. Modern impressions of colour have been influenced by Newtonian optics (and the colour spectrum) and by the advent of printing which employed black type on white paper thereby confirming these colours as opposites. Earlier perceptions could be different and, for example, the opposite of white was red [2]. While availability and cost were important factor in the value and limited use of some colours, medieval colour theories also played an important role in everyday use of colour. Tyrian Purple, for example, due to its rarity and the technical challenge associated with its production, was associated with nobility and prestige/power [3]. Even more recently the association of 'blue collar' and 'white collar' with specific roles in the workplace remains present although, in this case, it originally reflected the type of work done and the practicality of some colours of clothing for manual and non-manual workers. The classification of health status by employment type [4] and even the presumption of guilt [5] has been associated with certain types of employment classified on the basis of the colour of clothing worn. While the use of colour in art and textiles is an important societal aspect, the colours of organisms (including eye and hair colouration as well as skin colour) is an important example of environmental and genetic interplay. A detailed discussion of the factors that influence these characteristics is outwith the scope of this chapter, but it has produced some ideas and thoughts about the role of selection mediated by the environment on skin colour [6]. The importance of enzymes such as tyrosinase (the rate determining enzyme in cutaneous melanin production) has been noted [7]—as have the various polymorphisms that can influence pigmentation [8].

From a chemistry perspective, the gradual development of methods to create and effectively use pigments from natural sources was, at least in part, overtaken by the synthetic dyestuffs industry in the nineteenth and twentieth centuries. This in turn had a profound effect on industry and art (particularly the Impressionist movement). As natural colours increased in popularity again (with the growth in finding industrial alternatives to petrochemicals as a source of feedstocks) towards the end of the twentieth century, the application of biology to the controlled synthesis of complex dye molecules became an area of considerable interest and was an aspect of the growth of synthetic biology. This progression is reflected in the development of blue dye (indigo) from woad and other plants, the replacement of natural indigo with synthetic indigo and, most recently, the application of fermentative techniques to produce indigo from a non-plant, biological source. These aspects will be further illustrated.

5.1 Colours from Nature. Land and Sea

The importance of colour has already been discussed in the context of woad (*Isatis tinctoria*) and the development of woad growing and processing in medieval Europe (see Chap. 4). The use of valuable agricultural land (and the foul odours associated with woad processing) and the occurrence of regular famines led Elizabeth I to issue a proclamation 'against the sowing of woade' in October 1585. It is known that the growing tips of some *Brassica* plants are rich in indole compounds [9] including the indole glucosinolate, glucobrassicin so the practice of harvesting of woad leaves in a sequential manner rather than all at once may be a way of maximising the indole content that can be extracted from the younger rather than the more mature leaves [10]. In addition to the dyeing of textiles, the blue pigment extracted from woad was also used to colour the walls of medieval churches [11] and since East Anglia in general [12] and Norfolk in particular had a large concentration of churches with wall paintings and the arable land was not permitted to be used for growing woad, it is likely that imported, processed woad will have entered Norfolk at Kings Lynn or one of the other Hanseatic ports on the east coast from the Hanseatic League city of Erfurt (the centre of the medieval woad trade) [13]. In this case, the restriction on growing woad noted above would result in a lack of raw materials being available in Norfolk and this makes it likely that processed (or partially processed) woad rather than the technology for its processing from plant material will have been imported. The history of methods used for the production of the blue pigment from woad has been reviewed [14] based on collecting together information from a number of historical sources and carrying out experimental archaeology to confirm details of the processing used in the medieval times. The authors concluded that their experimental studies "validate the efficiency of the whole technological line of woad production and use as practiced in Europe since the Middle Ages." There is evidence for other technologies (for example for salt production) being imported to King's Lynn from the near continent [15] and the availability of technology for dye production in England influenced the use of blue colour in textiles and in other applications throughout society [16].

In order to better understand the importance of colour it is important, as with chemistry, to become familiar with the precise meanings of some of the terms that are used since an appreciation of these will help to explain the chemistry of the process. There are a number of physical mechanisms involved in the recognition of light and its wave/particle duality. This particular aspect was first demonstrated by Thomas Young in 1801 [17] and was very much in opposition to the corpuscular theory advanced by Newton and others. Young included a discussion of both light and colours. Another important principle associated with colour is that of perception and of additive and subtractive properties. Prior to the publication of Newtons work 'Optiks' [18], one of the most popular theories of light and colour was that black and white were 'extreme colours' and other colours were intermediates. This could be considered to fit with the 'tria prima' theory of Paracelsus in which matter was considered to be made up of sulphur, mercury and salt. The idea was that any material could be created from different proportions of mercury and sulphur. Colours could similarly be created by combining different proportions of black and white. Newton, although he was a strong believer in alchemical principles [19] proposed that the colours in a spectrum were continuous. The principle that all colours could be created from three primary pigments was already known, in a practical sense, to artists and this seemed to result in a conflict between the wave (continuous) and the corpuscular (atomist) theories of the nature of colour. The progression of ideas of colour (including the history of the apparently competing wave and corpuscular theories) has been described at length [20]. The evolution of the competing theories eventually give rise to the expression of a light (additive) theory and the subtractive (dark) theories of colour mixing in terms of colours being expressed in the context of wavelengths but also having distinct (corpuscular) properties. The additive theory of colour creation involves the addition of different proportions of the primary colours (red, green and blue) transmitted wavelengths of light to give all possible colours. There is clearly a corpuscular element to this since the proportions of the different colours present define the final colour achieved however, rather than being distinct particles, light is made up of a spectrum of different component colours. The ultimate proof of the continuous, rather than atomist theory is expressed when considering the addition of all possible wavelengths which creates white light. Removal of all wavelengths which, in an additive sense, is the absence of all colour results in the formation of black and can be considered to be the start point for additive colour creation. The principle of subtractive colour is, in some ways, the inverse of the additive theory outlined above. It involves the transmission, absorption, and reflection of wavelengths by different pigments. The starting point for the subtractive colour system are the three colours that are complementary to the three primary colours in the additive system noted above. The three subtractive primary colours are cyan, magenta and yellow. In the subtractive system, white light is the starting point and colours are created by the absorption of certain colours and reflecting back the wavelengths of the colours that remain. A colour that is a combination of cyan/magenta is created by the absorption of yellow and the reflection of the other colours. Eventually, if all three primary colours are absorbed then all wavelengths are absorbed and the absence of colour (black) is reflected back. Pigments and dyes use subtractive colour theory while additive colour is a feature of computer screens, projectors and other producers of transmitted light.

While the terms 'dye' and 'pigment' tend to be used interchangeably, they actually refer to different concepts and different types of chemicals. A dye is a soluble substance that imparts colour by being absorbed onto a surface. A pigment is an insoluble component that has to be dispersed in a binder or dispersed in a solvent. In general dyes are chemically bound to the material they are imparting colour to while pigments tend to produce, primarily, an effect on the surface of the material being coloured. This means that while dyes are often applied by immersion, pigments primarily are effective as a consequence of a surface coating effect.

A mordant is a substance (often a metal salt) that is used to facilitate the binding of a dye molecule to a textile. A mordant is a means of 'fixing' a dye such that it is stable and does not wash out. One of the problems with treatments involving mordants is that they are not 100% effective and the effluent from the mordant process often contains toxic metal ions. For this reason, a range of 'bio-mordants' have been developed [21] and are used in combination with natural dyes to create environmentally friendly materials [22]. Mordants are also used for biological tissue staining in histology [23] in order to identify particular cellular characteristics. The description of biological staining in terms of the different types of bonds that are formed between the biological material and the dye used has been carried out [24]. One of the most important biological staining techniques (which is used for bacterial classification) is the Gram stain which was developed by Hans Christian Gram in 1882 [25]. The first step in the process is the dyeing of the bacterium with crystal violet (Fig. 5.1).

If a bacterium irreversibly takes up the dye, it appears to be purple/brown under a microscope and is defined as Gram positive. Bacteria which fail to accumulate it appear to be red after counterstaining and are classified as Gram negative. The process combines both cell wall permeability and a chemical transformation. In the first instance all bacteria take up the dye. The sample is then treated with iodine and a crystal violet-iodine complex is formed. In effect the iodine is acting as a mordant.

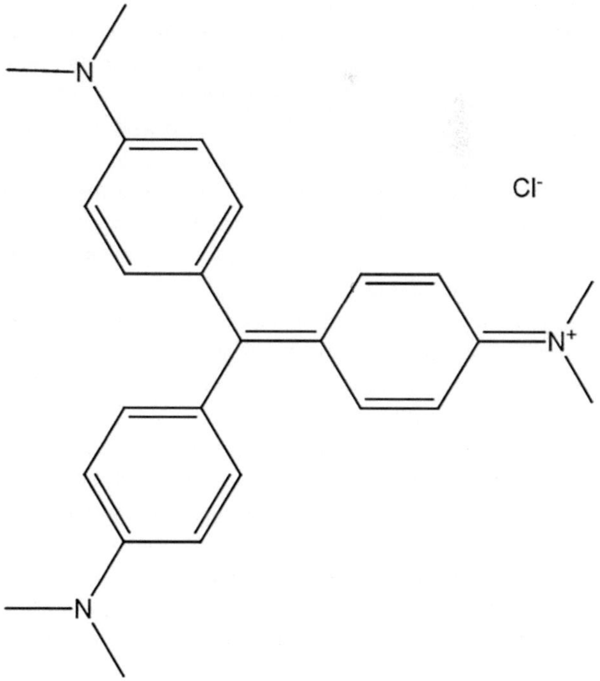

Fig. 5.1 The structure of crystal violet stain

A solvent (ethanol/acetone 50:50) is used and if the bacterium has a high lipid content in its cell wall content and a thin peptidoglycan layer, then the complex is easily washed out and the bacterium is Gram negative. If the bacterial cell wall has a high peptidoglycan content, then the complex is trapped in this layer and the solvent shrinks the cell wall and the coloured complex remains. The cells are then stained purple (the colour of the crystal violet-iodine complex). If a basic fuschin counterstain is then added, the Gram-negative bacteria are stained pink and are more easily identified. Gram staining is the basic method of typing of bacteria. The soil bacterium family, *Streptomyces*, which have been extensively studied in Norwich, are Gram positive bacteria and an important source of antibiotics. This has led to the development of procedures for their phenotypic visual and microscopic evaluation which includes Gram staining [26].

Another aspect of the practical use of pigments and dyes is the creation of Lakes. A Lake is the form of a colour that is used subsequent to prior dispersion using a mordant. A Lake pigment is, therefore, another name for a pigment that has been precipitated using a suitable mordant and is ready for dispersal, generally with an inert binder. Natural Lake pigments can be extracted from several sources including plants [27]. And some of these have been used traditionally following the empirical study of their suitability, light fastness etc. [28]. Such trial-and-error approaches are common in other areas of natural product chemistry before the development of spectroscopic methods. Pigments that are suitable for use as Lakes often have a negative charge and this enables them to chelate with metal cations. It is of interest (and also for validation of authenticity) for the lake pigments of older works of art to be characterised and this has been carried out using microspectrofluorimetry [29] and different pigments and mordants are able to be distinguished using an archive of certified materials as a reference source supplemented by spectroscopic and mass spectrometric techniques [30]. In addition to validation of works of art through verification of materials, it is also possible to determine the extent to which different colours may have degraded over time and how this may alter the appearance of the work. Colours which are not light fast are termed ‘fugitive’ and, once analysed and characterised, digital reconstruction of an original ‘lost’ painting can be carried out [31]. Not all Lake colours are equally easy to analyse and to quantify and techniques continue to be developed. While natural colours can be extracted and used for a range of purposes, another aspect of the use of natural colours is the change in colour (and pattern) in water and water-borne creatures as a consequence of changes in environmental factors. In these cases, colour change is being used as an *in situ* environmental indicator. This has been applied, among other areas, to the study of gene flow in water snakes in Lake Erie [32], to colour variation in cichlid fish [33] and to the link between colour and the iron content of boreal lakes [34]. Indeed, pigment-based chemotaxonomy has been suggested to be a potential tool in the investigation of phytoplankton populations [35]. Remote sensing data enables significant numbers of sites to be investigated in a temporal fashion [36] over a large number of comparable sites.

The isolation of colours was an important aspect of pre-industrial technology. A major source of earth-related pigments was rocks and minerals which could be mined or isolated from surface deposits. One of the most significant and earliest colours that was isolated and used was ochre [37], an iron-rich pigment that is a mixture of ferric oxides and varying amounts of clay to give a range of colours from yellow to deep orange and brown. The different colours have been found to be associated with the mineral constituents and not with particle size [38]. The darker red ochres contain predominantly haematite while the yellow ochre contains only goethite (named after Goethe). Earth pigments also include red ochre, purple ochre, sienna, and umber. They can also contain variable amounts of manganese which change the hues of the colouring material. In some cases, the ochres are heated to give 'burnt' (dehydrated) colours which often have a more pronounced red colour. A range of naturally occurring ochres from different sources have been examined by Raman spectroscopy and the content determined [39] and regional differences noted. In addition to cave excavations [40] there is evidence of mining for ochre [41]. In addition to its decorative use ochre also has some photoprotective effects and may, therefore, have been used (purposely or inadvertently as a protective agent against UV skin damage and affected human dispersal [42]. Techniques such as saturation isothermal remnant magnetisation (RT-SIRM), isothermal magnetisation sweeps (hysteresis loops) and zero-field-cooled/field-cooled (ZFC-FC) remanence enable the spread of ochre in ancient civilisations to be tracked and this is a surrogate of the spread of civilisations and technology [43]. In addition to the use of terrestrial minerals such as ochre, some other minerals have been used as colouring agents. For example, lapis lazuli was originally transported to Europe from Afghanistan and its intense blue colour was sought for painting. It was ground into a powder to make the pigment ultramarine. The analysis of the lapis lazuli (using techniques such as Raman microspectroscopy, scanning electron microscopy with energy-dispersive spectroscopy, electron probe microanalysis, and laser ablation inductively coupled plasma mass spectrometry) in ultramarine pigments in artefacts allows the geological origin of lapis lazuli in pigments to be determined [44]. The pigment ultramarine was synthesised in the nineteenth century after it was found to be formed as a by-product from glass manufacture but the method of synthesis was not disclosed for commercial reasons and was later investigated using modern analytical techniques [45] and materials available in the nineteenth century. A mixture of kaolin (100 parts), sodium carbonate (100 parts), bitumen emulsion (or a similar carbon source) (12 parts) and sulphur (60 parts) heated at 750 °C for 4 h gave an optimal colour. Industrial zeolite waste and rice husks have also been used to prepare ultramarine [46]. The development of marine pigments has been reviewed including sources, their exploitation and questions regarding sustainability [47]. Many marine pigments are derived from marine organisms (and therefore are 'biopigments') and this may limit both their availability and ease of procurement. An example of this will be shown below in the case of Tyrian Purple. In addition to their potential as pigments, many have ancillary properties as bioactive agents for treatment of a range of conditions and this provides further focus for their isolation and investigation. Compounds such as astaxanthin, which has a range of proposed biological activities [48] and is used traditionally as a colouring agent [49] (Fig. 5.2) fall into this category.

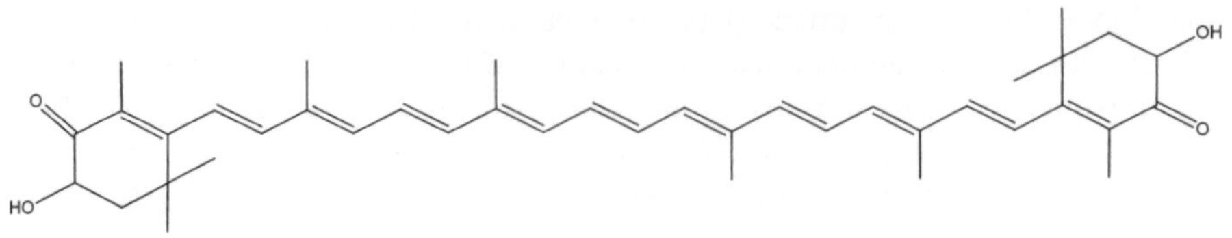

Fig. 5.2 The structure of astaxanthin

A range of health-promoting and colour compounds have been found in seaweed, algae and in lichen. In common with terrestrial ochres, some compounds from sea lichen are also UV protective [50] but, unlike ochres, they are organic rather than minerals and their biosynthesis may be induced by UV light [51]. One of the most significant colours from an aquatic source is Tyrian Purple. The colour Tyrian Purple derives from the secretions of a particular gland of the sea snail of the family *Muricidae*. The snails (and their secretions) are not coloured but post-harvest treatment gives rise to the major colour agents (6,6′-dibromo indigo), although other compounds (6-bromo indigo and 6,6′-dibromoindorubin, for example) have also been isolated. The proportions of the different compounds present and the particle size of the dyeing agent are responsible for the shade of purple obtained (Fig. 5.3).

These reactions have been investigated and the sequence of formation of intermediates determined by synthesis of final products and precursors [52]. The high cost of Tyrian Purple (and hence its limited use which was confined to the wealthy and those of high status) is a consequence of the limited source of starting materials (many thousands of snails are required to produce a small amount of the precursor) and the complexity of the process used to prepare the purple compound from its precursors. In addition to their use for the production of a rare colour, snails of the *Muricidae* family have also traditionally been used in medicine [53]. It has been suggested that some of these biological properties are a consequence of the formation of the bromo indoles that are the functional colour ingredients [54] although other aromatic compounds have also been implicated [53]. Other compounds from the same source (bromo indirubins) have also been investigated for their biological properties [55]. The isolation of other coloured materials from the sea (for example carotenoids) has a substantial history [56]. In addition to chemically derived visually coloured molecules, fluorescence has also been noted in some marine organisms [57]. The reasons and biological significance of fluorescence in marine organisms remain, in many cases, unclear [58] although behavioural and physiological effects have been suggested. Fluorescent diverse organic matter (FDOMs) is found extensively in marine environments and is thought to derive from picocyanobacteria [59] (see below for further discussion of marine bacteria) although other sources seem to also make a lesser contribution. From a functional standpoint, one of the factors suggested to be important in the development of fluorescence is the possibility that it may be a compensatory mechanism to counteract the loss of the longer wavelengths of light at greater depths [60]. Fungi are also a source of polyketide-derived fluorescent compounds [61]. In essence, fluorescence is a

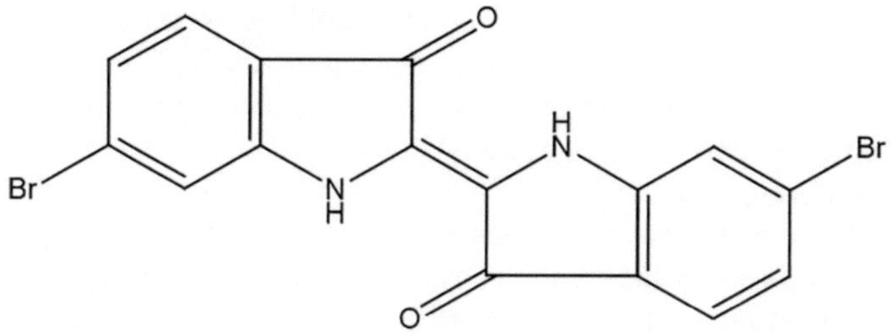

6,6'-Dibromo indigo

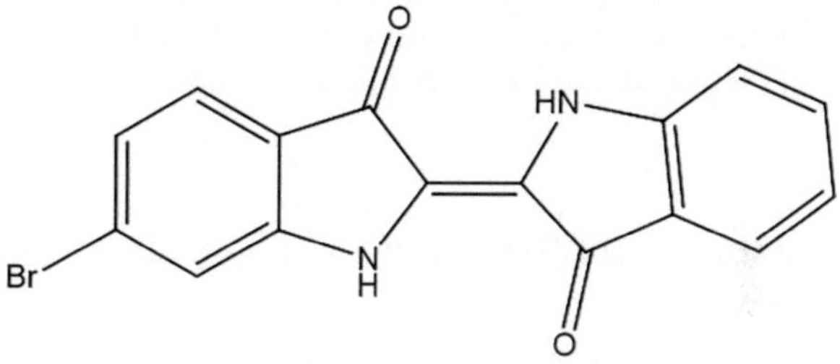

6-Bromo indigo

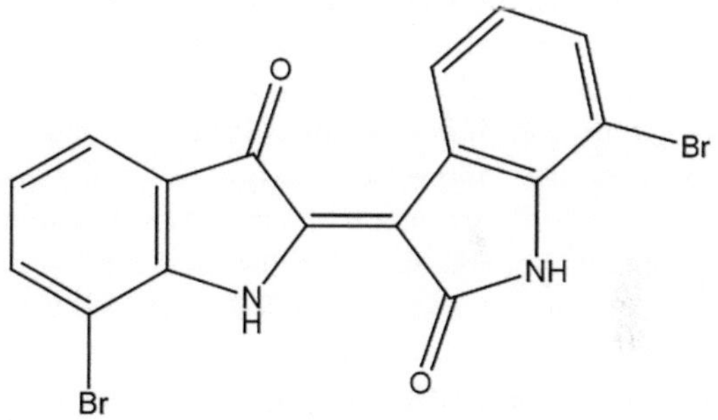

6,6'-Dibromo indirubin

Fig. 5.3 Bromo Indigo from sea snails

process whereby a molecule absorbs light (energy) and this excites an electron. When the electron falls back to its ground state, it emits energy in the form of a photon. This is then detected. Apart from fluorescence, as noted above in the case of Tyrian Purple, a range of colours are derived from marine sources, and these have an increasing number of applications [47] of which Tyrian Purple serves as an example. In the case of marine bacteria, for example, in addition to fluorescence, a range of pigments have been isolated and these may have a range of biological activities in addition to being coloured [62]. Many marine pigments have applications in food and health [63] however dye molecules from marine sources (including fungi, mentioned previously in the context of fluorescence and polyketide production) [64] have also found increased application for dying textiles. Algae have also been found to be a good source of dye molecules and are increasingly being applied [65]. The development of sustainable and naturally sourced pigments has become an important objective for industry however many of the technological advances that underpinned the industrial revolution in Europe arose as a consequence of the development of the synthetic dyestuffs industry in the nineteenth century. While the importance of textiles in Norfolk will be considered later (theme 3 Chap. 6), the development of technology to create coloured textiles (and other materials) which was imported from Europe lay at the core of industrial development in Norfolk [66]. The beginnings of the important and influential dyestuffs industry arose in England and was the result of a failed attempt to synthesise an important medicinal compound, quinine and this will be considered in Sect. 5.2.

5.2 Colours from Synthesis: The Quinine Story

The story of the discovery and subsequent use of quinine has been told—with apparently varying levels of accuracy—a number of times. Malarial fevers were a considerable burden on health in the eighteenth and nineteenth centuries—in part because of the occupation of tropical and sub-tropical parts of the Empire by the British. Malaria was also common in and around marshlands in Southern and Eastern England until Victorian times [67] and populations that lived in coastal and marshland areas were more likely to contract malarial-like diseases [68]. Common practices such as crop irrigation and improved transport links with increased transport of people and goods increased the incidence of malaria. Malaria is caused by the parasitic *Plasmodium* species with *P. vivax* and *P. malariae* being most common in northern climates while in Africa it is *P. falciparum*. The parasite is transmitted through the bite of a female mosquito (*Anopheles*) and the mosquito is, therefore, the vector by which the parasite is transmitted from infected to healthy individuals. Various possible causes of malarial fever were postulated before the cause was determined [69] however the surgeon Laveran first saw parasites in the blood of malarial patients [70]. While quinine (see below) was an effective anti-malarial whose mechanism of action has been suggested to be by interference, concerned with the ability of the parasite to break down haemoglobin [71], precise details of the mechanism of action of quinine and structurally related drugs is unknown [72] and this has contributed to the continued use of the administration of quinine as a treatment. The major source of quinine is the bark of a tropical tree. The discovery that the bark of the Cinchona tree contained an anti-malarial agent was somewhat serendipitous and, indeed, the bark has a number of biological activities that appear unrelated to its anti-malarial properties so its use specifically for malaria is largely a chance discovery [73]. The active principle from the bark of the Cinchona tree was isolated in 1820 by Pelletier and Caventou [74] by a relatively simple process of extraction and acidification leading to the formation of crystals of quinine—since quinine is less soluble than the other alkaloids of Cinchona bark in acidified water. The isolation of this active principle was particularly important since it allowed treatment to be standardised. This is particularly important in the case of bioactive substances like quinine where the level of dosage that can safely be administered between pharmacological efficacy and toxicity (the therapeutic window) is narrow. The Cinchona bark (and extracts of it) remained the only source of quinine (and the only significantly effective anti-malarial treatment) throughout the nineteenth century. A source of quinine became increasingly important as Europeans extended their colonial influence into tropical areas where malaria was endemic. One variety of Cinchona grown in Dutch Java (*Cinchona ledgeriana*) contained about 13% quinine in its bark and was, therefore, a good source of the medicinal agent. Attempts to circumvent South America as the sole source of Cinchona trees (and hence quinine) led to attempts to grow plantations of the quinine-producing tree in Java, India and Algeria however these were not all completely successful. By the mid-eighteenth century, the drive to find an alternative source of quinine had increased to a significant extent [75]. In the case of England, the success of the Dutch Cinchona plantations in Java was perceived as a threat to British expansionism in tropical areas [76]. The need to combat tropical diseases led (as will be explained below) to the synthesis of the first aniline dye by W H Perkin in London and to the development by Ehrlich and others of anti-microbial agents from dyes [77]. As noted previously, malaria was common in marshy areas of England (and Europe) and however improved drainage and the extension of arable land and the elimination of the marsh environment that led to a decrease in the occurrence of ‘fevers and agues’. In a similar way, the development of dichlorodiphenyltrichloroethane (DDT) (originally synthesised in 1874) as an insecticide (use discovered in 1939) led to a decrease in the population of the mosquito vector and the reduction in the number and severity of malaria outbreaks globally [78]. Despite the increase in the proportion of arable land in Norfolk and East Anglia, in general, and the reduction in indigenous incidences of water-borne diseases this did not impact on malaria in the tropics and there remained a perception that control over the supply of quinine was essential to the wealth and prosperity of the country. It was against this background (and the competition between the British and Dutch East India Companies) [79] that science in general and chemistry in particular operated. The nationalistic sentiment was a common feature of science is also demonstrated by the competing claims regarding the discovery of oxygen by Priestley and Lavoisier [80]. This was later reflected in the development of the pigments and dyestuffs industry in, particularly, Britain and Germany.

Although Britain was one of the leading pre-industrial countries, this was largely built upon the existing industries of agriculture and traditional goods manufacture including, as noted earlier, initiatives such as crop rotation and animal breeding. The work by Liebig at Giessen on the importance of nitrogen and minerals in agriculture was of central importance to the development of agronomic practice and Liebig visited Britain frequently in the nineteenth century where his ideas were enthusiastically adopted. A number of British chemists (and industrialists) studied with Liebig and two of them, Lloyd Bullock and John Gardner, proposed setting up a chemistry research institute in London and, as a result, the Royal College of Chemistry (RCC) was founded with private capital [81]. This later became the Chemistry Department of Imperial College. The popularity and reputation of German chemists led to the appointment of one of Liebig's former students, August Hoffman as the first Director of the RCC in 1845. Unlike Liebig's home institute at Giessen (a university), the RCC was oriented towards the production of chemists who could make an industrial rather than an academic contribution. Hoffman was interested in the possibilities offered by synthetic chemistry. This was, in part, fuelled by the work he did in Liebig's laboratory on the transformations of the compounds present in coal tar (a waste product from the production of coke and coal gas). Hoffman eventually returned to Berlin in 1865 attracted by the offering of a university Chair and significant financial incentives to continue his chemistry investigations. One of the challenges that were of interest to Hoffman (and those who invested in the RCC) was finding a synthetic route for the preparation of quinine. This would be both financially and nationally important. Hoffman had worked on the compounds of coal tar while in Liebig's laboratory (see above) and as early as 1849, he had speculated that quinine could be synthesised by oxidative condensation of two molecules of N-allyl toluidine (Fig. 5.4) and he gave this as a challenge to one of his students, W H Perkin.

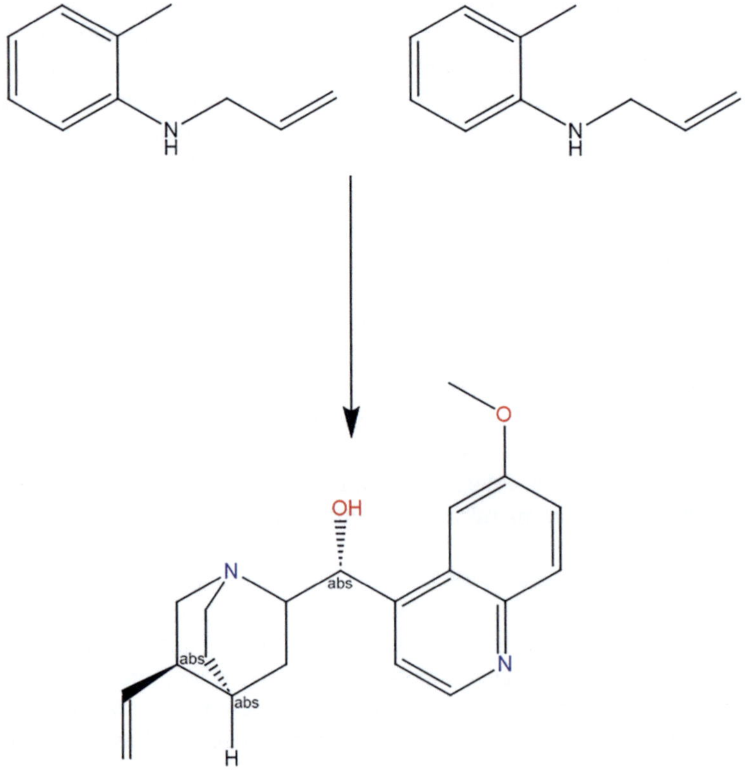

Fig. 5.4 The putative synthesis of quinine from N-allyl toluidine

Hoffman may also have been influenced by the work carried out by Ehrlich on the medicinal uses of aniline dyes derived from coal tar and their use for treating the parasitic cause of sleeping sickness, trypanosomes [82]. The supposition was supported by the relative empirical formulae of N-allyl toluidine ($C_{10}H_{13}N$) and quinine ($C_{20}H_{24}N_2O_2$) and the idea that the addition of two molecules of N-allyl toluidine to oxygen would, with the removal of water, give an identical empirical formula to that of quinine. However, in terms of the actual chemical structures (information that was not available to Hoffman or Perkin), this was clearly not a coherent plan. Perkin attempted the reaction but obtained a 'black intractable tar'. He then tried to carry out the reaction using (he thought) aniline which was a simpler starting material. In fact, his aniline contained impurities (ortho- and para-methyl anilines) and these reacted to give a distinctive colour, christened mauveine by Perkin. While he failed in his primary objective of synthesising quinine, Perkin recognised the potential of the stable purple colour, and he carried out work on mauveine at home rather than at Hoffman's laboratory at the RCC. This ensured that the patent for the synthesis and use of mauveine was wholly ascribed to Perkin [83].

The development of the European synthetic dyestuffs industry was highly influenced by German companies such as Hoechst, BASF, Bayer and Agfa which were created and grew in the 1860s on the basis of the production of aniline dyes based upon Perkin's original work. Indeed, the names of two of them incorporate a sense of their beginnings in aniline exploitation—AGFA (Gesellschaft Für Anilin-Fabrikation), BASF (Badische Anilin- und Sodafabrik) and all four began as aniline dye manufacturers. Diversification was an important factor in the growth of all four companies. For BASF, this was in part a consequence of the development of the Haber-Bosch process for the production of ammonia (and nitrate). Bosch was employed by BASF and the process, which was brought in house by BASF, was the first industrial synthesis to make use of high pressure. Bosch was instrumental in the setting up of IG Farben—an amalgamation of the major dyestuff manufacturers in Germany and, at one time, the largest chemical company in the world. By the turn of the twentieth century, BASF remained primarily a dye manufacturer (as, indeed did Hoechst and AGFA). The German companies gradually increased their interests in pharmaceuticals. There were commercial reasons for this but, in addition, German industry had strong links to Technical Universities and colleges (Technische Hochschulen) including those at Karlsruhe, Munich, Aachen, Braunschweig, Stuttgart, Darmstadt, Berlin, Hanover and Dresden [84]. These links gave German industry access to the leading chemists and biologists of the time such as Paul Ehrlich and led to the development of compounds such as the important treatment for syphilis, Salvarsan in 1910 [85] which was marketed by Hoechst. Other anti-trypanosome molecules were developed such as Trypan Red which had dual functionality and was used both for histological staining and medicinal purposes [86]. In addition to the formal links between industry and the German Technical Universities and Colleges, there was also an exchange of ideas between chemists (and, in particular pharmacists who operated within the alchemical tradition) and leading German philosophers of the eighteenth and nineteenth centuries [87]. Throughout Europe, the mixture of chemistry and philosophy was seen in the works of Stahl,

Boerhaave, Geoffroy and, in particular, Lavoisier and the new tradition he espoused [88]. In addition, in contrast to Germany, universities in England were limited to Oxford and Cambridge neither of which embraced science and technology. One of the first universities to be established outside Oxford and Cambridge took place with the founding of Durham University in 1832 (other English universities also lay claim to being the first ‘post-Oxbridge’ learned institution). The formation of another competing university in England was firmly opposed by Oxford and Cambridge. Despite this, in addition to the technical universities in Germany, medieval universities were set up across Europe and acted as centres of learning and scholarship [89]. The traditional and insular approach in England was a factor in the spread of the chemical industry in Europe and the loss of the lead created by Perkin’s fortunate experiment.

The role of dyes (particularly those used in cell staining) as pharmaceuticals was a significant factor in the development of synthetic chemistry for multiple uses in the nineteenth and twentieth centuries [90]—particularly in Europe. Drug discovery was also driven by the isolation and purification of natural products (particularly alkaloids) and this was supplemented by the synthesis of novel compounds—in many cases initially prepared as colouring agents as a consequence of the growth of the dyestuffs industry [91]. The impact of the new pigments and dyes in their primary target areas of art and manufacturing will be considered in more detail in Sect. 5.3 while the importance of the development of dyeing in Norfolk will be examined in Sect. 5.4.

5.3 New Colours in Art and Manufacture

The importance of the development of synthetic chemistry was significant in both art and manufacture. In addition to the new colours, many of the pigments that were developed had significant advantages in use (for example, light fastness). In addition, the compounds that give rise to colours that can be extracted from natural material are a relatively small proportion of the whole material and the residual plant material must be discarded. For example, in the case of indigo, the colour compounds are not present in the plant material and the glycoside precursors must be converted [92] into the corresponding aglycones and then further processed to give the coloured compounds (Fig. 5.5). Typically, the plant material has to be extracted into water, treated with calcium hydroxide and oxygen and the compound of interest precipitated.

The amount of indigo that can be isolated is variable and is dependent upon agronomic and harvesting characteristics. The production of synthetic indigo is more reproducible however it is highly dependent on the use of harsh petroleum-derived chemicals [93] and effluent that must be treated to remove toxic components—including residual dye [94]. More recently bio-fermentative approaches have been developed [95] and these will be discussed in more detail in Sect. 5.5.

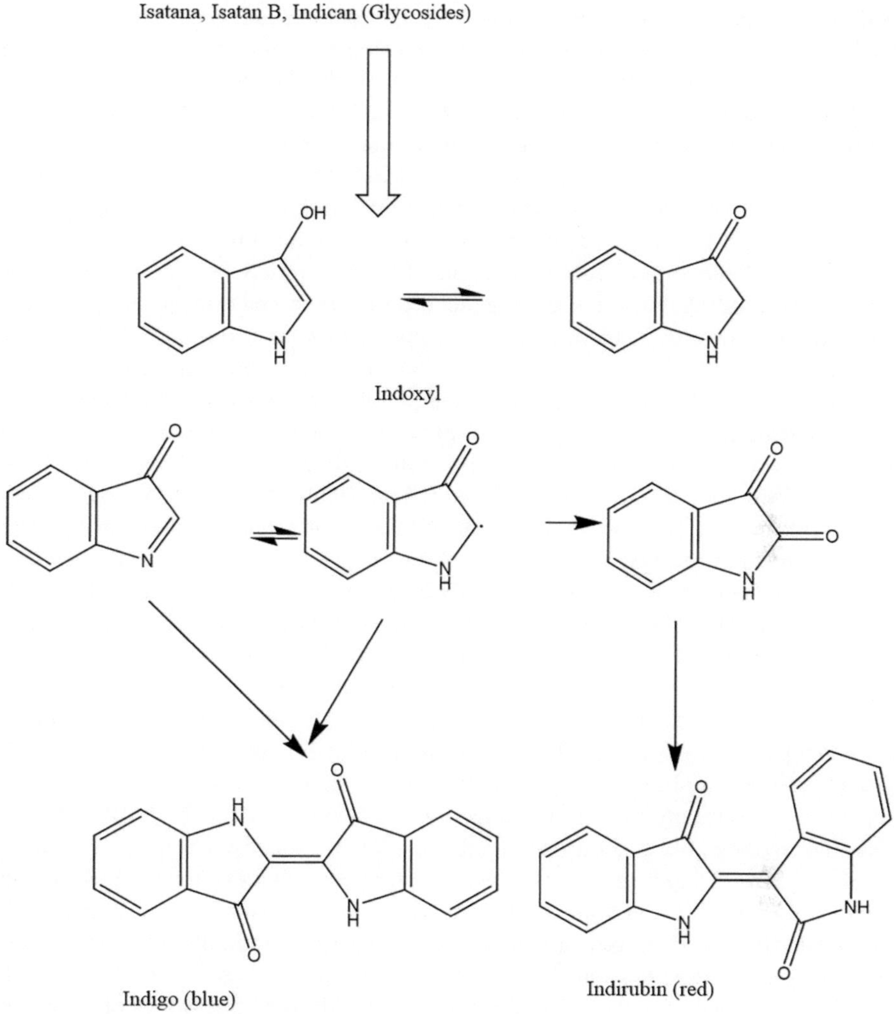

Fig. 5.5 The production of Indigo Blue and Indigo Red

The development of synthetic pigments and dyes was not confined to the work carried out by Perkin and others in the nineteenth century. Previously both the Egyptians and the ancient Chinese [96] produced blue pigments, and they were not alone in this endeavor [97]. The significant advances in the development of synthetic colours were of great importance to the industrial age in the 19th and twentieth centuries. The palette of colours available to impressionist artists included a number of synthetic colours which, as noted previously, had the dual advantages of being both novel and stable [98]. The palette available to impressionist artists was both vivid and able to be transported to be used outdoors (en plein air). The use of synthetic pigments was not restricted to European artists and many American (North and South) artists who trained in Europe brought colours and expertise in their use back to the Americas with them [99]. The development of new colours was also a factor in the establishment of the first provincial school of artists in Britain—The Norwich School which was founded in 1803. The Norwich School was, not surprisingly, influenced by Dutch art of the time (the similarity of the landscape subject matter has been noted) as a consequence (at least in part) of the trading and other links between Norfolk and the Low Countries. Although the synthetic colours that began with the invention of aniline pigments in the middle of the nineteenth century, were not available to the initial proponents of the Norwich School at the beginning of the nineteenth century, the third generation of artists painting in the Norwich style towards the end of the nineteenth century had access to much of the colour palette of the Impressionists (1873 onwards). Although some pigments were not available, the development of premixed and portable paints was a significant step in the increasing popularity of painting en plein air. In addition, a number of colours had been developed prior to the synthesis of the aniline dyes including Cobalt Blue (1802), Chromium Green Oxide (1809), Indian Yellow, Cadmium Yellow (1817) and Cerulean Blue (1821) and these were added to the largely Renaissance palette in the eighteenth century. The tradition of the Norwich School was continued into the twentieth century by artists such as Tuck (1876–1951) and Seago (1910–1974). Indeed, in their treatment of their subject matter, the earlier Norwich School prefigured the later rise of Naturism and Impressionism. The appreciation of colour and light was also apparent in aspects of the weaving and textiles businesses in Norfolk and this will be examined further later. The Norwich School painters were influenced by Dutch artists (particularly Hobbema) and, while both the Norwich School and Hobbema have been largely initially ignored, the rising reputation of the latter may, in time, have an invigorating effect on the appreciation of the former. At any rate, the cross-sea links to Europe had the result that a localised School of Painting in England owed a greater debt to Northern Europe art than to painters operating in the rest of England at that time. Indeed, among Cotman's list of works are paintings of Dutch ships and barges at Yarmouth. John Crome (another member of the Norwich School) and his son, John Berney Crome also painted scenes in Holland. One of the reasons for the relatively poor appreciation of the works of Norwich School painters such as Crome and Cotman was the extensive patronage of them by local wealthy individuals such as J J Colman and the subsequent exclusion of their work from a more national perspective. Colman will be discussed in more detail later (Theme Four, Chap. 8).

Another significant beneficiary of the development of new colours and new technologies was the car industry. Henry Ford famously made his Model-T available in any colour—so long as it was black. In fact, originally the Model T was available in a range of colours however whether it was because of processing advantages, rapid drying time of black paint compared to other colours and cost, for a time (1914–1925) the Model T was only available in black. The establishment of General Motors (GM) in 1908 increased the choice for customers and the competition for Ford. While Henry Ford was primarily an engineer, GM attached great importance to the design and colour of their cars. In this way they sought to create affordable cars with a degree of style and individuality. This was realised through the creation of an "Art and Colour Section" in 1927. A key figure was Harley Earl who had created a reputation for customising standard production models for discerning customers. He became the first head of the Design and Colour Section of GM. The concept of design and colour as key consumer drivers were an important part of the GM business model [100] and led to a greater appreciation of design. In the case of colour, a productive collaboration between the DuPont chemical company and GM resulted in the development of new automobile colours and finishes [101]. While the mass production of colours in response to industry (and ultimately consumer) demand led to the establishment of large factories and facilities for dye and pigment production, small scale production for specific industries continued and this was the case for the weaving industry in Norwich.

5.4 Colour Production and Sustainability: The Red Wensum

The concept (and actuality) of the red Wensum (a term commonly used in the nineteenth century) is not a political statement but is, rather, the signifier both of industry and of its environmental impact. Norwich sits on the river Wensum which is navigable from the port of Yarmouth. This was an important outlet for the transport of manufactured goods to Europe and supported a number of industries. One of these was the production of dyed cloth. Much of the wealth of Norwich and Norfolk was built upon the wool trade and worsted cloth was a major export. Some aspects of this were the combined impact of the growth of agriculture (which has already been mentioned) and the consequence of the importing of technology and people in the seventeenth century (which will be considered in more detail in Theme three, Chap. 6). The development of woad as a colouring material has already been mentioned. Since the plant itself was prohibited from being grown, the import of indigo derived from woad or partly processed woad through the Hanseatic League ports such as Kings Lynn likely became more common. Other colours that were commonly used in dyeing cloth and textiles were madder (red) and weld (yellow). The production of madder was particularly significant in Norwich and the river Wensum was severely polluted by waste from the production process and from the dye itself (see below).

Madder is produced from the roots of the madder plant *Rubia tinctorum* and are fixed to cloth using a mordant (commonly alum). The precursor of the red colour is a mixture of anthraquinone glycosides which, through a process involving fermentation or treatment with acid, gives the red pigment, alizarin [102] and the less stable purpurin (Fig. 5.6).

The colour of anthraquinones is pH dependent with a range of colours observed, yellow ($pH < 5$), red ($5 < pH < 10$) or purple ($pH > 10$). Alizarin was recognised as an important dyestuff with considerable commercial value and was synthesised by Perkin and by Grabe and Liebermann (working for BASF in Germany) who obtained a patent for their process [103] and this became a major source of income for the

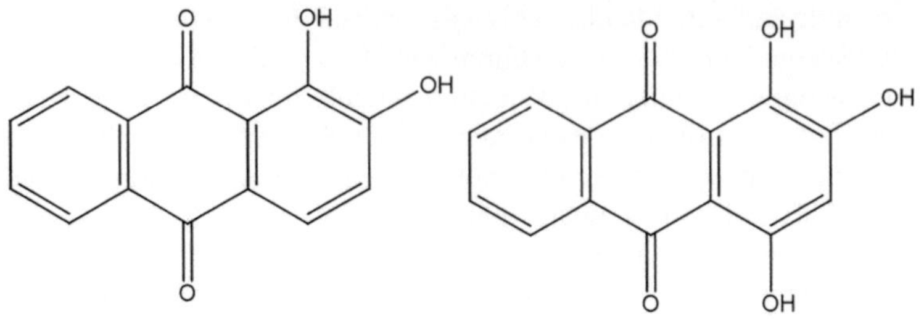

Alizarin Purpurin

Fig. 5.6 The formation of alizarin and purpurin

German company [104]. It was one of the first of the natural dyes to be synthesised and, indeed, its synthesis predated knowledge of its chemical structure (Fig. 5.4). Natural madder, in addition to alizarin, also contains a number of other compounds including quinalizarin, purpuro-xanthin, purpurin, pseudopurpurin, munjistin, ruberythric acid and other hydroxyl anthraquinone derivatives. The processes involved in the extraction and synthesis of alizarin have been reviewed [105] and these investigations have helped to delineate the properties of different madder extracts. Anthraquinones and purpurin in particular, in addition to being used as colouring agents, have significant biological activity [106] and a number of potential uses have been proposed. The levels of anthraquinones in plants of the *Rubia* genus are relatively small and the use of culture methods to increase secondary metabolite production has been investigated [107] and this creates the possibility of using bioreactor technology to produce good yields of commercially interesting compounds as well as developing more environmentally sustainable routes. Alizarin was one of a group of dyes that were of interest to Ehrlich in his studies of the specific biological properties of natural products [108] and of histological staining of biological material [109].

Alizarin binds to calcium and is preferentially extracted into hard water so the chalk spring-derived water of the river Wensum (and other similar rivers in Norfolk) will improve the yield of extracted colour. As noted previously, however, the levels of anthraquinones are relatively low and considerable plant material must be disposed of after extraction of the desired compounds and this reduces the sustainability of the process of extraction of colour from natural sources. In addition, alizarin will photodegrade with concomitant loss of colour and this is apparent in older textiles and paintings—in particular those which make use of alum in a madder lake pigment [110]. The mechanism of the fading reactions of alizarin have been proposed to include light-mediated oxidation of alizarin to give colourless small molecules through an ·OH radical-mediated route [111]. This may also have relevance to the biological properties *in vivo* of alizarin as a radical scavenger [112]. The photodegradation of alizarin has also been used to investigate photodegradation reagents [113].

Madder cultivation was principally carried out in the provinces of South Holland and Brabant [114] until the nineteenth century when it was grown elsewhere in Europe (in particular France and Germany) and in Asia. The strong trade links between Norfolk and the Low Countries (and the ingress of Flemish weavers in the sixteenth century) together with the ready availability of chalk spring water helped the dyeing industry become established. The effluent from the dyeing industry (and, in particular, the washing of cloth), led to the river Wensum being dyed red where it ran through Norwich. Interestingly the "Norfolk rebellion" in 1381 was called Lister's rebellion, because headed by John Lister, or Littester (which means someone who dyes cloth and textiles), a dyer of Norwich. There was, therefore, indirect historical evidence for dyeing and dyers being resident in Norfolk even in the fourteenth century and, in Norwich itself, part of the city is referred to as the "Madder Market" [115]. The production of dye was common both in Norfolk and Belgium and evidence for the growing of the dye plants gaude (*Reseda luteola*), madder (*Rubia tinctorum*) and woad (*Isatis tinctoria*) from the fifteenth century has been found in fluvial deposits in Brussels and in Mechelen [116]. As is the case in Norfolk, a supply of water was crucial for production of pigments from plants in Belgium, however finding remains of waste plant material in river deposits is relatively rare since they are normally rapidly decomposed in the environment. While madder could be grown in England, by the middle of the eighteenth century, it was difficult to grow and caused nutrient depletion of soil [117], and, although it was introduced to market gardens around London in the 1620s, madder was largely imported into England from Europe (particularly Holland) and was a significant aspect of trade between the countries in the centuries before synthetic alizarin became available.

5.5 The Principles of Synthetic Biology Applied to the Production of Colours

The development of more sustainable methods for the production of commercially important colours (particularly indigo and alizarin as examples) which do not involve the use of petrochemicals is a significant challenge. The relatively low levels of pigments in plants (generally less than 5%) means that harvesting and isolation from plants involves the production (and disposal) of a significant amount of waste plant material. An alternative based on 'natural' rather than synthetic processes will have considerable sustainability challenges and opportunities—removing the dependence on petrochemicals and the need to dispose of chemical waste. Because of this, biosynthetic approaches have been developed using the expertise gained from synthetic biology and fermentations. The first step in the development of such an approach is the elucidation of potential biosynthetic pathways and their incorporation into a suitable fermentative organism. The history of the development of colours provides a set of useful paradigms for the exploitation of naturally occurring chemicals.

Before the development of syntheses of colours, extraction and processing of natural material was the principal route to their isolation. A number of coloured compounds were also found to have biological properties therefore the two drivers for studying coloured molecules were their practical use as colouring agents and their potential as bioactive/medicinal agents. The isolation and purification of compounds was the first stage in the processes that were traditionally developed and adopted by chemists. The second aspect of structural confirmation was the synthesis of the isolated and purified compound [118]. This was by no means a trivial matter when few methods were available for structural determination. As noted previously, during the discussion of Perkin's attempted synthesis of quinine, empirical formulae were a useful but insufficient tool to ascertain not just what atoms were present but how they were arranged to give distinct functional groups and molecules. After having become wealthy as a consequence of his synthesis of the dye, mauveine [119] and his industrial successes, Perkin was able to retire from commerce to consider other ways to make use of the chemicals from coal tar as starting materials for chemical synthesis. One of these was the synthesis of coumarin in a one pot reaction that became known as the Perkin condensation [120]. This synthesis of a compound that was formerly available only as a natural product and was an important perfumery ingredient [121] as well as having potential as a medicinal product [122] was an important advance.

The growth in synthetic methodologies together with functional group analysis [123] was an important aspect of pre-spectroscopic chemistry since they enabled structures to be confirmed and built up the knowledge of which functional groups were located in specific molecules. However, even if synthetic methods could be devised (as they were, for example, for quinine [124]), the synthetic methodologies were often cumbersome or gave the product in such a low yield that biological extraction methods remained the route of choice and gaps in the knowledge of biosynthesis *in planta* had to be filled in order that chemical transformations occurring *in vivo* could be understood [125]. The driver for the development of a biologically derived synthesis of indigo is not a function of the complexity of the laboratory synthesis or the yield obtained. It is, rather, a sustainability challenge (to produce the dye in substantial amounts without the use of harsh chemicals and the production of waste chemicals that must be subsequently disposed of) and an efficiency challenge since, as noted previously, extraction from a natural plant source leaves considerable vegetative material that is then discarded. These have been the drivers for the development of 'green' fermentative approaches to the production of indigo [126]. The use of recombinant microbes to produce bio-indigo has been demonstrated to be feasible [127]. This analysis suggests that an annual output of 540 tonnes is achievable using fermentation technology and this should be compared to the annual production by chemical synthesis (using aniline which is toxic and creates an environmental burden) of 8000 tons per year. The bio solution may not be precisely biomimetic in terms of the existing plant biosynthetic pathway and, for example, one promising route involves metabolic engineering by modifying the tryptophan pathway in *E coli* to cause high-level indole production and then adding the *Pseudomonas putida* genes encoding naphthalene dioxygenase to give the indigo precursor, indoxyl and hence indigo [128].

The use of biotechnology to produce bio-alizarin has also been investigated as an alternative to madder production from the plant or alizarin production by synthesis. As noted previously, the commercial synthesis of alizarin was carried out almost simultaneously in Germany and England. Alizarin can be synthesised conveniently by a number of routes including from anthraquinone or from catechol (Fig. 5.7).

The bromination route from anthraquinone (1869) was developed in the year following the discovery that alizarin was a derivative of anthraquinone however it was not suitable for industrial use and in 1869, BASF (Carl Graebe and Carl Liebermann)

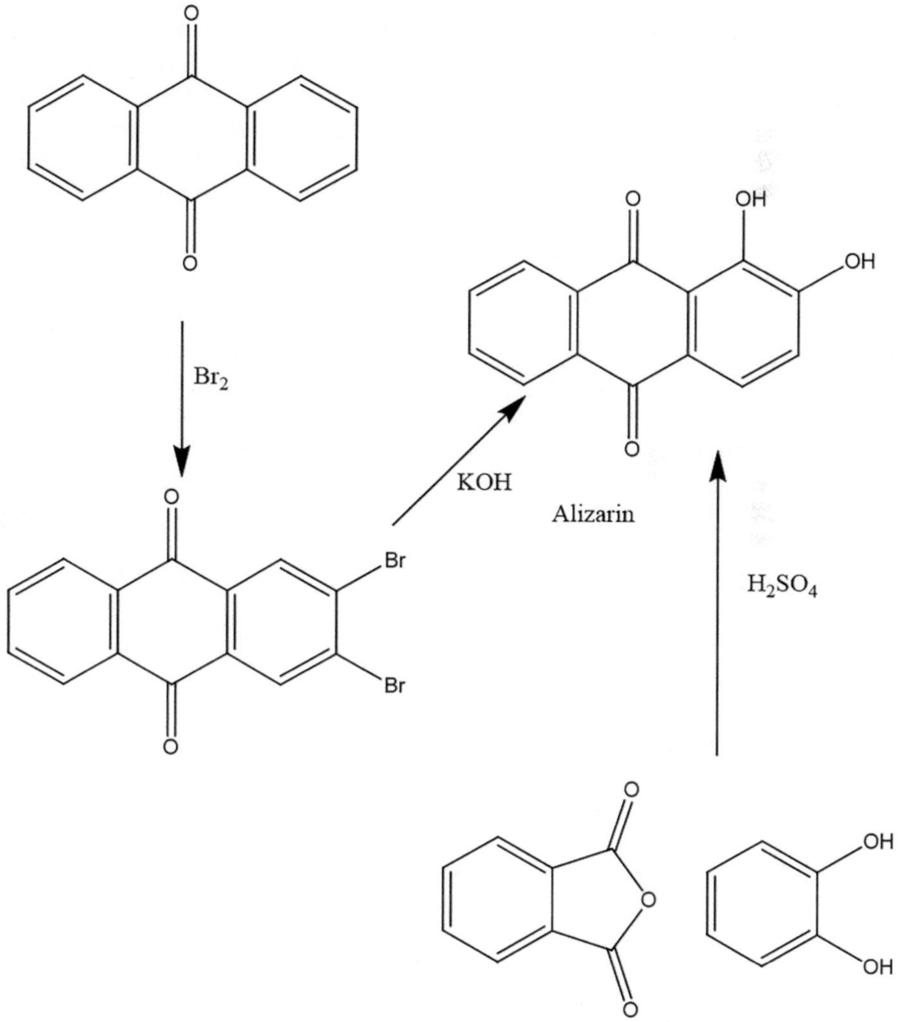

Fig. 5.7 Routes for the synthesis of alizarin

and William Henry Perkin independently discovered the same synthesis *via* the sulphonation of anthraquinone (Fig. 5.8) [104].

The biological profile of alizarin derivatives (including anti-tumour activity) has acted as a driver for the synthesis of a range of alizarin derivatives [129]. Because, in part, of its effect on cellular events such as those involved in reproduction and development, the environmental consequences of the release of alizarin has encouraged methods to be developed for its removal [130]. The concept of natural substances invariably being safe does not apply in this instance. Nonetheless, biological routes to anthraquinones such as alizarin have been developed. The starting point is knowledge of the sequence of enzyme-catalysed reactions that are used in nature. The biosynthesis of anthraquinones occurs through two routes—the polyketide and the shikimate pathways. In the case of alizarin-type anthraquinones, the shikimate pathway is the most significant. Prenylation of 1,4-dihydroxy-2-naphthoic acid (DHNA) is the first step in the pathway and a prenyl transferase has been found to be involved [131]. While the precise details of all the steps in alizarin (and other anthraquinones) biosynthesis has not been described in detail the physical and chemical parameters that have an impact on the biosynthesis have been studied [107]. In addition to its medicinal and use as a pigment, a putative role for alizarin

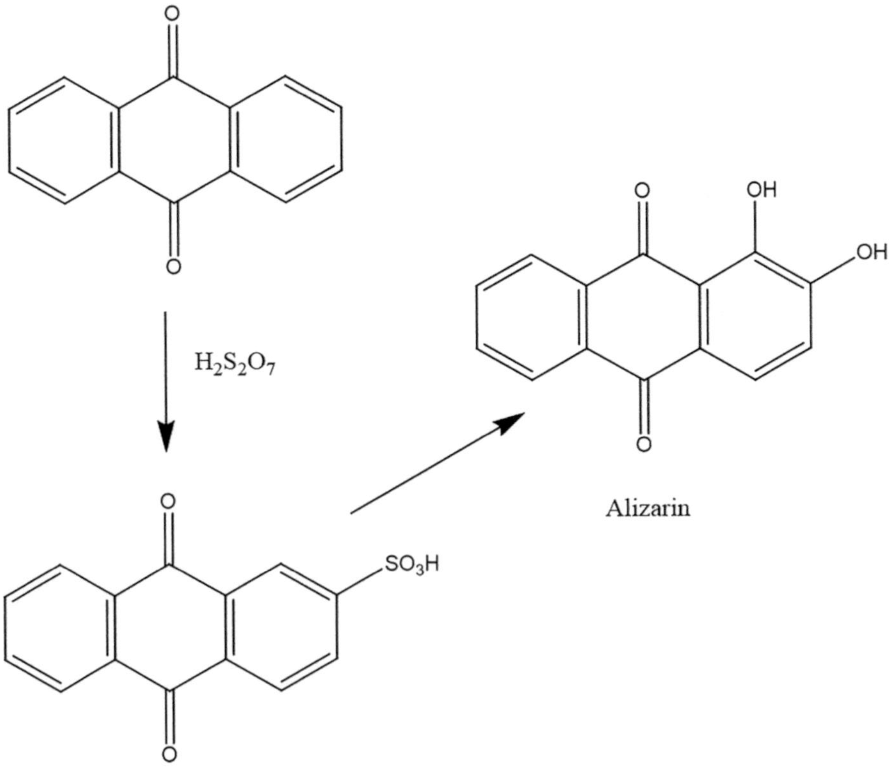

Fig. 5.8 Commercial route for the synthesis of alizarin via sulphonation

in plant defence strategies has been suggested [132]. The development of biotechnological approaches to the production of alizarin-type molecules has been achieved in a bioreactor [133]. *Oldenlandia umbellata* is a member of the *Rubiaceae* family of plants and is therefore related to *Rubia* species including *Rubia cordifolia*, *Rubia tinctorum*, and *Rubia akane*. All three are able to be grown in cell and organ cultures to produce anthraquinones [107]. Manipulation of genes using gene editing techniques such as CRISPR will result in the biosynthesis of new anthraquinones with novel physical and biological properties however such methods will have to compete with the growing development of 'green' synthetic chemistry [134].

References

1. Atanasov AG, Zotchev SB, Dirsch VM, et al. Natural products in drug discovery: advances and opportunities. Nat Rev Drug Discov. 2021;20:200–16. https://doi.org/10.1038/s41573-020-00114-z.
2. Huxtable M, O'Donnell R. Medieval colour. In: Gerrard C, Gutiérrez A, editors. The Oxford handbook of later medieval archaeology in Britain. Oxford: Oxford Academic; 2018. https://doi.org/10.1093/oxfordhb/9780198744719.013.57.
3. Cooksey C. Tyrian purple: the first four thousand years. Sci Prog. 2013;96(Pt 2):171–86. https://doi.org/10.3184/003685013X13680345111425.
4. Clougherty JE, Souza K, Cullen MR. Work and its role in shaping the social gradient in health. Ann N Y Acad Sci. 2010;1186:102–24. https://doi.org/10.1111/j.1749-6632.2009.05338.x.
5. Marriott L. The construction of crime: the presumption of blue-collar guilt and white-collar innocence. Soc Policy Soc. 2017;16(2):237–51. https://doi.org/10.1017/S1474746416000063.
6. Sturm RA, Duffy DL. Human pigmentation genes under environmental selection. Genome Biol. 2012;13(9):248. https://doi.org/10.1186/gb-2012-13-9-248.
7. Bae-Harboe YS, Park HY. Tyrosinase: a central regulatory protein for cutaneous pigmentation. J Invest Dermatol. 2012;132(12):2678–80. https://doi.org/10.1038/jid.2012.324.
8. Manga P, Loftus S. Genetics of skin, hair, and eye color in human pigmentation disorders. Ann Hum Genet. 2025;89(5):305–20. https://doi.org/10.1111/ahg.70003.
9. Hanley AB, Heaney RK, Fenwick GR. Improved isolation of glucobrassicin and other glucosinolates. J Sci Food Agric. 1983;34:869.
10. Kokubun T, Edmonds J, John P. Indoxyl derivatives in woad in relation to medieval indigo production. Phytochem. 1998;49:79–87. https://doi.org/10.1016/S0031-9422(97)01069-8.
11. Montana G, Giarrusso R, D'Amico R, Di Natale B, Vizzini MA, Ilardi V, Mulone A, Randazzo L, Bordenca CV. Multi-analytical study of the medieval wall paintings from the rupestrian church Grotta del Crocifisso at Lentini (eastern Sicily): new evidence of the use of woad (Isatis tinctoria). Archaeol Anthropol Sci. 2022;14(9):183. https://doi.org/10.1007/s12520-022-01656-6.
12. Champion M. Devotion, pestilence and conflict: the medieval wall paintings of St Mary the virgin, Lakenheath. In: Heslop TA, Mellings E, Thøfner M, editors. Art, faith and place in east anglia: from prehistory to the present. Woodbridgem: Boydell & Brewer; 2012. p. 88–104.
13. Stade H, et al. Erfurt: eine Stadt im Wandel. Leipzig: Edition Leipzig; 2015.
14. Cardon D, Koren ZC, Sumi H. Woaded blue: a Colorful approach to the dialectic between written historical sources, experimental archaeology, chromatographic analyses, and biochemical research. Heritage. 2023;6:681–704. https://doi.org/10.3390/heritage6010037.
15. Clarke G. Salt-winning on the Lyn: Anglo-Saxon and medieval industry at Gaywood's north marsh, King's Lynn. Barnsley: East Anglian Archaeology; 2023. p. 180.

16. Hayward M. Dressed in blue: the impact of Woad on English clothing, c. 1350–c. 1670. Costume. 2015;49(2):168–85. https://doi.org/10.1179/0590887615Z.00000000074.
17. Young T. The Bakerian lecture: on the theory of light and colours. Philos Trans R Soc Lond. 1802;92:12–48.
18. Newton I. Opticks. London: Innus; 1704.
19. Newman WR. Newton the alchemist: science, enigma, and the quest for nature's "secret fire". Princeton: Princeton University Press; 2019.
20. Lee BB. The evolution of concepts of color vision. Neurociencias. 2008;4(4):209–24.
21. Benli H. Bio-mordants: a review. Environ Sci Pollut Res Int. 2024;31(14):20714–71. https://doi.org/10.1007/s11356-024-32174-8.
22. Eyupoglu S, Eyupoglu C, Merdan N. Investigation of dyeing Behaviors of bio-mordanted wool fabric dyed with natural dye. Fibers Polym. 2025;26:707–21. https://doi.org/10.1007/s12221-025-00858-8.
23. Horobin RW. Biological staining: mechanisms and theory. Biotech Histochem. 2002;77(1):3–13.
24. Dapson RW. Dye-tissue interactions: mechanisms, quantification and bonding parameters for dyes used in biological staining. Biotech Histochem. 2005;80(2):49–72. https://doi.org/10.1080/10520290500219982.
25. Bartholomew JW, Mittwer T. The gram stain. Bacteriol Rev. 1952;16(1):1–29. https://doi.org/10.1128/br.16.1.1-29.1952.
26. Bennett JA, Kandell GV, Kirk SG, McCormick JR. Visual and microscopic evaluation of streptomyces developmental mutants. J Vis Exp. 2018;139:57373. https://doi.org/10.3791/57373.
27. Szadkowski B, Kuśmierek M, Śliwka-Kaszyńska M, Marzec A. Structure and stability characterization of natural lake pigments made from plant extracts and their potential application in polymer composites for packaging materials. Materials. 2022;15(13):4608. https://doi.org/10.3390/ma15134608.
28. Clementi C, Melo MJ, Romani A, Seixas de Melo JS. Organic dyes and lake pigments. In: Springer series on fluorescence. Cham: Springer; 2024. https://doi.org/10.1007/4243_2024_50.
29. Nabais P, Melo MJ, Lopes JA, et al. Microspectrofluorimetry and chemometrics for the identification of medieval lake pigments. Herit Sci. 2018;6:13. https://doi.org/10.1186/s40494-018-0178-1.
30. Sabatini F, Degano I. Analysis of natural and synthetic organic lakes and pigments by chromatographic and mass spectrometric techniques. In: Colombini MP, Degano I, Nevin A, editors. Analytical chemistry for the study of paintings and the detection of forgeries. Cultural heritage science. Cham: Springer; 2022. https://doi.org/10.1007/978-3-030-86865-9_9.
31. Fieberg JE, Knutås P, Hostettler K, Smith GD. "Paintings fade like flowers": pigment analysis and digital reconstruction of a faded pink lake pigment in Vincent van Gogh's undergrowth with two figures. Appl Spectrosc. 2017;71(5):794–808. https://doi.org/10.1177/0003702816685097.
32. King RB. Color-pattern variation in Lake Erie water snakes: prediction and measurement of natural selection. Evolution. 1993;47(6):1819–33. https://doi.org/10.1111/j.1558-5646.1993.tb01272.x.
33. Maan ME, Sefc KM. Colour variation in cichlid fish: developmental mechanisms, selective pressures and evolutionary consequences. Semin Cell Dev Biol. 2013;24(6-7):516–28. https://doi.org/10.1016/j.semcdb.2013.05.003.
34. Xiao Y, Riise G. Coupling between increased lake color and iron in boreal lakes. Sci Total Environ. 2021;767:145104. https://doi.org/10.1016/j.scitotenv.2021.145104.
35. Tamm M, Freiberg R, Tõnno I, Nõges P, Nõges T. Pigment-based chemotaxonomy–a quick alternative to determine algal assemblages in large shallow eutrophic lake? PLoS One. 2015;10(3):e0122526. https://doi.org/10.1371/journal.pone.0122526.

36. Topp SN, Pavelsky TM, Dugan HA, Yang X, Gardner J, Ross MRV. Shifting patterns of summer lake color phenology in over 26,000 US lakes. Water Resour Res. 2021;57(5):e2020WR029123. https://doi.org/10.1029/2020WR029123.
37. Velliky EC, d'Errico F, van Niekerk KL, Henshilwood CS. Unveiling the multifunctional use of ochre in the middle stone age: specialized ochre retouchers from Blombos cave. Sci Adv. 2025;11(26):eads2797. https://doi.org/10.1126/sciadv.ads2797.
38. Mortimore JL, Marshall LJ, Almond MJ, Hollins P, Matthews W. Analysis of red and yellow ochre samples from clearwell caves and Catalhöyük by vibrational spectroscopy and other techniques. Spectrochim Acta A Mol Biomol Spectrosc. 2004;60(5):1179–88. https://doi.org/10.1016/j.saa.2003.08.002.
39. Froment F, Tournié A, Colomban P. Raman identification of natural red to yellow pigments: ochre and iron-containing ores. J Raman Spectrosc. 2008;39:560–8. https://doi.org/10.1002/jrs.1858.
40. Velliky EC, Porr M, Conard NJ. Ochre and pigment use at Hohle Fels cave: results of the first systematic review of ochre and ochre-related artefacts from the upper palaeolithic in Germany. PLoS One. 2018;13(12):e0209874. https://doi.org/10.1371/journal.pone.0209874.
41. MacDonald BL, Velliky EC, Forrester B, Riedesel S, Linstädter J, Kuo AL, Woodborne S, Mabuza A, Bader GD. Ochre communities of practice in stone age Eswatini. Nat Commun. 2024;15(1):9201. https://doi.org/10.1038/s41467-024-53050-6.
42. Rifkin RF, Dayet L, Queffelec A, Summers B, Lategan M, d'Errico F. Evaluating the photoprotective effects of ochre on human skin by in vivo SPF assessment: implications for human evolution, adaptation and dispersal. PLoS One. 2015;10(9):e0136090. https://doi.org/10.1371/journal.pone.0136090.
43. Crombie M, et al. A new protocol for fingerprinting cultural ochre sources using mineral magnetism. J Archaeol Sci. 2025; https://doi.org/10.5194/egusphere-egu24-7855.
44. Schmidt CM, Walton MS, Trentelman K. Characterization of lapis lazuli pigments using a multitechnique analytical approach: implications for identification and geological provenancing. Anal Chem. 2009;81(20):8513–8. https://doi.org/10.1021/ac901436g.
45. Hamerton I, Tedaldi L, Eastaugh N. A systematic examination of colour development in synthetic ultramarine according to historical methods. PLoS One. 2013;8(2):e50364. https://doi.org/10.1371/journal.pone.0050364.
46. Wang H, Zhang S, Hu S, Zhen Z, Gomez MA, Yao S. A systematic study of the synthesis conditions of blue and green ultramarine pigments via the reclamation of the industrial zeolite wastes and agricultural rice husks. Environ Sci Pollut Res Int. 2020;27(10):10910–24. https://doi.org/10.1007/s11356-020-07624-8.
47. Muñoz-Miranda LA, Iñiguez-Moreno M. An extensive review of marine pigments: sources, biotechnological applications, and sustainability. Aquat Sci. 2023;85(3):68. https://doi.org/10.1007/s00027-023-00966-8.
48. Donoso A, González-Durán J, Muñoz AA, González PA, Agurto-Muñoz C. Therapeutic uses of natural astaxanthin: an evidence-based review focused on human clinical trials. Pharmacol Res. 2021;166:105479. https://doi.org/10.1016/j.phrs.2021.105479.
49. Nishida Y, Berg PC, Shakersain B, Hecht K, Takikawa A, Tao R, Kakuta Y, Uragami C, Hashimoto H, Misawa N, Maoka T. Astaxanthin: past, present, and future. Mar Drugs. 2023;21(10):514. https://doi.org/10.3390/md21100514.
50. Solhaug KA, Gauslaa Y, Nybakken L, Bilger W. UV-induction of sun-screening pigments in lichens. New Phytol. 2003;158:91–100. https://doi.org/10.1046/j.1469-8137.2003.00708.x.
51. Leksin I, Shelyakin M, Zakhozhiy I, Kozlova O, Beckett R, Minibayeva F. Ultraviolet-induced melanisation in lichens: physiological traits and transcriptome profile. Physiol Plant. 2024;176(5):e14512. https://doi.org/10.1111/ppl.14512.
52. Cooksey CJ. Tyrian purple: 6,6'-dibromoindigo and related compounds. Molecules. 2001;6(9):736–69. https://doi.org/10.3390/60900736.

53. Nongmaithem BD, Mouatt P, Smith J, et al. Volatile and bioactive compounds in opercula from Muricidae molluscs supports their use in ceremonial incense and traditional medicines. Sci Rep. 2017;7:17404. https://doi.org/10.1038/s41598-017-17551-3.
54. Benkendorff K, Rudd D, Nongmaithem BD, Liu L, Young F, Edwards V, Avila C, Abbott CA. Are the traditional medical uses of Muricidae molluscs substantiated by their pharmacological properties and bioactive compounds? Mar Drugs. 2015;13(8):5237–75. https://doi.org/10.3390/md13085237.
55. Vougogiannopoulou K, Skaltsounis AL. From Tyrian purple to kinase modulators: naturally halogenated indirubins and synthetic analogues. Planta Med. 2012;78(14):1515–28. https://doi.org/10.1055/s-0032-1315261.
56. Fox DL. Carotenoids and other lipoid-soluble pigments in the sea and in deep marine mud. Proc Natl Acad Sci USA. 1937;23(6):295–301. https://doi.org/10.1073/pnas.23.6.295.
57. Macel ML, Ristoratore F, Locascio A, Spagnuolo A, Sordino P, D'Aniello S. Sea as a color palette: the ecology and evolution of fluorescence. Zool Lett. 2020;10(6):9. https://doi.org/10.1186/s40851-020-00161-9.
58. Poding LH, Jägers P, Herlitze S, Huhn M. Diversity and function of fluorescent molecules in marine animals. Biol Rev Camb Philos Soc. 2024;99(4):1391–410. https://doi.org/10.1111/brv.13072.
59. Zhao Z, Gonsior M, Luek J, Timko S, Ianiri H, Hertkorn N, et al. Picocyanobacteria and deep-ocean fluorescent dissolved organic matter share similar optical properties. Nat Commun. 2017;8:15284. https://doi.org/10.1038/ncomms15284.
60. Bähr S, van der Meij SE, Terraneo T, Oury N, Michiels NK, Ogg S, Marchese F, Benzoni F. Integrative phylogenomics sheds light on the diversity and evolution of fluorescence in coral-dwelling gall crabs. Proc Biol Sci. 2025;292(2042):20242403. https://doi.org/10.1098/rspb.2024.2403.
61. Ding W, Dong Y, Xu R, Liang D, Yuan X, Zhou L, Sun C, Zhang H, Li M, Ma J, Ju J, Li Q. Seven cyclization patterns of type II PKS yield fluorescent compounds, including fungal-type folded Globismycin a. ACS Chem Biol. 2025;20(8):1916–27. https://doi.org/10.1021/acschembio.5c00299.
62. Nawaz A, Chaudhary R, Shah Z, Dufossé L, Fouillaud M, Mukhtar H, Haq IU. An overview on industrial and medical applications of bio-pigments synthesized by marine bacteria. Microorganisms. 2020;9(1):11. https://doi.org/10.3390/microorganisms9010011.
63. Negi A, Aggarwal S, Kathuria D, Barthwal R, Singh N. An overview on natural marine pigments: extraction, potential health benefits and stability in food applications. Food Rev Int. 2025; https://doi.org/10.1080/87559129.2025.2528956.
64. Venil CK, Velmurugan P, Dufossé L, Devi PR, Ravi AV. Fungal pigments: potential coloring compounds for wide ranging applications in textile dyeing. J Fungi. 2020;6(2):68. https://doi.org/10.3390/jof6020068.
65. Salman M, Adeel S, Naseer M, Haider MZ, Anjum F. Modern approach toward algal-based natural pigments for textiles. In: Thomas S, Jose S, Mathew SS, Samant L, editors. Plant biomass derived materials. Hoboken: Wiley; 2024. https://doi.org/10.1002/9783527839032.ch9.
66. Gidney T. The world and a small place in England: Norwich's textile industry from the 'middle ages' to 'industrial revolution. World Hist Connect. 2023;15(1) https://doi.org/10.13021/whc.v15i1.3818.
67. Dobson MJ. Malaria in England: a geographical and historical perspective. Parassitologia. 1994;36(1-2):35–60.
68. Hutchinson RA, et al. Malaria and deaths in the English marshes. Lancet. 2006;367:1947–51.
69. Molyneux ME. History of discoveries, malaria. In: Hommel M, Kremsner P, editors. Encyclopedia of malaria. New York: Springer; 2013. https://doi.org/10.1007/978-1-4614-8757-9_115-1.
70. Tan SY, Ahana A. Charles Laveran (1845-1922): nobel laureate pioneer of malaria. Singapore Med J. 2009;50(7):657–8.

71. Foley M, Tilley L. Quinoline antimalarials: mechanisms of action and resistance. Int J Parasitol. 1997;27(2):231–40. https://doi.org/10.1016/s0020-7519(96)00152-x.
72. Achan J, Talisuna AO, Erhart A, Yeka A, Tibenderana JK, Baliraine FN, Rosenthal PJ, D'Alessandro U. Quinine, an old anti-malarial drug in a modern world: role in the treatment of malaria. Malar J. 2011;24(10):144. https://doi.org/10.1186/1475-2875-10-144.
73. Yip A, Sivarajah S, Natkunarajah J. H22 barking up the right tree: history of quinine. British J Dermatol. 2023;188(4):ljad113.304. https://doi.org/10.1093/bjd/ljad113.304.
74. Pai-Dhungat JV. Caventou, Pelletier–history of quinine. J Assoc Physicians India. 2015;63(3):58.
75. Steere-Williams J. Essay review: the fictive history of Victorian science and empire. Br. J Hist Sci. 2024;19:1–3. https://doi.org/10.1017/S0007087424001249.
76. van der Hoogte A, Pieters T. Science in the service of colonial agro-industrialism: the case of cinchona cultivation in the Dutch and British East Indies, 1852–1900. Stud Hist Philos Biol Biomed Sci. 2014;47(Pt A):12–22. https://doi.org/10.1016/j.shpsc.2014.05.019.
77. Greenwood D. The quinine connection. J Antimicrob Chemother. 1992;30(4):417–27. https://doi.org/10.1093/jac/30.4.417.
78. Ray D. Organochlorine and pyrethroid insecticides. In: Comprehensive toxicology. Amsterdam: Elsevier; 2010. p. 445–57. https://doi.org/10.1016/B978-0-08-046884-6.01327-0.
79. Neal L. The Dutch and English East India companies compared: evidence from the stock and foreign exchange markets. In: Tracy JD, editor. The rise of merchant empires: long distance trade in the early modern world 1350–1750. Studies in comparative early modern history. Cambridge: Cambridge University Press; 1990. p. 195–223.
80. Rodwell G. Lavoisier, Priestley, and the discovery of oxygen. Nature. 1882;27:8–11. https://doi.org/10.1038/027008d0.
81. Roberts GK. The establishment of the royal college of chemistry: an investigation of the social context of early-victorian chemistry. In: Historical studies in the physical sciences, vol. 7. Berkeley: University of California Press; 1976. p. 437–85.
82. Bentivoglio M, Kristensson K. Tryps and trips: cell trafficking across the 100-year-old blood-brain barrier. Trends Neurosci. 2014;37(6):325–33. https://doi.org/10.1016/j.tins.2014.03.007.
83. Perkin WH, Patent AD (1856) No 1984
84. Anderson RD. 'Germany: academic Golden age', European universities from the enlightenment to 1914. Oxford: Oxford Academic; 2010. https://doi.org/10.1093/acprof:oso/9780198206606.003.0010.
85. Vernon G. Syphilis and salvarsan. Br J Gen Pract. 2019;69(682):246. https://doi.org/10.3399/bjgp19X702533.
86. Wainwright M. Dyes, trypanosomiasis and DNA: a historical and critical review. Biotech Histochem. 2010;85(6):341–54. https://doi.org/10.3109/10520290903297528.
87. Cluskey JE. Goethe and chemistry. J Chem Educ. 1951;28(10):536. https://doi.org/10.1021/ed028p536.
88. Beretta M. Lavoisier and the history of chemistry. Ambix. 2024;71(2):209–24. https://doi.org/10.1080/00026980.2024.2324618.
89. de la Croix D, Docquier F, Fabre A, Stelter R. The academic market and the rise of universities in medieval and early modern Europe (1000–1800). J Eur Econ Assoc. 2024;22(4):1541–89. https://doi.org/10.1093/jeea/jvad061.
90. Manavi MA, Salehi M, Mohammad Jafari R, Dehpour AR. From dyes to drugs: the historical impact and future potential of dyes in drug discovery. Arch Pharm. 2024;357(11):e2400532. https://doi.org/10.1002/ardp.202400532.
91. Jones AW. Early drug discovery and the rise of pharmaceutical chemistry. Drug Test Anal. 2011;3:337–44. https://doi.org/10.1002/dta.301.

92. Frignani E, D'Eusanio V, Grandi M, Pigani L, Roncaglia F. A continuous extraction protocol for the characterisation of a sustainably produced natural indigo pigment. Life. 2024;14:59. https://doi.org/10.3390/life14010059.
93. van Bommel R, de Keijzer M. The long industrial road to synthetic indigo. In: Bright colours from the past. Cultural heritage science. Cham: Springer; 2025. https://doi.org/10.1007/978-3-031-88664-5_21.
94. Castillo-Suárez LA, Sierra-Sánchez AG, Linares-Hernández I, Martínez-Miranda V, Teutli-Sequeira EA. A critical review of textile industry wastewater: green technologies for the removal of indigo dyes. Int J Environ Sci Technol. 2023;27:1–38. https://doi.org/10.1007/s13762-023-04810-2.
95. Linke JA, Rayat A, Ward JM. Production of indigo by recombinant bacteria. Bioresour Bioprocess. 2023;10(1):20. https://doi.org/10.1186/s40643-023-00626-7.
96. Berke H, Wiedemann HG. The chemistry and fabrication of the anthropogenic pigments Chinese blue and purple in ancient China. EASTM. 2000;17:94–120. https://doi.org/10.1163/26669323-01701006.
97. Berke H. Chemistry in ancient times: the development of blue and purple pigments. Angew Chem Int Ed. 2002;41:2483–7.
98. Kalba LA. Color in the age of impressionism: commerce, technology, and art. University Park: Penn State University Press; 2017. https://doi.org/10.5325/j.ctv14gp14d.
99. Vermeulen M, Miranda ASO, Tamburini D, et al. A multi-analytical study of the palette of impressionist and post-impressionist Puerto Rican artists. Herit Sci. 2022;10:44. https://doi.org/10.1186/s40494-022-00683-9.
100. Gartman D. Harley earl and the art and color section: the birth of styling at general motors. Des Issues. 1994;10(2):3–26.
101. Clarke S. Managing design: the art and colour section at general motors, 1927-1941. J Des Hist. 1999;12(1):65–79.
102. dos Santos TC. Alizarin. In: Shamey R, editor. Encyclopedia of color science and technology. Cham: Springer; 2023. https://doi.org/10.1007/978-3-030-89862-5_438.
103. Graebe C, Liebermann C (1869) Improved process of preparing alizarine. U.S. Patent no. 95,465
104. van Bommel R, de Keijzer M. Copying nature: synthetic alizarin. In: Bright colours from the past. Cultural heritage science. Cham: Springer; 2025. https://doi.org/10.1007/978-3-031-88664-5_20.
105. Do KL, Mushtaq A, Zhao F, Su M. Alizarin old and new: extraction techniques for coloration, advances in detection methods for historical textiles and novel applications as a functional dye. Color Technol. 2024;140(1):5–29. https://doi.org/10.1111/cote.12711.
106. Singh J, Hussain Y, Luqman S, Meena A. Purpurin: a natural anthraquinone with multifaceted pharmacological activities. Phytother Res. 2021;35(5):2418–28. https://doi.org/10.1002/ptr.6965.
107. Murthy HN, Joseph KS, Paek KY, Park SY. Anthraquinone production from cell and organ cultures of Rubia species: an overview. Meta. 2022;13(1):39. https://doi.org/10.3390/metabo13010039.
108. Travis AS. Chemical modeling: from Paul Ehrlich's dyes to β-blockers-a brief history. J Comput Biol. 2019;26(7):726–34. https://doi.org/10.1089/cmb.2019.0092.
109. Travis AS. Models for biological research: the theory and practice of Paul Ehrlich. Hist Philos Life Sci. 2008;30(1):79–97.
110. Grazia C, Clementi C, Miliani C, Romani A. Photophysical properties of alizarin and purpurin Al(III) complexes in solution and in solid state. Photochem Photobiol Sci. 2011;10(7):1249–54. https://doi.org/10.1039/c1pp05039g.
111. Rehman R, Waheed-Uz-Zaman, Raza A, Noor W, Batool A, Maryem H. Photocatalytic degradation of Alizarin Red S, Amaranth, Congo Red, and Rhodamine B dyes using UV light modified reactor and ZnO, TiO2, and SnO2 as catalyst. J Chem. 2021;9:6655070. https://doi.org/10.1155/2021/6655070.

112. Zhang J, Chi Y, Feng L. The mechanism of degradation of alizarin red by a white-rot fungus Trametes gibbosa. BMC Biotechnol. 2021;21(1):64. https://doi.org/10.1186/s12896-021-00720-8.
113. Raj ANP, Bennie RB, Antony AJ, et al. Visible light driven effective photodegradation of alizarin red S using zirconium doped metastable hexagonal MoO3 Nanorods. J Clust Sci. 2023;34:2079–91. https://doi.org/10.1007/s10876-022-02371-x.
114. Schot J, Hornix WJ, Bud R, Schot J. Technology in decline: a search for useful concepts. Br J Hist Sci. 1992;25(1):5–26. https://doi.org/10.1017/S0007087400045301.
115. Bolingbroke LG. St John madder market, Norwich: its streets, lanes, and ancient houses, and their old-time associations. Norfolk Archaeol. 2020;20(2):215–39. https://doi.org/10.5284/1077651.
116. Speleers L, Vanden Berghe I, Ghesquière V, et al. Colourful rivers: archaeobotanical remains of dye plants from urban fluvial deposits in the southern low countries (Belgium). Veget Hist Archaeobot. 2025;34:291–306. https://doi.org/10.1007/s00334-024-01009-4.
117. Science Museum Group. The method of cultivating madder, as it is now practised by the Dutch in Zealand. 1959–82 Science Museum Group Collection Online. https://collection.sciencemuseumgroup.org.uk/objects/co482409/the-method-of-cultivating-madder-as-it-is-now-practised-by-the-dutch-in-zealand.
118. Newman WR. Introduction: traditions of analysis and synthesis. In: Newman WR, Schickore J, editors. Traditions of analysis and synthesis. Archimedes, vol. 73. Cham: Springer; 2025. https://doi.org/10.1007/978-3-031-76398-4_1.
119. Cova TFGG, Pais AACC, Seixas de Melo JS. Reconstructing the historical synthesis of mauveine from Perkin and Caro: procedure and details. Sci Rep. 2017;7:6806. https://doi.org/10.1038/s41598-017-07239-z.
120. Gupta D, Guliani E, Bajaj K. Coumarin—synthetic methodologies, pharmacology, and application as natural fluorophore. Top Curr Chem. 2024;382:16. https://doi.org/10.1007/s41061-024-00462-z.
121. Boisde PM, Meuly WC. Coumarin. In: Kirk-Othmer, editor. Encyclopedia of chemical technology. Hoboken: Wiley; 2000. https://doi.org/10.1002/0471238961.0315211302150919.a01.
122. Singh S, Kaushik N, Paliwal A, Sengar MS, Paliwal D. Biological activity and therapeutic potential of Coumarin derivatives: a comprehensive review. Curr Drug Discov Technol. 2025; https://doi.org/10.2174/0115701638374546250323152450.
123. Vershinin VI. Methodological aspects of the group analysis of organic substances. J Anal Chem. 2023;78:162–74. https://doi.org/10.1134/S1061934823020144.
124. Seeman JI. The Woodward-Doering/Rabe-kindler total synthesis of quinine: setting the record straight. Angew Chem Int Ed Engl. 2007;46(9):1378–413. https://doi.org/10.1002/anie.200601551.
125. Trenti F, Yamamoto K, Hong B, Paetz C, Nakamura Y, O'Connor SE. Early and late steps of quinine biosynthesis. Org Lett. 2021;23(5):1793–7. https://doi.org/10.1021/acs.orglett.1c00206.
126. Chandel N, Singh BB, Dureja C, et al. Indigo production goes green: a review on opportunities and challenges of fermentative production. World J Microbiol Biotechnol. 2024;40:62. https://doi.org/10.1007/s11274-023-03871-2.
127. Hu H, Li Z, Chen R, Lu R, Zhou D, Zhu Y, Mu W. Enhancement of indigo production by a metabolically engineered Escherichia coli MG1655 using membrane engineering and two-stage cultivation. J Agric Food Chem. 2025;73(16):9782–92. https://doi.org/10.1021/acs.jafc.5c01400.
128. Berry A, Dodge TC, Pepsin M, Weyler W. Application of metabolic engineering to improve both the production and use of biotech indigo. J Ind Microbiol Biotechnol. 2002;28(3):127–33. https://doi.org/10.1038/sj.jim.7000228.
129. Ye MY, Yao GY, Pan YM, Liao ZX, Zhang Y, Wang HS. Synthesis and antitumor activities of novel α-aminophosphonate derivatives containing an alizarin moiety. Eur J Med Chem. 2014;18(83):116–28. https://doi.org/10.1016/j.ejmech.2014.02.067.

130. El-Hallag I, Al-Owais A, El-Mossalamy ES. Kinetic and thermodynamic investigation of the removal of alizarin red dye using silica-supported nanoscale zero-valent iron particles. Sci Rep. 2025;15:31461. https://doi.org/10.1038/s41598-025-15233-z.
131. Liu C, Wang R, Wang S, Chen T, Lyu C, Kang C, Wan X, Guo J, Li Q, Huang L, Guo L. A prenyltransferase participates in the biosynthesis of anthraquinones in Rubia cordifolia. Plant Physiol. 2024;195(4):2860–76. https://doi.org/10.1093/plphys/kiae171.
132. van Tegelen LJP, Bongaerts RJM, Croes AF, Verpoorte R, Wullems GJ. Isochorismate synthase isoforms from elicited cell cultures of Rubia tinctorum. Phytochemistry. 1999;51:263–9.
133. Aswanilal LV, Siril EA. Enhanced production of alizarin type anthraquinone dye achieved through biotransformation of Oldenlandia umbellata L. using different strains of agrobacterium rhizogenes. Plant Cell Tissue Organ Cult. 2025;160:19. https://doi.org/10.1007/s11240-024-02956-6.
134. Madje BR, Shelke KF, Sapkal SB, Kakade GK, Shingare MS. An efficient one-pot synthesis of anthraquinone derivatives catalyzed by alum in aqueous media. Green Chem Lett Rev. 2010;3(4):269–73. https://doi.org/10.1080/17518251003776877.

Part III
Chemistry and Manufacturing

Chapter 6
Chemistry and Manufacturing. New Technologies

Abstract The importance of change and transformation in chemistry. The creation of value by the manipulation of existing material and the creation of novel and new substances.The building of metabolic iterations and relationships to create a deeper understanding. The development of contacts between Norfolk and Northern Europe through trade routes and the exchange of technical and scientific knowledge. The importance of centres of excellence such as Leiden to the acquisition of knowledge. The shortage of labour as a consequence of outbreaks of plague and wars. The movement of second and third generation immigrants to study at European universities. The continuation of macro weaving methodologies (textiles) to produce Covalent Organic Frameworks (COFs) with distinct molecular properties. The acquisition of knowledge about biological manipulation and the development of synthetic biology – engineering applied to biology. The increased understanding of macromolecular events such as crop rotation in the context of molecular and micrological interactions. The substantial importance of immigration and the two way flow of people, skills and knowledge between individuals, institutions and communities.

Keywords Manufacturing · Reductionism · Empiricism · Affinities · Observation · Huguenot · Weavers · Knowledge exchange

One of the most important aspects of chemistry is that instigates change and thereby creates new opportunities. Chemistry is rationalised in the context of being concerned with the interactions between substances (or parts of substances) at a molecular level. The reactivity of compounds is a consequence both of the molecules themselves and the reaction conditions. For example, phenol is also called carbolic acid. In aqueous conditions, it dissociates to give a phenoxide ion and a hydrogen ion however the equilibrium favours the undissociated molecule. The measurement of the acid dissociation is called the pKa and is defined as the negative logarithm of the acid dissociation constant. The lower the value, the more acidic (dissociated) a compound is. The pKa of phenol in water is 9.95. The value for trichlorophenol is 6.3. This means that the reactivity of trichlorophenol will be different to that of

B. Hanley, *The Chemistry of a County*,
https://doi.org/10.1007/978-3-032-21491-1_6

phenol. These types of interactions means that atoms that are distant from the putative reaction centre may have an impact on reaction rates and may impact upon in the formation of new molecules and new interactions. In brief, chemistry is concerned with the structure and transformation of substances and the consequences of increasing the understanding of the molecular interactions that lie at the core of reactions. The consequences of molecular interactions are visible on a macromolecular scale. While Theme Two was concerned primarily with chemistry and commercialisation in broad terms (isolating and purifying something useful and of value and the consequences of that process), Theme Three deals with a particular aspect of the link between the commercial realisation of chemistry and the way in which this is achieved—specifically by the creation (manufacture) of something of value. Chemistry and manufacture examines specific examples of adding value rather than the more generic commercialisation process. There are, of course, overlaps between manufacture and commercialisation however a distinction can be drawn between processes that increase the overall commercial viability of a system (for example, crop rotation) and those where the increase in value is more easily understood on a molecular level. While previous chapters have explained how commercialisation can be applied to existing (mainly agricultural) processes and how the optimisation of these, even if the molecular chain of causality is not fully understood at the time, demonstrates a clear chemical conceptual lineage from crops primarily for consumption to the added value from the creation of dyestuffs and medicinal products, the theme of chemistry and manufacture will demonstrate the synthesis between existing resources (wool, textiles, even dyestuffs) and imported technology (the macromolecular weaving to create cloth and textiles) to manufacture products with economic and aesthetic value. This will then lead (Chap. 7) to an exposition of manufacturing based upon an extension of weaving to create textiles to the macromolecular manipulation of metals and other materials. In both cases, the importance of creating products with added value is an essential component and is the ultimate objective. The discussion about the unique nature of chemistry compared to physics on the one hand and biology on the other is one that continues to create debate [1]. One important factor is that the development of chemistry comes both from the manipulation of known materials and the creation of new substances. These in turn create a new starting point for the next transformation. In order to achieve both these objectives (manipulation and creation), it is necessary to define the content of materials—how does a chemist define what a substance actually comprises. Atomic and elemental theory provides one approach. These suggest that matter (and hence a given substance) can be divided into component parts but only up to a point. Infinite divisibility is not possible. Once a substance cannot be further subdivided (analysed) then what remains is its elements. The elements may not be recognisable as components of the substance because a substance is not a blend of the properties of its constituent parts. An alternative approach is to hypothesise the existence of a multi-component mixture and to define an element as each of the parts that, when combined, make up the complete multi-component mixture (compound). The first definition is empirical and reductionist and involves removing components successively until what remains is recognisable while the second is perhaps more theoretical and additive as each component is added until the whole is achieved. These operations are analogous to the analysis/synthesis paradigm that has been used to define and confirm substances. While analysis has a logical, reductionist endpoint—the confirmed structure of the molecule once it has been separated into recognisable components—the synthetic paradigm is essentially limitless since it can be used to design new molecules with whatever properties are desired [2] as well as create a specific, known structure. Rather than being exclusively a structure-driven paradigm, it has also a functional component. Whichever approach is used, each can define compounds in terms of components (by addition or subtraction) and, by defining them in those terms, we arrive at the empirical formula—the proportions of each element in the substance as a whole. It has, however, been clear to chemists that an empirical formula is not invariably sufficient to unambiguously define a substance (see below). This is particularly noticeable for organic compounds. Two substances may have the same relative proportions of carbon, hydrogen and oxygen molecules but may be very different in terms of their properties and characteristics. Molecules that have this property are isomeric forms of the same empirical formula. An additional function of isomers which relates to the transformative aspect of chemistry, is that, as Perkin discovered, rearranging atoms to convert one molecule to another is not a trivial task. This led to the development of theories about organic molecules being made up of functional groups as the fundamental critical component and the connections between molecular components being formed of bonds which required energy to sever them. Rather than an atomic-driven principle, molecules were defined by the complementary properties of their components (functional groups) rather than being the macromolecular summation of their micro-level, identical constitutive units (atoms). Such an approach, when combined with an understanding of the biological properties of a compound, is the basis of quantitative structure activity relationships (QSAR)—the main paradigm of a significant part of medicinal chemistry [3]. The concept of QSAR is an extension of the complexity of chemistry. Atoms of a particular element are identical units that cannot be divided further. Molecules are composed of arrangements of atoms and have properties that are different to the component atoms—they are not simply a blend of the most significant atomic properties or an average of them. The functions and properties of molecules are dictated by how the atoms and groups of atoms are arranged in the molecule. QSAR measures the effect of particular molecular arrangements when they interact with living systems. The progression in complexity is from atoms to molecules to systems with a gradually diminishing level of certainty.

Unlike observational studies, the practical and empirical aspects of chemistry lie at the core of the discipline and of the chemistry-oriented approach to the world. Chemistry proposes hypotheses and then devises ways of testing these. In this way a framework is built up based on molecules and how they interact. By extrapolating from the molecular to the systemic, it is possible to build up a world view of integrated parts. It is possible, then to identify two distinct approaches to the laws of nature and how they operate depending on how chemistry is interpreted. The first approach suggests that laws of nature are derived from summaries of the experiences and observations carried out by chemists (and other scientists). This in, in some ways, the chemistry equivalent of the statement by Descartes 'Cogito ergo sum' however the basis of chemistry lacks the certainty of the statement by Descartes. As such, the whole structure of chemistry could collapse. Even the existence of atoms, molecules and bonds is a convenience that explains empirical observations but has no core of certainty and may, indeed be nothing more than an illusion. At a systems level, chemistry avoids the uncomfortable uncertainty and instead considers the accumulation of evidence as populating a sort of Tabula rasa which has been used to develop theories about artificial states which can be created from an initial zero state [4] and built up by metabolic iterations. An alternative view, which is also grounded in chemical uncertainty, is that nature is built up of relationships between its constituent essential parts. These essential parts are also unproven concepts however they do not have to have a provable existence for the consequences of their interaction to be observed. This approach also encompasses microstructural essentialism and carries with it the concept that it is the microstructural characteristics of a substance that create its macromolecular (and observable) properties [5]. This does not, however, imply that macromolecular properties are a blend of the summation of the most indivisible micro structural properties and this is an important distinction. Both approaches can be tested empirically however the first of them, since it is observational only, does not allow for prediction of outcomes. A theory based on relationships and the observable consequences of these rather than purely on observation of specific (but unproven) entities enables predictions about outcomes to be made and tested. Nonetheless, there remains an important aspect of structuralism that is implied in chemistry—namely that structure depends on size and scale and the type of theory that is most useful is dependent upon the scale that is being considered (atomic, molecular, systemic). At one level, the structure of a substance can be hypothesised in terms of electrons and bonds. As the wave theories of the nature of the chemical bond have developed, these have enabled chemists and physicists to envisage structures and bonds as distinct energetic entities [6]. The combination of mathematical rigour combined with the practical observation of properties as observed by chemists has been a goal of this approach. It creates a theoretical framework that, at best, can describe the subatomic basis for the observed macromolecular properties of substances. It does not, however, create an empirical or theoretical underpinning for predicting the outcome of interactions between molecules or, indeed, of individual molecules. It is based upon a hypothesis rather than on a certain foundation. At a second level, substances can be considered in terms of the atoms present and both the proportions present found and their relative orientations to one another within the molecule ascertained. This is a reductionist approach that takes molecules and their components as a starting point and enables empirical formulae (the relative proportions of the different constituents) to be determined by degradation The determination of the proportions of different elements present can enable prediction of properties and, to an extent, inter and intra molecular interactions, to be ascertained. An example of the consequences of this is the concept of 'affinities' and these became popular in the eighteenth century and were tabulated by Geoffroy in 1718. The basis of affinities was that some elements had a greater affinity for a specific element than others however it was largely observational and, in the absence of hypotheses about how bonds were formed, its predictive power was limited [7]. It provided a partial explanation for the observation that displacement reactions occur in which an element of higher affinity could displace one of lower affinity from a compound. In addition, the concept of affinities also lies at the core of biological systems and changes in affinities is significant evidence for changes in biological systems [8]. While the concept of displacement of one atom by another has been mentioned and forms part of the basis of aspects of inorganic chemistry, there are also some examples at the junction of organic and inorganic chemistry in which affinity is important and selective, including the use of DNA aptamers to probe inorganic nanomaterials [9]. While such examples are an interesting application of the principles of affinity-guided chemistry, the traditional use of affinities and the reductionism of molecules to single elements can, as noted above, be problematical for organic compounds which, because in particular of the tetravalent nature of carbon atoms, are able to exist in multiple forms (isomers) with the same empirical formula. Even for the simplest inorganic molecules, the idea that a substance can be defined as a single entity at a fixed point and that a substance is composed of sets of identical units is overly simplistic. For example, water has the molecular formula of H_2O however, even a pure sample of water will be partially dissociated into H_3O^+ and OH^- ions and, for individual molecules, this is a dynamic process [10]. In the case of organic molecules, the situation is even more complex. The possibility of the existence of isomers means that compounds with the same empirical formulae may have very different macromolecular properties. In addition, as will be considered later, molecules may exist in different forms. This suggests that total reductionism to the constituent atomic parts may not be the most insightful approach. Identification of specific functional groups (acids, alcohols etc.) allows molecules to be classified according to microscopic functionality and macroscopic properties. This does not provide an absolute connection between substances and, even for very simple compounds, the presence of different environments and other components in the molecule may have a significant effect. For example, phenols can be considered to be a distinct class of molecules (a benzene ring in which one of the H atoms is replaced by an OH group). However, the parent compound dissociates in a water environment, into an anion ($C6H5O^-$) and a positively charged cation (H^+). If the aromatic ring is substituted with chlorine atoms then the level of dissociation (as measured by the dissociation constant—pKa) is reduced [11]. In conclusion, the use of subatomic, elemental or functionality-guided principles to probe chemical structures is very dependent upon the type of investigation being carried out. The paradigm of observation as a sole way of investigating chemistry has some value however, whether measured on a molecular or a macromolecular level, the chemistry of substances (and their interactions) is a dynamic process and should, therefore, be tested empirically over time rather than just being a single moiety frozen at a particular timepoint. Chemistry is about how things change as well as how they exist. The same type of thinking can be applied to the impact of chemistry-related changes to manufacturing and processes. The dynamic processes that are governed by chemistry apply in many different circumstances and therefore predicting what will happen (even if the foundations are insecure) can be arrived at by extrapolation. There is a consistency to the application of chemistry in manufacturing in different areas. This will be examined from the context of technological advances and the consequences of these.

Some of the new technologies that were developed (improvements to agronomic practice and the extraction and synthesis of colours) have already been discussed. In many cases, improvements were able to be noted and quantified even if an understanding of the nature of the causality between actions and consequences were not understood. While these actions still had a significant impact on technology in the eighteenth century and an impact on manufacturing processes, the driver for some of the development of new technology resulted from improvements in manufacturing processes that were developed in the seventeenth century or earlier and were, in part, only made possible by the development of chemistry as a distinct discipline that was separate from alchemy (and, in particular, more reproducible). Some of the technologies that were developed had unexpected consequences and were impacted by, for example, traditional trade routes. In this case the strength of trading links throughout Northern Europe and the subsequent development of trade through to Asia (including along the traditional Silk Road) enabled goods and technology to be transported including, in particular textiles such as silk and finished goods. With its strong links through the Hanseatic League and the traditional interchange of agronomic expertise (for example, crop rotation), the Eastern counties of England had particular geographical and agricultural trade advantages that facilitated the adoption of goods and technologies from Asia *via* Northern Europe and the subsequent creation of wealth. Indeed Norwich was more a product of its region (and its links with Northern Europe extended that region) than of England as a whole. The similarities and links between the extremities of the Silk Road (Norwich and Nara, Japan) has been noted [12]. The focus towards Europe (and beyond) and the popularity and effectiveness of sea trading routes may explain why railway development between Norwich and London did not start until December 1845 and was preceded by the establishment of the line from Norwich to Yarmouth (April 1844). This was in contrast to the development of railways linking the major cities of the Northwest (Manchester, Liverpool, Leeds and Birmingham) with each other and with London in the 1830s. Prior to the coming of the industrial revolution that changed the focus of wealth creation in England from East to West (and facing the New World rather than Europe), new technologies were imported and established which increased the economic power of Norfolk and Norwich in the sixteenth and seventeenth centuries.

6.1 New Technology in the Sixteenth and Seventeenth Centuries

The importation of immigrants and technology into Norfolk occurred in distinct waves in response both to a push (based largely on religious persecution on mainland Europe) and to the pull of technological and economic advantage and the availability of raw materials (particularly wool) from the arable land in East Anglia. The source of the majority of these immigrants was Flanders (in the eleventh century) and a combination of Flemish and Walloon weavers in the sixteenth century then an influx of Huguenots fleeing religious persecution in France in the seventeenth century. Each of these groups brought particular skills which not only had a direct effect on the local production and manufacture of goods but also facilitated the development of adjacent trades and skills. Many of these additional crafts were underpinned by chemistry which was emerging from alchemy [13] and from attempts to transmute base metal into gold using alchemical methods [14]. Although the terms 'alchemy' and 'chemistry' were both in use in the seventeenth century, they are not conceptually or philosophically interchangeable, but a degree of linguistic confusion was inevitable [15]. In this section, some of the new technologies that were imported from Europe will be revealed and the impact these had on the development of chemistry-related manufacturing processes will be considered. Where appropriate, the influx of new technologies and skills will be contextualised specifically with regard to England as a whole however the primary focus will be on Norfolk and Norwich since there is a (sometimes rather obscure) lineage that can be discerned and this is important for the successful implantation of new technologies. This is apparent in the history of the Flemish and Walloon populations, in particular, so these will be described in more detail than the earlier (low countries) or later (Huguenot) influxes—both of which brought distinct advantages to the pre-industrial revolution economy and way of living in Norfolk and, indeed, England as a whole [16].

As noted previously, Norfolk was an important agricultural area from the time of the Norman Conquest onwards. The growth in agricultural expertise and the increased yields led to substantial time and resources being spent on the production of goods from agriculture and within this were sown the seeds of later developments in the processing of raw materials (grain, meat, fish etc.). One of these key industries was a consequence of the woollen trade. Climatic conditions were also important and there was frequent flooding in near Europe. The spread of disease, including the black death, along trade routes [17] affected Norfolk particularly. This led to a shortage of people with the skills necessary for taking advantage of the output from the highly productive arable land of East Anglia and the frequent translocation of people from the low countries to Norfolk was well established from the eleventh century onwards. By the twelfth century, weavers from Flanders were encouraged to settle in areas of Norfolk. They made use of the long-haired sheep that had been developed in England (compared to the short-haired sheep) and used this (after combing to remove short fibres—see below) to make a particular type of yarn that could be woven into cloth which kept its shape and so was particularly sought after for some types of clothing. The centre of this industry was the village of Worstead and the woven cloth became known as 'worsted' cloth. The process by which worsted cloth is made begins with carding. This is a process whereby the fibres in wool are disentangled. The process was originally carried out using a spiny plant (teasel—*Dipsacus fullonum*). Short fibres were removed and longer fibres were then spun into yarn that was used to make cloth and it was this cloth that was used to make worsted clothing. While this cloth was in demand and was exported, it was more expensive (and heavier) than cloth from Flanders and so the market or worsted cloth gradually deteriorated until, by the sixteenth century, a period of decline was underway. By the middle of that century various attempts were made (some endorsed by Elizabeth I) to encourage Flemish and Walloon weavers to settle in England and to bring with them the technology for the production of lighter woven cloth. In addition to the 'pull' from England which needed new technology and a replacement workforce due to intermittent outbreaks of the Black Death, there was both an economic and religious persecution 'push' that encouraged the emigration of weavers (and others) both from the Low Countries and, later, from France. A number of this latter group (Huguenots) arrived and settled in and around Norwich bringing technology and new insights. One of the key elements that catalysed the departure of the protestant Huguenots from France was the Massacre of Vassy in 1562 when 300 Huguenots who were holding a service outside of Vassy, were attacked by French soldiers. Various attempts were made to defuse the situation however in 1572, the St Bartholomew's Day Massacre resulted in the mass murder of around 70,000 Huguenots across France. This (and subsequent persecutions) led to the emigration of Huguenots from France to elsewhere in Europe and to England. While many of the Huguenots had a background in the textile industry and they brought these skills with them, they were also, in many cases, highly educated and became key figures in a number of professions. This not only created considerable expertise and wealth in England but also depleted France of an important part of their workforce. There was, however, some resistance from the indigenous English population as immigrants encroached on trades adjacent to textile production. It was estimated that up to 50,000 Huguenots left France and settled in England. By 1600 the population of England was about five million so an influx of 50,000 French would have made a considerable impact—particularly in areas where their settlements were concentrated. This was in addition to the Flemish/Walloon immigrants and their descendants who comprised about a third of the population of Norwich (see Sect. 6.2). While immigrants became integrated into English society, to an extent, the relatively poor provision of universities in England compared to Europe encouraged second and third generation immigrants to study abroad. By 1500 there were over 50 universities in Europe [18] including many significant seats of learning. Since the teaching was carried out in Latin, there was not a significant language barrier across Europe. The development and spread of the Enlightenment in the eighteenth century did not particularly include England, which remained somewhat isolated from European thought, but was prominent in the development of Scotland and, in particular, Edinburgh—in part due to its close links to Europe [19].

In brief, Norwich and Norfolk benefitted significantly from the immigrant populations from Europe and this enabled the creation of wealth based upon technological advances. In addition, following the development of the worsted cloth industry, there was an increased demand for dyes from woad and madder. As noted previously, the outbreaks of famine in England led to a ban on the growing of non-food agricultural crops such as woad therefor importation from Europe was an essential requirement. Dye itself was more often applied to the yarn than to the completed garment. The development of dye derived from madder is commemorated by the church of St John, Maddermarket and the naming of a particular textile dye shade as 'Norwich Red'. Madder red could only be harvested once a year while indigo from woad could be extracted from leaves picked at various times. Because of this, red was a more valuable colour than blue. Prior to the development of synthetic dyes in the nineteenth century, different colours were only available from natural sources by extraction. These included, significantly, the following.

red—madder root, *Rubia tinetorum*, kermes or grana from insects.
blue—woad leaves, *Isatis tinctoria.*
violet—orchil from lichen.
crimson—brasilwood from the East India tree, *Dalbergia latifolia.*
purple—brasilwood from the East India tree.
yellow—weld, *Reseda luteola*, dyers' rocket, turmeric, saffron, onion skin, marigold, chamomile.
green—indigo, weld, turmeric.
brown—walnut shells, bark.

As noted previously, the chemicals that were responsible for colouring agents were often derived from secondary metabolites and were, therefore, present in relatively small amounts. This presented a problem when disposing of waste material. In addition, a proportion of the dye molecules were either not extracted or not taken up by the material being subjected to the dyeing process. One of the most attractive aspects of dye molecules is their persistence and stability (including light fastness). This contributes to their long environmental half-life since they are not broken down or easily decomposed. Dyeing is a highly water intensive process and this, together with the polluting effects of the dye molecules themselves (mentioned above) mean that the potential environmental impact is significant. It is possible to use QSAR to model the ecotoxicity of dye molecules and this, together with information about solubility and binding to biological materials and cellular location of accumulation [20], may provide a way of assessing the appropriateness of use of particular dyes for particular purposes [21]. Extensive exposure to dye-related chemicals could give rise to occupation-related illness and this has been extensively studied in modern populations [22]. In addition, occupational diseases in textile workers specifically have been reviewed [23]. It is likely that exposure to plant material during extraction of colours could also result in allergic dermatitis in medieval dyers and some information related to occupational health in dyers in medieval Europe is available [24]. In a related area, the history of disease associated with bleaching has been considered [25]. As noted previously, rose madder contains alizarin and purpurin while the component responsible for the colour extracted from woad is indigo. Alizarin is reported to be mutagenic and carcinogenic due to its effect on catalase [26] while its potential environmental effect has also been investigated as a consequence of its effect on microalgae [27]. The toxicity and tumourigenicity of purpurin has also been investigated [28] however, in contrast to alizarin, it may also be protective against oxidative damage thereby preventing neurological damage [29]. Indigo (and dyes derived from it) may be carcinogenic and teratogenic [30] however the use of *Indigo naturalis* in traditional Chinese medicine (TCM) has been advocated [31]. Overall, the toxicity of indigo is considered to be low [32] and therefore it most probably did not represent an occupational hazard to medieval dyers or other textile workers—particularly when compared to other pigments commonly used such as white lead, Naples yellow (which contains lead and antimony) and others.

6.2 Flemish Weavers and Integration into Norfolk

The importance and context of the immigration of Flemish weavers into Norwich in the sixteenth century has already been noted. A range of documents regarding the Flemish and Walloon weavers and their settlement in Norfolk are available from the Norwich Record Office (NRO) [33]. In addition to the communities in Norwich, other exile migrants settled elsewhere in England subsequent to arrival including in London or, in particular, in Kent at Canterbury. While these migrations were in part a consequence of religious persecution of the non-Catholic population—particularly in France and the Low Countries, there was also an economic driver with the lighter fabrics produced by the immigrants being favoured in the textile market over the heavier English (worsted) products. In addition, the effects of plague had led to a shortage of workers and to increasing levels of unrest [34].

The materials introduced by the immigrants were termed ‘says and bays’ and the techniques for their production were particularly adopted by those who settled in Colchester in East Anglia. Bays and says were approximately equivalent to baize and serge cloth. While textiles were important protection against the weather, the cloth produced by the immigrants had other functional properties since they were lighter in weight. This can be seen as a precursor of the current interest in ‘wearable technology’. For example, the interaction of cloth of various types with the skin microbiome has become an area of increasing interest with the growth of ‘functional clothing’ [35]. Optimisation of the microbiome and, more generally of the microbiology of textiles has become an important objective [36] and this is affected both by the conditions used in the production of textiles and in the environmental exposure while wearing [37]. While the Flemish/Walloon community in England was mainly centred on London, by the early 1560s, there was also a presence in Norwich. In addition, a number of weavers based in London expressed a wish to move to Norwich and to be established there. This fitted with a local and national policy of distributing immigrants in various places to prevent overcrowding and the build up of resentment in a single particular place. A proportion of exiles moved to

Sandwich (Kent) and to Colchester and Kings Lynn. In addition there were a number of immigrants who arrived in Sandwich directly from Flanders. The communities established in Colchester and Norwich did not intermingle with each other to any significant degree while there was an interaction from both London and Sandwich to Colchester. This may be a reflection of the relative ease and history of travel by sea to Colchester from the two southern ports and the strong trading links for cloth and other material from Colchester to London. Additionally, the difficulty of travelling by land from Colchester to Norwich (2 days travel over difficult terrain) or by sea *via* Great Yarmouth was likely also a factor. It was only the advent of improvements to the main roads and the development of horse-drawn coaches in the seventeenth and eighteenth centuries that changed the emphasis of travel to Norwich and Norfolk from marine to land.

Although they grew largely independently, the European weaver communities in Norwich and Colchester soon comprised 30% and 20% of the respective populations of the two towns. The beginnings of this can be seen in 1565 when there were official attempts to establish two 'alien' immigrant communities in Norwich (one Flemish and one Walloon). A key figure in this was the mayor of Norwich, Thomas Sotherton. The economic rationale that lay behind this establishment was the gradual decline in the Norwich worsted wool industry and the successful establishment of lighter cloth production (says and bays) in Colchester and Sandwich. This encouraged Sotherton to attempt to attract foreign weavers to Norwich. In addition, the importation of cloth from Flanders was becoming more difficult due to local challenges to the weavers in Flanders. In principle, therefore, the argument for the settlement of Flemish and Walloon weavers was compelling and received Royal assent in 1565. This stated that the weavers had to be born outside England (i.e. could not be second generation individuals translocated from Colchester, Sandwich or elsewhere). Although the official declaration was signed by Sotherton in 1566, he was approaching the end of his time as mayor and there was opposition in Norwich to the acceptance of an immigrant community [16]. This opposition was based around several points—first of all it was felt that Sotherton personally would gain commercially from the revival of the Norwich cloth trade. Added to this, there was a degree of anti-alien (and anti-Protestant) feeling among some of the citizens of Norwich and an impression that aliens were responsible for bringing disorder to the local settled community. In total, eventually 24 Flemish-speaking and 6 French-speaking (Walloon) masters and their households were reported to have settled in Norwich. Following further disturbances in Europe (called the *Beeldenstorm*) in 1566, more textile workers and merchants arrived in Norwich in 1567 and 1568. Some of the incomers continued to be involved in uprisings in Flanders and, despite this, were able to remain in Norwich. Economic prosperity continued to benefit Norwich and Norfolk and weaving also became established in other places such as Thetford, King's Lynn and Great Yarmouth. The establishment of weaving in Norfolk also impacted upon trade with dyestuffs and raw materials being imported and finished goods and cloth being exported. The diversity of trades in the urban centre of Norwich in the sixteenth and seventeenth centuries has been noted [38] and this is reflected in the gradual relocation of weavers and ancillary trades from villages and towns in Norfolk (particularly in the East of the county) into Norwich through the seventeenth century. This is reflected in the goods that were exported. In 1640 (near the height of worsted woollen manufacture), the majority (>60%) of goods exported from England was woollen cloth. This declined with the advent of other materials however in 1700 woollen cloth still accounted for about half the exports—the majority of it sold to Europe. This continued emphasis on cloth allowed East Anglia as a whole to remain wealthy since the 'New Draperies' that replaced worsted cloth which came to England from the Low Countries, came first to East Anglia. As noted above, Norwich added a miscellany of cloths referred to as 'stuffs' to its traditional worsted cloth thereby retaining a significant market share in textiles. Indeed, the importance of woven cloth and the industry that it supported in Norfolk and the rest of East Anglia was only diminished when the Industrial Revolution took hold in England and manufacturing shifted to mill towns with a commercially useful water supply in Lancashire and other manufacturing enterprises with access to coal and iron in centres such as Birmingham and Sheffield [39]. Against this background, the assimilation of a significant population of aliens who brought new technology, wealth and prosperity and delayed the demise of the East Anglian textile industry is significant.

6.3 The Chemistry of Weaving and Molecular Weaving

Chemistry is the science of molecular interactions however these interactions are surmised from the accumulation of indirect evidence rather than being directly observed. While the idea that the macromolecular properties of substances is simply an extrapolation of their micromolecular structures is not necessarily and invariably valid, the use of similar paradigms to describe the interrelationship between bulk and molecular events can be useful. For example, the structure of benzene has been mythologised to be suggested by a snake biting its tail [40]. The reinterpretation of an imagined creature as a representation of a molecular structure is insightful however it should be remembered that the benzene ring is, itself, an imagined structure that fits the available evidence rather than being an actual proven physical entity. This also applies to the principle of delocalised electrons which is a key thought experiment in the consideration of an explanation of the properties of aromatic molecules. In a similar way, the discussion of molecular weaving may borrow some concepts from the macromolecular physical representation of weaving as applied to cloth but ultimately it is representational in nature and built upon current understanding of the properties of atoms and molecules and the bonds that exist between them.

Weaving textiles is, in principle, a simple process. It involves interlacing vertical threads (warp) with horizontal threads (at 90°) (weft) to create fabric. The main piece of equipment that is used is a loom which is a device that holds the warp threads in place while the weft thread is woven through them. This basic design is used for the creation of multiple cloths. The earliest looms were vertical however horizontal looms were introduced during the medieval period. As more complex patterns could be created so weaving was recognised as a highly skilled operation. Before the advent of mechanisation, looms were developed which allowed weavers to control multiple warp threads with their feet. A detailed discussion of the technology of weaving is outwith this chapter however some of the key features will, be described. One of the important factors in the development of weaving is the spinning of yarn. The use of longer fibres has already been mentioned as being the basis of worsted cloth. The competition with the remaining Flemish weavers in Flanders was one reason for fluctuations in the numbers of weavers (and quantity of cloth) produced in Norfolk however since England was a major producer of wool, a shortage of raw materials also affected the Flemish weavers and this may have been a factor (together with religious persecution) in their relocation from Europe to Norfolk. Another reason was the development of the dyestuffs industry in Norwich. As noted previously, Elizabeth I placed some restrictions on the production of blue colours from woad. Despite this, the necessary materials for the production of blue pigment could be imported *via* the Hanseatic ports to King's Lynn and Norwich. This enabled weaving to continue to be an important industry and with the development of lighter cloth production (Sect. 6.2), Norwich was positioned to take advantage of advances in related technologies. One of these was the optimisation of the extraction of madder to give a consistent red colour for both the wool weft and the silk warp (called 'Norwich Red') by the Scottish chemist (based in Norwich), Michael Stark, who learned the trade of dyeing in London, in the eighteenth century. Starks son, James, was one of the Norwich School of artists and the knowledge of colours was adapted for a different use to dyeing textiles. The process developed by Stark used the combination of madder, a tin mordant and the chalk stream-derived waters of the Wensum which were rich in calcium carbonate. Calcium is known to deepen the red colour of madder root and this is the basis of its use for histochemical staining [41]. The use of water from a chalk spring which is naturally rich in calcium (as is the case for the Wensum in Norwich) will, therefore deepen the pigment and provide uniformity of colour for different materials (for example, wool and silk). Wool keratin and silk fibroin will both complex Ca^{2+} ions [42] and this may also affect madder dye (alizarin) uptake. The importance of weaving using materials other than wool and the production of goods other than textiles will be described in Chap. 7.

Molecular weaving can be used to describe the structure of biological macromolecules such as DNA, proteins and amphipathic bilayers. The concept of using molecular weaving to define polymer structures has attracted significant attention [43]. The use of metal complexes has enabled the creation of 2-D structures (for example 'chain-link type molecules') and these Covalent Organic Frameworks (COFs) have been described and manipulated [44]. The potential for a link between molecular structure and macro molecular properties has encouraged researchers to consider applications of COFs including the separation of gas molecules [45] and a range of biomedical [46] and industrial [47] applications. Textile weaving and the interactions of substances on a macromolecular scale can provide the inspiration for some of these molecular manipulations. A number of researchers based in Belgium and the Netherlands (in regions that have their origin in Flanders) are active in the field of COFs and in the development of materials with novel properties based upon their molecular structure for a range of applications. From macromolecular applications in weaving to micromolecular construction of COFs, a flow of thought and conceptual awareness exists. Apart from their use in the industrial and biomedical applications mentioned above, specifically designed COFs have also been proposed as offering an opportunity for enabling specific protein refolding [48]. This is an important application since misfolding of proteins *in vivo* has been implicated in a number of disease conditions including, notably prion disease [49] and Alzheimer's disease [50]. The use of COFs or similar structures may offer the possibility of correcting the conformation of the misfolded protein on a molecular level thereby alleviating the diseases associated with protein misfolding.

6.4 Science and Wealth Creation

While exploring the nature of the world has been a prime motivation for science, the development of ways of making a profit from ingenuity has also been an important aspect of the development and application of science. A major motivation for alchemists, for example was the quest for the means to achieve the transmutation of base metals into gold. In ancient times (and also more recently) scientists such as Archimedes, while trying to understand and explain the world [51], also enjoyed support through patronage [52] and the benefits of wealth. This was not, however, a purely philanthropic exercise. The weapons of war developed by Archimedes had a very practical use [53] for the defence of the city he resided in and, later, were known to Leonardo da Vinci. Indeed Leonardo designed a number of weapons—some of which were derived from original ideas of Archimedes [53]. Leonardo da Vinci is perhaps significantly better known for his contributions to knowledge of the anatomy of the human figure [54]. While royal, aristocratic or religious patronage of science has become less obvious more recently and funding comes (often with commercial application of innovation being favoured) from national science bodies who claim to prioritise innovation [55] however the more practical impact of industry on research has been acknowledged [56]. While industry requires science to be honestly and truthfully carried out, the relative importance of some areas compared to others in terms of potential economic impact is frequently a decisive factor in choosing which scientific studies to prioritise. In some cases, this is apparent from the choice of study target while in others the questions being framed and prioritised are designed to respond to specific industry needs. In addition, research funding is known to have impacts beyond the direct effects on the funded researchers [57]. While traditionally research funding has been perceived to be awarded to outstanding individuals, the importance of research teams has been increasingly appreciated [58]. In addition, the importance of commercial applications has been acknowledged over many years even after the direct application of transmutation by alchemists was abandoned. For example, one of the achievements of Albert Einstein was the development of a refrigerator for the Elecrolux Servel Corporation (granted in 1930)— [59]. This work was carried out several years after his groundbreaking publications in 1905 [60].

Other examples of the development of science as a means of creating wealth are known. One of the most significant (and well known of these) was the fortuitous synthesis of the first aniline dyes by W H Perkin in the nineteenth century. With the advent of technology that enabled chemists to measure the relative proportions of different elements in a substance, chemistry began to be a discipline that was increasingly driven by metrology. The most significant problem, perhaps, arose with organic chemistry. Organic chemistry can be defined as the chemistry of carbon-containing substances, and this lies at the core of the challenge of understanding organic chemistry. Before examining the practical and applied aspects of organic chemistry (which is closely nuanced by the way that carbon atoms are surmised to form bonds), it is important to consider the philosophical underpinning of organic chemistry as a reductionist and empirical exercise and how it relates to the theoretical drivers of quantum mechanics [61]. There is a considerable literature regarding the philosophical basis of quantum mechanics and the interested reader is referred to that area of study. The question of whether chemistry in general and organic chemistry in particular can be considered to be a proper scientific discipline is one that Immanuel Kant considered [62]. This has led philosophers to consider chemistry as an empirical, applied science rather than a 'proper' scientific subject whose rules can be wholly derived from nature. Such an interpretation implies that chemistry exists for a set of specific and defined purposes and its formative function must be to support those purposes. From this standpoint, it is reasonable to suggest that one of the purposes of chemistry is the creation of wealth and, as such, activities that do not actually or potentially create wealth fall outside the remit of chemistry. Is the subdivision of chemistry that includes organic chemistry a scientific discipline or is it actually the consequence of the application of technology to particular types of molecules? Since Synthetic Biology has been developed as the application of engineering and technology principles to biology then the question of the role of organic chemistry is posed. Does organic chemistry arise from the application of technology to living systems or does it create the knowledge and conditions that enable the manipulation of living systems? An important aspect of organic chemistry is that it relies on models and these models, it can be argued, do not increase understanding at a basic level but rather enable the chemist to manipulate molecules and, through this, to impact upon systems that rely on molecules. The question that still arises is what is the nature of organic chemistry? Does it only exist, in a purely applied sense, only to manipulate molecules or does it also have a cognitive, explanatory aim? Will a better understanding of organic chemistry (even if it is based upon an unprovable set of models) provide insights? In part, the answer to this depends on what question is asked and how it is posed. As we have seen previously (see above) physical science and technology as studied by Archimedes, Leonardo da Vinci and even Einstein can have distinct practical applications. In a similar way, the models that enable organic chemistry to be envisioned also facilitate a greater understanding of the interactions of biological systems that are so dependent upon the molecular relations between carbon-containing molecules. These same models are crucial to the organic chemist who wishes to manipulate organic molecules (and biological systems) in a predictable way. The important caveat to this is that chemistry does not (unlike physics) define a particular state but rather it tracks the transitions that result from molecular interactions within particular environments. This does not displace the role of organic chemistry, in particular, in wealth creation, but it is first of all necessary to understand the particular attributes of the chemistry of carbon atoms and how they combine to make molecules and how such combinatory rules can be used to manipulate molecules. The increased understanding and development of models of increasing levels of sophistication are important aspects of the gradual development of chemistry from a 'black box' whose manipulation is by trial and error to a predictive model in which the path of outcomes can be anticipated and benefits achieved. Models can also be used to better understand the natural world and how it responds. Some examples, at different scales and at different levels of understanding, will illustrate some of these principles. For example, in the case of Norfolk, as noted previously, arable land became increasingly available in part due to developments in drainage technology (including mills) imported from the Netherlands [63]. While this is certainly evidence of human intervention, and the changing conditions altered the chemistry of the system, the significant advance in the optimisation of arable production occurred with the introduction of crop rotation between the sixteenth and eighteenth centuries. The optimisation was to a large extent, a process of trial and error and details of the science that underpins specific crop rotations did not emerge until later with the development of statistical models [64] and a more detailed understanding of the consequences of different crop rotations on ecologically significant microbial populations [65]. Even with such advances, crop rotation is not an exact process, and the results are dependent upon a number of factors that may combine to affect eventual outcomes [66]. The process then starts with an external change (removal of water) continues with a planned intervention (crop rotation) in which the consequences (increased crop yield, less disease) could be measured on a macro level even if not fully understood. More complete data analysis enables correlation to be understood better and some level of prediction employed. Further reductionism defines one particular factor (microbial population) and brings a partial understanding of causality and finally the impact of multiple factors on a number of outcomes simultaneously is able to be computed.

In the case of organic chemistry, an understanding of the system can be constructed from a consideration of the interactions at a molecular level and these interactions provide much of the information necessary for the prediction of the outcomes of manipulations based on model systems. The empirical formula of a compound is the simplest, whole number representation of the proportions of the different atoms in a molecule. In the case of many inorganic molecules, the empirical formula is a definitive guide to the molecular structure. There are, however, some exceptions. For example, ionic compounds may have an empirical formula that does not reflect the arrangement of atoms and some inorganic macromolecules (for example SiO_2) may comprise a number of repeating units so the empirical formula does not fully reflect the molecular formula. In the case of carbon (which is the prime constituent of organic molecules) the situation is further complicated by the existence of different bonding properties. Atoms are proposed to combine by the formation of bonds between them. The concept of chemical bonds has progressed through various iterations but it is important to remember that, although the nature of the chemical bond has evolved, it remains a model that describes the behaviour of atoms and molecules [67]. One of the important principles developed by the ancient Greeks was that substances were composed of atoms and the spaces between them (voids). One way of considering atoms is that they are the basic building blocks that cannot be subdivided further. This leads to the concept that atoms do not contain voids—they are continuous species. The concept that all matter was composed of atoms was recognised as a philosophical principle however a further development that removed the necessity for the existence of atoms was finally summarised by Aristotle who proposed that all matter was composed of earth, air, fire and water with a fifth component (ether) which is found in the heavens. The progression of thought about matter from the continuous nature espoused by Aristotle which was popular until the eighteenth century, to atomism has been extensively examined [68] and will not be further discussed. The idea of spaces between atoms (voids) is an essential prerequisite for chemical reactions to take place and so fits with the concept of chemistry as a dynamic discipline. The development of thought regarding bonds and bonding did not occur in isolation since other observations regarding the nature of matter were also being investigated. The properties of electricity and the concepts of anions and cations and attraction and repulsion were also active areas of investigation [69]. The acceptance of the concept of an electron was crucial to an understanding of how bonds are formed between atoms to create molecules. In essence this gave rise to the concept of ionic bonds (where an electron is substantively transferred from one atom to another). The idea that a bond could exist as a consequence of sharing electrons (covalent bonding) arrived later [70]. This was expressed in the concept of ionic and covalent compounds—the detailed consideration of which are beyond the scope of this chapter. Some approaches suggest there is no definitive differences between different types of chemical bond [71] while more conventional definitions of bonds focuses on electronegativity and polarizability, and thus to the classification of polar and non-polar bonds. From an organic chemistry standpoint, the most important aspect of bond theory is the nature of the bonding that can be undergone by carbon atoms since this is responsible for the development of ideas of the importance of functional groups rather than the absolute ratios of different atoms (empirical formulae). The nature and functional consequences of the tetravalency and the importance of functional groups in organic molecules has been extensively described [72]. Without having a detailed exposition of the nature of organic chemical bonds as outlined by Pauling, [73], it is sufficient to observe that carbon atoms can form single, multiple and aromatic bonds both between themselves and some other atoms (Fig. 6.1).

The concept of aromaticity is one of the most counter-intuitive aspects of organic chemistry since it does not conform to the pairwise duality of chemical bonding and it is defined in different ways [74]. The properties of aromatic compounds are specifically different to those of other carbon-containing molecules. Perhaps the most

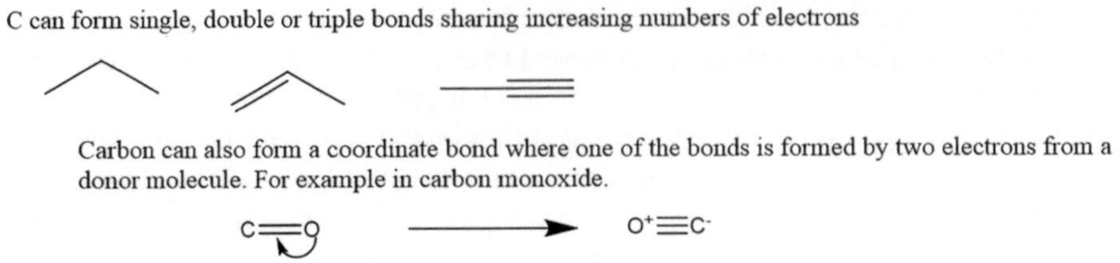

Fig. 6.1 Different types of carbon bonds

significant aspect of aromatic molecules is a consequence of the stability of the bond system (π-electrons). This means that aromatic systems are thermodynamically stable beyond that of hybrid bonds that lie between single and double bonds [75] and the maintenance of aromaticity as a consequence of the existence of aromatic stabilisation energy (ASE) has a particular effect on the reactivity of aromatic compounds [76] and breaking aromaticity is, therefore, energetically disfavoured. These effects are reflected in the types of compounds that are formed. These properties also affect the biological properties of the molecules and their resistance to chemical reactions and to biotransformations.

As mentioned previously, one distinct feature of organic molecules is the capacity of the same empirical formula to form different molecules, and this can be reflected in the chemical and biological properties (and value) of the isomers. One example of this is the isomers of menthol (Fig. 6.2).

The most valuable of these is L-menthol which exhibits a clear minty taste and is, therefore, of particular value as a flavouring compound. The other isomers are computed to possess varying degrees of off flavours [77]. Another example is carvone (Fig. 6.3).

in which different isomers not only have a different flavour profile [78] but also have different pharmacological profiles [79]. One of the best known consequential biological differences occurs between the different isomeric forms of thalidomide in which there are differences in toxicity between the R and S forms [80]. Some understanding of the potential molecular mechanisms that underlie this difference have been developed [81]. The target protein, cereblon, regulates the expression, assembly, and activities of other special proteins related to cell proliferation and metabolism [82]. There are a number of other examples of substances whose isomeric forms exhibit differential activity when interacting with living systems. Some of these will be mentioned in the context of mint and mustard (Theme Four Chap. 8). The wealth generated by both is highly dependent upon the organoleptically important compounds contained within the plants—although other physiological attributes are also of some interest. In more general terms, the link between science and wealth creation has presented a challenge for all interested parties [83]. In many cases, the main initiative for action has come from trying to find a use for a particular scientific discovery rather than finding science that can solve a real-world problem or benefit from an opportunity. The interaction between science and wealth creation *via* market pull is considered by some participants to be a sequential and

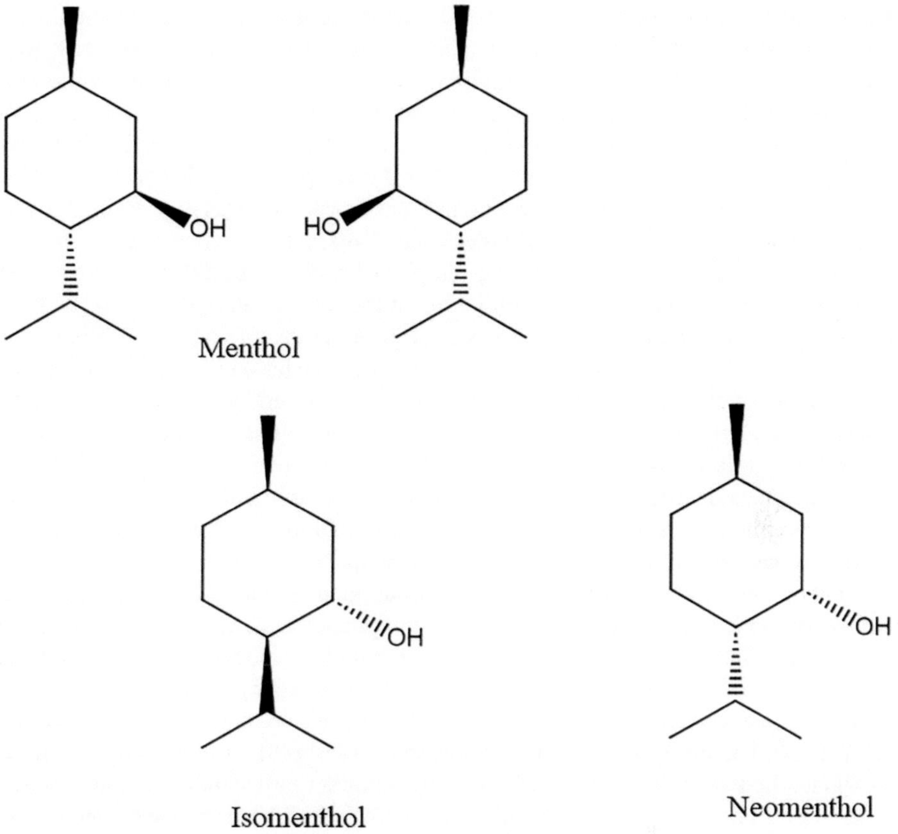

Fig. 6.2 The isomeric forms of menthol

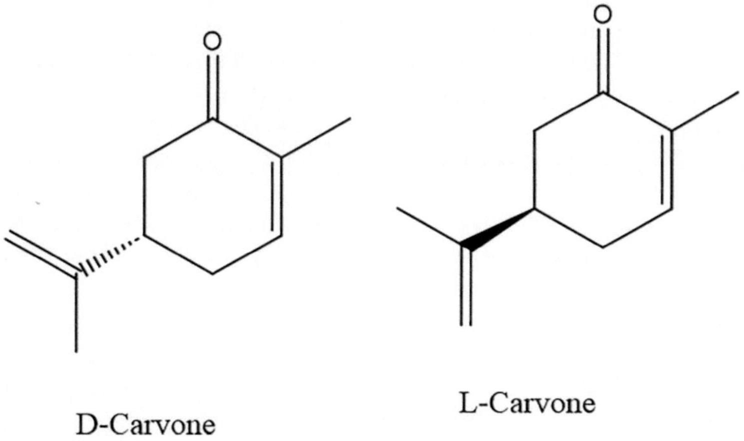

Fig. 6.3 Structures of L and D carvone

largely harmonious process. In the case of pharmaceuticals, eventual clinical development of new products is chiefly financed by free market pull mechanisms [84]. After the identification of a need and an opportunity, the starting point is the creation of a scientific insight however the insight may not be advised by enough mechanistic information to be confirmed as a solution and testing for efficacy will be required. The scientist together with clinical colleagues may then see the potential for this advance to be applied to the particular problem. It is then further developed and the solution begins to experience a market pull that enables further development and optimisation to take place. There are a number of stages in the process towards the creation of wealth. One of the most crucial of these is the flow of ideas that have the potential to be adapted to a marketable output. While this is essential, it also creates an intellectual barrier. The translation of scientific concepts into practical outcomes requires, first and foremost, an understanding of scientific concepts and language and secondly, an appreciation of what is required to create a market pull and, perhaps more importantly, what attributes of a potential solution will make the scientifically validated solution unworkable in the marketplace. This not only applies in the pharmaceutical area—although this is a high reward environment and has been extensively studied. One example in a different area is the development of a breath-freshening proposition. Endogenous oral malodour (halitosis) is primarily caused by the generation of thiol compounds and, in particular, hydrogen sulphide and methane thiol [85]. Alleviation of halitosis is most frequently achieved by masking the odour with a strong flavour such as mint [86] or removing the malodourous substances using zinc or lactic acid or a combination of the two together with an antibacterial preparation [87] or metal ions [88] or even some agent to modulate the bacterial population [89]. From a market pull standpoint, the problem with microbial modulation is that the treatment has to be continued over a long period of time in order to be effective and the agents used to remove the odourous organosulphur compounds (addition of metal ions, lactic acid or oxidising agents) impart a negative taste sensation. One approach that overcomes some of these barriers is the use of small amounts of allyl isothiocyanate which not only removes hydrogen sulphide and other thiols (Fig. 6.4) [90] but also has intrinsic antimicrobial activity [91]. The problem is that the solution (addition of allyl isothiocyanate) requires the introduction, in a factory environment, of significant amounts of a highly flammable pungent compound which is also an irritant and this is not a solution favoured by development engineers or those responsible for marketing the product.

The purely scientific solution for removal of compounds that cause halitosis was, therefore, only of very limited value in the marketplace and the scientifically less

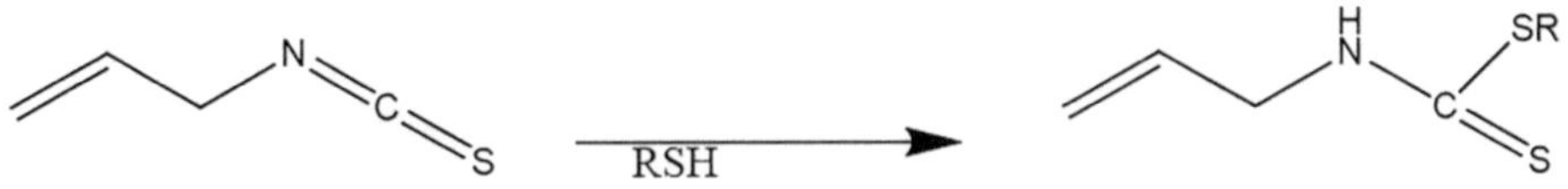

Fig. 6.4 The removal of thiols with allyl isothiocyanate

effective solution of using masking agents remained the commercial solution of choice and the newer technology was not adopted.

Another aspect of the development of science and technology and its link to wealth creation is the development (and later disintegration) of technological monopolies. An example of this is the development (and eventual demise) of the typewriter. The original type writing machine was patented in 1866 [92]. At first there were a number of manufacturers and this stimulated the development of many different aspects of the machine however, eventually, the market consolidated and common features such as the order of letters on the keyboard were adopted as industry-wide standards. The demise of typewriters resulted from the development of computers but, more specifically, the adjacent technology that gave rise to printers. The transition from mechanical to electronic office machines spelt the end of the typewriter as an essential office tool and changed, not only office practices, but also the nature of the work being carried out in offices and the skills required [93]. In a similar way, the development of weaving in Norwich and Norfolk as an occupation for an individual weaver was, at first, greatly advantageous to the working population and to the creation of wealth. The natural advantages of the county for the weaving process from the production of wool because of an abundance of arable land to raise sheep all the way to the preparation of dyestuffs aided by the nature of the chalk streams in Norfolk and the proximity to the European markets were all significant factors. The loss of the wealth generating capacity of Norfolk and Norwich started as early as the eighteenth century because of two factors—the loss of the worsted industry to Northern England and the impact of war in Europe on trade in cloth [94]. Equally significantly, the loss of the weaving industry in Norfolk was exacerbated by the spread of mechanisation and the increased scale made possible by the establishment of factories in England which reduced costs and made the individual weaver redundant and uncompetitive. In addition, the creation of synthetic pigments removed the functional advantage created by the chalk rivers of Norfolk in the production of dyes from natural sources. In the case of weaving and cloth production, as with typewriting, a change in technology in two adjacent areas—in this case the development of steam power and the production of synthetic pigments hastened the demise of the traditional weaving industry. The aspect of weaving that created value—particularly the limit on the amount of material that could be produced by an individual weaver and hence an increase in its scarcity value—became a significant weakness when cloth could be produced mechanically more rapidly and cheaply.

6.5 Trade and Knowledge Exchange

The importance of trade and the acquisition of knowledge and technical expertise has been mentioned previously. One of the most significant weaving towns in Flanders in the sixteenth century was Leiden which was also a centre of clinical medicine. Leiden is currently home to the Rijksmuseum Boerhaave which is the

Dutch museum for the history of science and medicine and was named after Herman Boerhaave, a significant figure in European science. Herman Boerhaave was active in the seventeenth and eighteenth centuries and was significantly involved in experiments on plants from his botanic garden [95]. In Leiden (and elsewhere in the region) small 'chemical' factories manufactured soap, wax, starch, paper, litmus, pigments, compounds of mercury, fat, and aromatic oils and these formed the basis of science-led wealth creation. Boerhaave published one of the first (and most significant) chemistry textbooks (Elementa chemiae). This volume was translated from Latin into other languages (including English) and this placed Boerhaave at the forefront of the dissemination of knowledge throughout Europe. In addition, the textile research centre is situated in Leiden and links between Leiden and Norwich have been noted and was a consequence both of the textile trade and the medical studies undertaken by Sir Thomas Browne of Norwich in Leiden [96]. The university at Leiden (founded in 1675), like the other European institutions inaugurated at a similar time [97] was designed to spread scientific and medical knowledge. These centres—and particularly those in the Netherlands, were an important cultural and scientific resource [98]. This, together with the trade links from Norfolk through links to the Hanseatic League enabled Norfolk to benefit from the Northern European association and trade links [99]. The cultural, technological and economic links to Northern Europe facilitated wealth generation in Norfolk and created ties which persisted. Some of these are reflected in records held (in Flemish) at the Norwich Record Office [100] while other pieces of evidence show the continued transnational nature of the influence of exiles in Norfolk contributed to the development of the Netherlands [101]. It was not only weaving that linked Norwich with Northern Europe. The development of printing in Norwich by one particular immigrant showed evidence of his Dutch origins [102] and this is further evidence of the strong associations that developed.

References

1. Chang H. What history tells us about the distinct nature of chemistry. Ambix. 2017;64(4):360–74. https://doi.org/10.1080/00026980.2017.1412135.
2. Gao W, Coley CW. The synthesizability of molecules proposed by generative models. J Chem Inf Model. 2020;60(12):5714–23. https://doi.org/10.1021/acs.jcim.0c00174.
3. Muratov EN, Bajorath J, Sheridan RP, Tetko IV, Filimonov D, Poroikov V, Oprea TI, Baskin II, Varnek A, Roitberg A, Isayev O, Curtarolo S, Fourches D, Cohen Y, Aspuru-Guzik A, Winkler DA, Agrafiotis D, Cherkasov A, Tropsha A. QSAR without borders. Chem Soc Rev. 2020;49(11):3525–64. https://doi.org/10.1039/d0cs00098a.
4. Kruszewski G, Mikolov T. Emergence of self-reproducing metabolisms as recursive algorithms in an artificial chemistry. Artif Life. 2022;11:1–23. https://doi.org/10.1162/artl_a_00355.
5. Hendry RF. Structure, essence and existence in chemistry. Ratio. 2023;36:274–88. https://doi.org/10.1111/rati.12387.
6. Náray-Szabó G, Mezey PG. Int J Quantum Chem. 2022;122(8):e26686. https://doi.org/10.1002/qua.26686.

7. Nikolaychuk PA. How the concept of chemical affinity evolved and how it contributed to chemical science. ChemTexts. 2024;10:2. https://doi.org/10.1007/s40828-023-00185-6.
8. Jarmoskaite I, AlSadhan I, Vaidyanathan PP, Herschlag D. How to measure and evaluate binding affinities. elife. 2020;6(9):e57264. https://doi.org/10.7554/eLife.57264.
9. Zhou Y, Huang Z, Yang R, Liu J. Selection and screening of DNA aptamers for inorganic nanomaterials. Chemistry. 2018;24(11):2525–32. https://doi.org/10.1002/chem.201704600.
10. Debenedetti PG, Klein ML. Chemical physics of water. Proc Natl Acad Sci U S A. 2017;114(51):13325–6. https://doi.org/10.1073/pnas.1719350115.
11. Murray JW, Gordon NE. The dissociation constants of some chlorophenols. J Am Chem Soc. 1935;57:110–1.
12. Kaner S, Hutcheson A. Nara to Norwich: the arrival of belief, religion and archaeology at the extremities of the silk roads AD 500-1000. Abstract from theoretical archaeology group conference, London, United Kingdom. 2019.
13. Ball P. From alchemy to chemistry. Nature. 2007;449:788. https://doi.org/10.1038/449788a.
14. Szydło ZA. The beginnings of chemistry: from ancient times until 1661. Pure Appl Chem. 2022;94(7):869–88. https://doi.org/10.1515/pac-2022-0203.
15. Newman WR, Principe LM. Alchemy vs. chemistry: the etymological origins of a historiographic mistake. Early Sci Med. 1998;3(1):32–65. https://doi.org/10.1163/157338298x00022.
16. Joby C. Flemish and Walloon exile communities in sixteenth-century Norwich: a case study of local and national responses to large-scale migration from the low countries. Immigr Minorities. 2024; https://doi.org/10.1080/02619288.2024.2392831.
17. Lenz K, Hybel N. The black death: its origin and routes of dissemination. Scand J Hist. 2016;41(1):54–70. https://doi.org/10.1080/03468755.2015.1110533.
18. Wiesner-Hanks M. Early modern Europe, 1450–1789. Cambridge: Cambridge University Press; 2006.
19. Emerson RL. The Scottish enlightenment and the end of the philosophical Society of Edinburgh. Br J Hist Sci. 1988;21(68 Pt 1):33–66. https://doi.org/10.1017/s0007087400024377.
20. Horobin RW. Where do dyes go inside living cells? Predicting uptake, intracellular localisation, and accumulation using QSAR models. Coloration Technol. 2014;130:155–73. https://doi.org/10.1111/cote.12093.
21. Funar-Timofei S, Ilia G. QSAR modeling of dye ecotoxicity. In: Roy K, editor. Ecotoxicological QSARs. Methods in pharmacology and toxicology. New York: Humana; 2020. https://doi.org/10.1007/978-1-0716-0150-1_18.
22. Nemery B. Occupational diseasesl overview. In: Encyclopedia of respiratory medicine. Amsterdam: Elsevier; 2006. p. 186–91. https://doi.org/10.1016/B0-12-370879-6/00269-6.
23. Sargunar CG, S, R. Occupational diseases in textile dyers—a brief review. J Res Biol. 2012;2(7):626–33.
24. Kroustallis S. Medieval art technology and medical praxis: occupational health for painters, illuminators, and dyers. In: Technical knowledge in Europe, 1200–1500 AD. Cambridge: Cambridge Scholars Publishing; 2022.
25. Torén K, Blanc P. The history of pulp and paper bleaching: respiratory health effects. Lancet. 1997;349:1316–8.
26. Hu S, Yuan D, Liu Y, Zhao L, Guo H, Niu Q, Zong W, Liu R. The toxic effects of alizarin red S on catalase at the molecular level. RSC Adv. 2019;9(57):33368–77. https://doi.org/10.1039/c9ra02986a.
27. Moorthy AK, Shukla SP, Govindarajan RB, Kumar K, Bharti VS. Application of microalgal physiological response as biomarker for evaluating the toxicity of the textile dye alizarin red S. Bull Environ Contam Toxicol. 2022;109(2):401–8. https://doi.org/10.1007/s00128-022-03525-3.
28. Mori H, Ohnishi M, Kawamori T, Sugie S, Tanaka T, Ino N, Kawai K. Toxicity and tumorigenicity of purpurin, a natural hydroxanthraquinone in rats: induction of bladder neoplasms. Cancer Lett. 1996;102(1-2):193–8. https://doi.org/10.1016/0304-3835(96)04159-6.

29. Kim W, Kwon HJ, Jung HY, Hahn KR, Yoon YS, Hwang IK, Choi SY, Kim DW. Neuroprotective effects of purpurin against ischemic damage via MAPKs, Bax, and oxidative stress cascades in the gerbil hippocampus. Mol Neurobiol. 2022;59(4):2580–92. https://doi.org/10.1007/s12035-021-02642-0.
30. Porkodi M, Brahmane MP, Pathan MA, Poojary N, Singh S, Harshavarthini M, Nagpure NS. Indigo dyes: toxicity, teratogenicity, and genotoxicity studies in zebrafish embryos. Mutat Res Genet Toxicol Environ Mutagen. 2024;896:503752. https://doi.org/10.1016/j.mrgentox.2024.503752.
31. Qi-yue Y, Ting Z, Ya-nan H, et al. From natural dye to herbal medicine: a systematic review of chemical constituents, pharmacological effects and clinical applications of indigo naturalis. Chin Med. 2020;15:127. https://doi.org/10.1186/s13020-020-00406-x.
32. Ferber KH. Toxicology of indigo. A review. J Environ Pathol Toxicol Oncol. 1987;7(4):73–83.
33. Meeres F. Records relating to the strangers at the Norfolk record office. Dutch Crossing. 2014;38(2):132–53. https://doi.org/10.1179/0309656414Z.00000000055.
34. Senn MA. English life and law in the time of the black death. Real Property Probate Trust J. 2003;38(3):507–88.
35. Broadhead R, Craeye L, Callewaert C. The future of functional clothing for an improved skin and textile microbiome relationship. Microorganisms. 2021;9(6):1192. https://doi.org/10.3390/microorganisms9061192.
36. Deaja S, Amy G, Dunn RR. A review of clothing microbiology: the history of clothing and the role of microbes in textiles. Biol Lett. 2021; https://doi.org/10.1098/rsbl.2020.0700.
37. Whitehead K, Eppinger J, Srinivasan V, Ijaz MK, Nims RW, McKinney J. Microbiome of clothing items worn for a single day in a non-healthcare setting. Microbiol Res. 2023;14:948–58. https://doi.org/10.3390/microbiolres14030065.
38. Pelling M. Occupational diversity: barbersurgeons and the trades of Norwich, 1550-1640. Bull Hist Med. 1982;56(4):484–511.
39. Coleman DC. Growth and decay during the industrial revolution: the case of East Anglia. Scand Econ Hist Rev. 1962;10(2):115–27. https://doi.org/10.1080/03585522.1962.10407622
.
40. Rocke AJ. It began with a daydream: the 150th anniversary of the Kekulé benzene structure. Angew Chem Int Ed Engl. 2015;54(1):46–50. https://doi.org/10.1002/anie.201408034.
41. Norton SA. Useful plants of dermatology. IV. Alizarin red and madder. J Am Acad Dermatol. 1998;39(3):484–5.
42. Wurm F, Rietzler B, Pham T, Bechtold T. Multivalent ions as reactive crosslinkers for biopolymers-a review. Molecules. 2020;25(8):1840. https://doi.org/10.3390/molecules25081840.
43. Zhang ZH, Andreassen BJ, August DP, et al. Molecular weaving. Nat Mater. 2022;21:275–83. https://doi.org/10.1038/s41563-021-01179-w.
44. Han X, Neumann SE, Nannenga BL, Wang K, Li KK, Mirzaei S, Yao X, Zhu C, Gao MY, Zhang YB, Cui Y, Yaghi OM. Directing molecular weaving of covalent organic frameworks and their dimensionality by angular control. J Am Chem Soc. 2023;145(42):22885–9. https://doi.org/10.1021/jacs.3c09691.
45. Tian X, Cao L, Zhang K, Zhang R, Li X, Yin C, Wang S. Molecular weaving towards flexible covalent organic framework membranes for efficient gas separations. Angew Chem Int Ed Engl. 2025;64(4):e202416864. https://doi.org/10.1002/anie.202416864.
46. Shi Y, Yang J, Gao F, Zhang Q. Covalent organic frameworks: recent progress in biomedical applications. ACS Nano. 2023;17(3):1879–905. https://doi.org/10.1021/acsnano.2c11346.
47. Freund R, Zaremba O, Arnauts G, Ameloot R, Skorupskii G, Dincă M, Bavykina A, Gascon J, Ejsmont A, Goscianska J, Kalmutzki M, Lächelt U, Ploetz E, Diercks CS, Wuttke S. The current status of MOF and COF applications. Angew Chem Int Ed Engl. 2021;60(45):23975–4001. https://doi.org/10.1002/anie.202106259.
48. Guo J, Sun X, Wang J, et al. Precise modulation of protein refolding by rationally designed covalent organic frameworks. Nat Commun. 2025;16:4122. https://doi.org/10.1038/s41467-025-59368-z.

49. Anand U, Patra S, Sekar RV, Garen CR, Woodside MT. Different folding mechanisms in prion proteins from mammals with different disease susceptibility observed at the single-molecule level. Proc Natl Acad Sci U S A. 2025;122(29):e2416191122. https://doi.org/10.1073/pnas.2416191122.
50. Uddin MS, Al Mamun A, Rahman MA, Behl T, Perveen A, Hafeez A, Bin-Jumah MN, Abdel-Daim MM, Ashraf GM. Emerging proof of protein Misfolding and interactions in multifactorial Alzheimer's disease. Curr Top Med Chem. 2020;20(26):2380–90. https://doi.org/10.2174/1568026620666200601161703.
51. Chattopadhyay A, Kaufman B. Archimedean principles and mathematical heritage: a synthesis. Axiomathes. 2021;31:145–55. https://doi.org/10.1007/s10516-020-09484-w.
52. Wietzke J. Speaking wit to power: reconsidering Royal Patronage through Archimedes' sand-reckoner. Class Antiq. 2022;41(1):129–79. https://doi.org/10.1525/ca.2022.41.1.129.
53. Simms DL. Archimedes' weapons of war and Leonardo. Br J Hist Sci. 1988;21(2):195–210. https://doi.org/10.1017/S0007087400024766.
54. Wolach B, Wolach O. Science, anatomy, and art: revisiting Leonardo da Vinci, Pioneer of modern anatomy. Isr Med Assoc J. 2021;23(10):676–80.
55. Packalen M, Bhattacharya J. NIH funding and the pursuit of edge science. Proc Natl Acad Sci U S A. 2020;117(22):12011–6. https://doi.org/10.1073/pnas.1910160117.
56. Fabbri A, Lai A, Grundy Q, Bero LA. The influence of industry sponsorship on the research agenda: a scoping review. Am J Public Health. 2018;108(11):e9–e16. https://doi.org/10.2105/AJPH.2018.304677.
57. Sattari R, et al. The ripple effects of funding on researchers and output. Sci Adv. 2022;8:7348. https://doi.org/10.1126/sciadv.abb7348.
58. Wuchty S, Jones BF, Uzzi B. The increasing dominance of teams in production of knowledge. Science. 2007;316:1036–9.
59. Einstein A, Szilárd L. Refrigeration (Appl: 16 December 1927; Priority: Germany, 16 December 1926) U.S. patent 1,781,541, 11 November 1930.
60. Rigden JS. Einstein 1905: the standard of greatness. Cambridge: Harvard University Press; 2005.
61. Goodwin W. Quantum chemistry and organic theory. Philos Sci. 2013;80(5):1159–69. https://doi.org/10.1086/673734.
62. Kant I. Kant's prolegomena and metaphysical foundations of natural science. London: George Bell & Sons; 1883. p. 138.
63. Martin-Jones S. A study of the drainage Mills of the Norfolk Broadlands. Norfolk Archaeol. 1970;35(1):152–3.
64. Upcott EV, Henrys PA, Redhead JW, Jarvis SG, Pywell RF. A new approach to characterising and predicting crop rotations using national-scale annual crop maps. Sci Total Environ. 2023;860:160471. https://doi.org/10.1016/j.scitotenv.2022.160471.
65. Magurno F, Sasvári Z, Posta K. From monoculture to the Norfolk system: assessment of arbuscular mycorrhizal fungi communities associated with different crop rotation systems. Symbiosis. 2014;64:115–25. https://doi.org/10.1007/s13199-014-0309-8.
66. Kim DH, Wade T, Brym Z, Ogisma L, Bhattarai R, Bai X, Bhadha J, Her Y. Assessing the agricultural, environmental, and economic effects of crop diversity management: a comprehensive review on crop rotation and cover crop practices. J Environ Manag. 2025;387:125833. https://doi.org/10.1016/j.jenvman.2025.125833.
67. Constable EC, Housecroft CE. Chemical bonding: the journey from miniature hooks to density functional theory. Molecules. 2020;25(11):2623. https://doi.org/10.3390/molecules25112623.
68. Benn AW. The origin of the atomic theory. Mind. 1911;20(79):394–8.
69. Kinnersley E. New experiments in electricity: in a letter from Mr. Ebenezer Kinnersley, to Benjamin Franklin. Ll D F R S Philos Trans R Soc Lond. 1763;53:84–97.
70. Lewis GN. The atom and the molecule. J Am Chem Soc. 1916;38:762–85. https://doi.org/10.1021/ja02261a002.

71. Walton JR, Rivera-Rivera LA, Lucchese RR, Bevan JW. Is there any fundamental difference between ionic, covalent, and others types of bond? A canonical perspective on the question. Phys Chem Chem Phys. 2017;19(24):15864–9. https://doi.org/10.1039/c7cp02407j.
72. Renslo A, Koltun D. The nature of bonding in organic molecules. In: Renslo A, editor. The organic chemistry of medicinal agents. Columbus: McGraw-Hill Education; 2016. https://accessbiomedicalscience.mhmedical.com/content.aspx?bookid=1731&sectionid=116668954. Accessed 9 September 2025.
73. Pauling L. The nature of the chemical bond and the structure of molecules and crystals. Nature. 1941;148:677. https://doi.org/10.1038/148677c0.
74. Krygowski TM, Szatylowicz H. Aromaticity: what does it mean? ChemTexts. 2015;1(3):12. https://doi.org/10.1007/s40828-015-0012-2.
75. Jirásek M, Rickhaus M, Tejerina L, Anderson HL. Experimental and theoretical evidence for aromatic stabilization energy in large macrocycles. J Am Chem Soc. 2021;143(5):2403–12. https://doi.org/10.1021/jacs.0c12845.
76. Pino-Rios R. Aromatic stabilization energies in excited states at the multiconfigurational level: assessment in archetypal organic rings. RSC Adv. 2024;14(46):33741–6. https://doi.org/10.1039/d4ra05147e.
77. Wang M, Wen F, Zhang L, Sun B, Xie J, Sun S, Zhang Y. Decoding the molecular mechanisms of menthol isomer perception based on computational simulations. Foods. 2025;14(14):2494. https://doi.org/10.3390/foods14142494.
78. Ye J, Hua S, Liu S, Tian F, Ji X, Li Y, Hou M, Xu W, Meng L, Sun L. Enantioselective effects of chiral fragrance carvone (L- and D-carvone) on the physiology, oxidative damage, synthesis, and release of microcystin-LR in Microcystis aeruginosa. Sci Total Environ. 2022;20(853):158631. https://doi.org/10.1016/j.scitotenv.2022.158631.
79. Souza FV, da Rocha MB, de Souza DP, Marçal RM. (−)-Carvone: antispasmodic effect and mode of action. Fitoterapia. 2013;85:20–4. https://doi.org/10.1016/j.fitote.2012.10.012.
80. Fabro S, Smith RL, Williams RT. Toxicity and teratogenicity of optical isomers of thalidomide. Nature. 1967;215(5098):296. https://doi.org/10.1038/215296a0.
81. Mori T, Ito T, Liu S, Ando H, Sakamoto S, Yamaguchi Y, Tokunaga E, Shibata N, Handa H, Hakoshima T. Structural basis of thalidomide enantiomer binding to cereblon. Sci Rep. 2018;8(1):1294. https://doi.org/10.1038/s41598-018-19202-7.
82. Shi Q, Chen L. Cereblon: a protein crucial to the multiple functions of immunomodulatory drugs as well as cell metabolism and disease generation. J Immunol Res. 2017;2017:9130608. https://doi.org/10.1155/2017/9130608.
83. Richmond M. Science and wealth creation. Nature. 1993;362:584. https://doi.org/10.1038/362584a0.
84. Matthey MA, Hollis A. Pull me - push you? The disparate financing mechanisms of drug research in global health. Glob Health. 2024;20(1):14. https://doi.org/10.1186/s12992-024-01019-x.
85. Wu J, Cannon RD, Ji P, Farella M, Mei L. Halitosis: prevalence, risk factors, sources, measurement and treatment - a review of the literature. Aust Dent J. 2020;65(1):4–11. https://doi.org/10.1111/adj.12725.
86. Bradshaw DJ, Perring KD, Cawkill PM, Provan AF, McNulty DA, Saint EJ, Richards J, Munroe MJ, Behan JM. Creation of oral care flavours to deliver breath-freshening benefits. Oral Dis. 2005;11:75–9. https://doi.org/10.1111/j.1601-0825.2005.01098.x.
87. Winkel EG, Roldán S, Van Winkelhoff AJ, Herrera D, Sanz M. Clinical effects of a new mouthrinse containing chlorhexidine, cetylpyridinium chloride and zinc-lactate on oral halitosis. A dual-center, double-blind placebo-controlled study. J Clin Periodontol. 2003;30(4):300–6. https://doi.org/10.1034/j.1600-051x.2003.00342.x.
88. Young A, Jonski G, Rölla G, Wåler SM. Effects of metal salts on the oral production of volatile sulfur-containing compounds (VSC). J Clin Periodontol. 2001;28:776–81. https://doi.org/10.1034/j.1600-051X.2001.280809.x.

89. Feres M, Figueiredo LC, Faveri M, Guerra MC, Mateo LR, Stewart B, Williams M, Panagakos F. The efficacy of two oral hygiene regimens in reducing oral malodour: a randomised clinical trial. Int Dent J. 2015;65(6):292–302. https://doi.org/10.1111/idj.12183.
90. Tian M, Hanley AB, Dodds MW, Yaegaki K. Chewing gum containing allyl isothiocyanate from mustard seed extract is effective in reducing volatile sulfur compounds responsible for oral malodor. Am J Dent. 2013;26(4):180–4.
91. Romeo L, Iori R, Rollin P, Bramanti P, Mazzon E. Isothiocyanates: an overview of their antimicrobial activity against human infections. Molecules. 2018;23(3):624. https://doi.org/10.3390/molecules23030624.
92. Current RN. The original typewriter Enterprise 1867-1873. Wis Mag Hist. 1949;32(4):391–407.
93. Webster G. Martyn Lyons: the typewriter century—a cultural history of writing practices. Pub Res Q. 2021;37:510–2. https://doi.org/10.1007/s12109-021-09828-3.
94. Sugden K. Clapham revisited: the decline of the Norwich worsted industry (c. 1700–1820). Contin Chang. 2018;33(2):203–24. https://doi.org/10.1017/S0268416018000140.
95. Klein U. Experimental history and Herman Boerhaave's chemistry of plants. Stud Hist Philos Sci Part C Stud Hist Philos Biol Biomed Sci. 2003;34(4):533–67.
96. Hughes JT. The medical education of sir Thomas Browne, a seventeenth-century student at Montpellier, Padua, and Leiden. J Med Biogr. 2001 May;9(2):70–6. https://doi.org/10.1177/096777200100900202.
97. Otterspeer W. The University of Leiden: an eclectic institution. Early Sci Med. 2001;6(4):324–33.
98. Simon S. The glorious revolution and medicine in Britain and The Netherlands. Notes Rec R Soc Lond. 1989;43:167–90.
99. Fink A. The Hanseatic league and the concept of FOCJ. Kyklos. 2012;65:194–217. https://doi.org/10.1111/j.1467-6435.2012.00534.x.
100. Joby C. Early modern records in Dutch at the Norfolk record office. Dutch Cross. 2012;36(2):132–40. https://doi.org/10.1179/0309656412Z.0000000010.
101. Joby C. The Norwich exile community and the Dutch revolt. History. 2024;109:59–91. https://doi.org/10.1111/1468-229X.13387.
102. Feakes R. A familiar sight: 'Dutch type' and the first printer of Norwich. History. 2025;110:215–26. https://doi.org/10.1111/1468-229X.13441.

Chapter 7
Chemistry and Manufacturing. Manufacturing From Molecules to Macro Structures

Abstract The importance of institutional and societal awareness of science. The duality implicit in the analysis/synthesis approach to natural products. Theories about atoms (corpuscular nature of matter) and the development of paradigms regarding valence and bonding. The microscopic nature of matter and its relationship to the measurable, macroscopic properties as envisioned by the Greeks. The synthesis of mauveine. The development of electricity and its adoption to fuel general purpose technologies (GPTs). The application of GPTs in different areas leading to gradual progress. The importance of certain materials (wood and metal) and their impact upon ancillary areas (leather working, glass manipulation). The application of the principles of weaving from textiles to metal (wire). The Green Revolution and the importance of the Haber-Bosch process. The establishment of Norfolk and Norwich as a centre for the study of agriculture, ancillary areas such as soil microorganisms and food research.

Keywords Molecules · Structures · Materials · Analysis · Synthesis · Steam · Electricity · Haber-Bosch

The importance of chemistry to manufacturing has been explored (Chap. 6) and this will be investigated further in the following chapter. The three stages of the incorporation of chemistry—at individual, institutional and societal levels has already been mentioned and the following chapter will explore, in particular, the two latter views. Manufacturing in Norwich and Norfolk in the nineteenth and twentieth centuries was less built on the foundation of individual skill sets (agronomy, dyeing, textile production) and more on the institutional and societal awareness of scientific and technical skills and the opportunities that these would allow to be realised. There will, therefore, be less of an emphasis on specific skills and instead a more general consideration of the attitudes and atmosphere of the city and county during the period being considered. In addition to indigenous manufacturing itself, the historical aspects of the establishment of particular attitudes are an important factor in the development of technologies and their growth in the particular environment that was prevalent in Norwich and Norfolk. It is however the case that the position of

B. Hanley, *The Chemistry of a County*,
https://doi.org/10.1007/978-3-032-21491-1_7

Norfolk and its orientation towards Northern Europe rather than towards the rest of England, had a significant impact on the viability of manufacturing industries. The advent of the North Sea oil and gas industries in the twentieth century had the potential to create a new series of collaborations with Europe however at that time the UK was gradually growing more insular—particularly in its attitudes to the retention of natural resources. Later, the decision to leave the EU (which 58% of the population of Norfolk and over 70% of the population of Great Yarmouth supported) was historically (given the longstanding links with Northern Europe) and economically illiterate. The establishment of immigrant communities in Norfolk in the sixteenth century was a response to the need for new skills and the loss of the indigenous population as a consequence of outbreaks of the Black Death [1] and the shortage of labour. Unlike the indigenous population, who primarily lived close to where they were born, there is evidence that the Flemish and Walloon immigrants travelled more extensively both back to Europe to visit relatives or to seats of learning and even within England [1].

The reductionist approach that developed during the nineteenth century with a greater emphasis on attempting to understand the properties of substances in the context of what was discernible about their molecular structures continued into the twentieth century. The early chemical duality of analysis and synthesis continued to be a significant influence on the mindset of chemists—even if synthetic methods were often primitive and of limited applicability. As atomism became predominant, chemists were increasingly preoccupied with developing structural theories however these were not universally adopted [2] and remained controversial. In addition, atomism brought with it ideas about valence and bonding [3]. There were, therefore, three linked concepts—atomism, bonding and structure. These did not, however, just appear as a consequence of empirical observation but rather were part of a tradition of considering substances at a macromolecular (observable) and micromolecular (postulated) level. The concept that the properties of substances were reflected in their molecular structures was a continuation of philosophical thought about the nature of matter. The ancient Greeks, although their concept of atomism was different to current perceptions, were exponents of the idea that the microscopic and macroscopic were related therefore the granular structure of materials was reflected in their macroscopic properties. This led to the idea, for example, that the microscopic structure of iron must consist of hard entities while that of water must be slippery [4]. There are a number of problems with the microscopic/macroscopic duality of substances. The first of these is that it does not predict the properties of the various forms that a substance can take. For example, a piece of copper can be shaped into wire which is flexible and able to be bent into various shapes or it can be a slab of metal which requires considerably more force to deform. In both cases the elementary microscopic structure is the same (copper atoms) but the macroscopic properties are very different. For the duality of substances to be correct, the atoms in a copper wire would have to have different properties to those in a slab of copper. Similarly, atoms of water would have to have different properties depending on the macroscopic phase (solid, liquid or gas) of the substance. A second problem of the duality paradigm is that it fails to reconcile the transformative nature of chemistry. Changes at a molecular (or atomic) level are not necessarily reflected in macroscopic properties. There are, however, instances in which a molecular change can have a significant visible (and measurable) effect. A very simple example of this using a single element is the anaesthetic effect of xenon. Studies have suggested that radical pairs of electrons with entangled spins may be involved in the differential anaesthetic effects of different isotopes of xenon and hence in consciousness [5]. In this case, a simple (albeit quantum mechanical) mechanism is used to explain a macroscopic, biological property and the importance of electron spin is confirmed [6]. On a molecular level, receptor binding (and subsequent signal transduction effects) has been demonstrated to be influenced by the isomeric configuration of a given ligand [7] therefore even a small change in the configuration of a molecule can have significant macromolecular effects. Differences between the biological properties of enantiomers is perhaps one of most important examples of the impact of a relatively small micromolecular change on a macroscopic property. Other examples of these will be considered later and in subsequent chapters.

7.1 Introduction

The growth in manufacturing in the nineteenth and twentieth centuries had a profound effect on the application of science in society. The advent of the steam age was seen as being transformative however, apart from certain specific sectors (cotton and mining) the impact of the acquisition of significantly increased amounts of mechanical power was thought to be limited [8] however this has been challenged lately [9] and the significance of steam power across the whole industrial landscape emphasised. It remains challenging to confirm the actual impact of the adoption of steam—particularly in areas such as agriculture however the preference for steam power over water power in manufacturing (even when a good supply of water was available—although the multiple requirements for water suggest it was not as universally available as previously thought [10]) was growing in the second half of the nineteenth century. The introduction of steam had a particularly significant effect in areas such as Norfolk because of its introduction in manufacturing centres in weaving and in transport. The downstream effects of this were felt on the opening of new (particularly transatlantic) trade routes and the rapid transport of materials and finished goods. An example of this was the technological developments that reduced transatlantic crossing times from the time of the first steam paddle ship (The Great Eastern in 1858 which crossed the Atlantic in 15.5 days but with limited cargo space since it had to carry enough fuel (coal) to enable it to complete its voyage) to the advent of iron hulled ships and compound steam engines (by the 1860s, a significantly reduced crossing times of about 8–9 days could be achieved). Steam powered transportation enabled global trade routes to be developed and this facilitated the acquisition of materials and technology [11]. Just as the opening of trade routes to Europe facilitated the spread of the Black Death other infectious diseases became more globally prevalent in the nineteenth century [12]. In addition to enabling the more rapid spread of goods and disease, the demand for steam-powered transport was a key driver in the development of industrial and engineering technology [13]. One of the requirements for the development of ship building was access to deep water dockyards. While the west coast of Britain had advantages (not least with the growing importance of the transatlantic trade), there are some examples of east coast ports that were constructed to bolster trade with Europe [14] and it has been argued that coastal shipping remained an important aspect alongside the railways even into the twentieth century [15]. The tendency of ports in Norfolk such as Great Yarmouth to accumulate silt and thus be unusable and the lack of effective internal lines of communication (railways were, as noted previously, fairly late to reach Norfolk) meant that east coast ports remained relatively small and significant building of ships with global reach did not take place. However the transport of lower value, bulk products such as coal remained part of the coastal sea transport tradition [16]. Indeed, while Great Yarmouth, for example, developed when the railway reached the town in 1844, it was as a tourist destination and Wellington Pier was built in 1853 and Britannia Pier opened in 1858. Norfolk had benefited from its geographical features including an abundance of arable land, links to the Hanseatic League and near-European technology for weaving and cloth making and even chalk-derived streams to facilitate the development of the natural dyestuffs industry. With the advent of wars in Europe in the nineteenth century (Napoleonic war, Hundred Days war, Greek war of Independence, French and Belgian Revolutions etc.) and the developing interest in the Americas, the trade and manufacturing advantages that Norfolk had were gradually eroded. Norfolk lost out on many fronts almost simultaneously and even the agricultural revolutions of the 18th and 19th centuries [17] and the concomitant increase in agricultural productivity could not arrest the ongoing decline completely [18]. In addition, the application of Free Trade principles from the 1850s onwards gave American agriculture producers access to the British market and led to a significant fall in grain prices [19]. In addition, cloth production was increasingly mechanised and carried out in factories rather than being a home or small-scale enterprise and concentrated in the Northwest of England (where water and later steam was available to power production) and in Ireland [20]. In addition, by the middle of the nineteenth century, after increased levels of investment [21] significant amounts of wool were being imported from Australia—facilitated by the improved maritime trading links and the development of farming skills. By a combination of new trade links, new sources of raw materials and new technology (mechanisation), Norfolk's advantages in terms of production of both raw materials (wool) and finished products (cloth) was lost. A further challenge came from the development of the synthetic dyestuffs industry that started with Perkin's synthesis of mauveine in 1856 and the subsequent expansion of industrial processes to produce dyes and pigments [22]. Since chemical dyes were widely reported and people became familiar with the concepts involved [23], the production of dyes from coal tar became a rapidly expanding industry. In the 70 years after the production of mauve by Perkin in 1856, more than 1200 synthetic dyes were made [24]. Many of these were significantly superior to the natural alternatives [25] and, while they made use of petrochemical-derived starting materials, the stability of the pigments [26], the toxicity of natural pigments and the avoidance of the problem of disposing of large amounts of post-extractive material.

While steam power appeared to offer significant advantages and this was (literally) fuelled by the almost limitless supply of coal (and of the waste product, coal tar which was used as a raw material by the developing chemical industry), in fact the age of primacy of steam was relatively short. Steam power was overtaken, first by oil and gas and later by electricity. The production of oil from shale was prominent in parts of Scotland (particularly West Lothian—but also elsewhere [27]) and it pre-dated (1851) the widespread use of steam technology [28] but shale oil was primarily used for heating and lighting purposes and for lubrication of machinery [29]. In the case of electricity, some of the earliest work was carried out by Volta and published by the Royal Society in London in 1800 [30] following the creation of the electric battery in 1799 [31]. It was only a short period after the invention of the first battery that the first electric motor was invented and the first electric vehicle was created in 1839 by Robert Davidson in Scotland. Davidson was also a chemist whose business in Aberdeen made and supplied chemicals and developed dyes. Although his electric automobile did not lead to a commercial product, electric vehicles were being produced by the end of the nineteenth century. Batteries were large and unreliable, however and did not reach mass production. The first internal combustion engines were produced, slightly later, in 1880, and the mass-produced Ford Model T appeared in 1908 [32]. Prior to that, Nikolaus Otto invented the four-stroke engine in 1876 and this first example of the use of the four stroke process formed the basis of the subsequent development of the internal combustion engine [33]. Some of the technical challenges that have been overcome in the development of internal combustion engines may be of value in designing electric (and other motive force) vehicles [34]. The development of different types of technology (and the timing of these) has had a significant impact upon industrial success and the adoption of preferred technologies. These have included (as will be demonstrated) technologies that are directly relevant to certain types of knowledge acquisition and those which are tangential to a core base of knowledge, but which can illuminate a particular aspect of the manufacturing process.

7.2 Manufacturing in an Industrial Age: Molecules and Materials

The importance of molecules has been mentioned—as has the impact of these on macromolecular structures. Some specific examples will be considered as illustrations of the general principles involved. As noted previously, many of the design principles of machinery designed to realise the energetic potential of three sources of power (coal, petrochemicals and electricity) are similar and the thermodynamics of combustion have been compared [35]. The basic energetic principles of compression, expansion and subsequent power generation operate in steam engines, internal combustion engines and electric engines. The adaptation of technical insights to different types of energy release indicates that engineers and scientists work within technological constraints and are, therefore, applied practitioners of already determined sets of solutions (so-called, general-purpose technologies (GPTs)). Such thinking implies that innovation is an iterative process and that application of successive GPTs, as they are developed, can lead to acceleration and deceleration of growth. These occur as different GPTs are assimilated and are applied to different extents [36]. This diffusion of knowledge will occur no matter what basic technology is being applied. This means that, as noted above, the development of steam, electricity and the internal combustion engine can be compared and will be based upon the same (or very similar) paradigms. In the absence of a catastrophic change (for example, the complete loss of a technologically essential feedstock, a fundamental flaw leading to a severely impactful accident or a dramatic change in priorities), the progress enabled by the adoption of GPTs is, in essence, irreversible. Since adoption of GPTs leads to the improvement of a process, (otherwise why should they be adopted) then it follows that processes are in a state of continual and ongoing (if sometimes very gradual) improvement. This improvement may not be immediately obvious in the context of the wider economic picture and it may be challenging to demonstrate. Even taking a simple and easily understood metric such as productivity, it can be difficult to explain fluctuations in productivity in the context of the adoption of GPTs. Before the nineteenth century industrial revolution, growth rates of 0.25% per annum were normal and were representative of incremental improvements in systems and processes. The integration and adoption of multiple GPTs (even if they arose from the same core technology) across different technological areas in a short period of time—as occurred in the nineteenth century—has often led to a rapid increase in growth rates. These have been accompanied by a later cyclical retrenchment and decline in the rate of growth [37] on a periodic basis. Despite the adoption of GPTs as part of a gradual improvement, linear growth is generally the exception rather than the rule—except in times of industrial revolution. In part this is a consequence of innovation being an iterative process, so acceleration and deceleration of growth are to be expected. The analysis of trends suggests that short contractions followed by long, slow recovery is frequently the usual pattern [38]. This means that, while the adoption of GPTs leading to the industrial revolution in England had an impact both on growing economic areas in the North and on declining areas such as Norfolk, these changes were gradual and occurred over a period of time. While innovation (and the technology that accelerates its adoption and implementation) are important drivers for manufacturing, it is also the case that less direct factors such as improved infrastructure (for example, improvements in communication and transport) are also contributory factors [39] although they may be more difficult to quantify. In Norfolk, the loss in relevance of international trade routes to Northern Europe and the comparative lack of a highly developed national transport infrastructure exacerbated the impact of increased competition from factories in the North of England. As noted previously, changes in modes of transport which were brought about by the application of the same core technology (the application of steam power) as in factories changed trade routes and created economic opportunities. Adjacent technology application can stimulate the growth made possible by the adoption of new GPTs. In the case of Norfolk, competition from home (the North of England in cloth manufacture) and abroad (the increase in availability of imported agricultural products and raw materials at a competitive price) effectively short circuited the economy and led, eventually, to the adoption of smaller scale industry including manufacture and an increased emphasis on technology and the knowledge economy rather than on the relatively lower value agricultural (and marine industries) and craft working enterprises such as cloth making and weaving. The economic decline in Norfolk was the consequence of a range of factors and was, therefore, a gradual process with the sequence of short contractions followed by long, slow recovery which was mentioned above as a usual economic pattern being, in essence, reversed. There were short economic expansions followed by a long, slow contraction. Some of these will be considered in the following sections.

7.3 Wood and Metal Working: Boats, Wire, Cars and Technology

While the traditional industries of wool manufacture (and processing), cloth and finished garment manufacture and agriculture were gradually replaced, small scale manufacture continued in Norfolk and Norwich. Two of these industries, that relied upon the use (or adaptation) of local skills, were shoe making and boat building. These will be briefly discussed.

The development of Norfolk as a fishing county has already been mentioned. The fact that the river Wensum was navigable from Great Yarmouth to Norwich is reflected in street names in Norwich such as Quayside and Fishergate. Archaeological investigations of the former showed the presence of debris dating from the twelfth century derived from leatherworking, shellfish landing and basket weaving [40]. The link between the archaeology of Norwich and that of other parts of northern Europe has been studied [41]. In addition to debris related to specific industries, the area near to quayside also contained a number of warehouses and stone-built Merchants' houses. The discovery and analysis of waste material and tools suggests that, at various places within the city walls of Norwich, up to sixty-eight distinct crafts, industries, trades and services were being practiced in medieval times. In addition to the production of goods for local use, weaving, glass making and some other industries (for example, bell making [42]) had national (and perhaps international) significance. There are signs of a two-way trade with Germany and the Low countries through the discovery of distinct pottery deposits and the discovery of Baltic-derived timber in Norwich. A twelfth century wharf which was used to land shellfish was excavated at Quayside and the subsequent development of the site is reflected in the history of a building which became the New Star Inn (at Number 8 Quayside) and incorporated two previous buildings as follows:

House (Medieval—1066 AD to 1539 AD).
Pile (Medieval—1066 AD to 1539 AD).
Warehouse (Medieval—1066 AD to 1539 AD).
House (Post Medieval—1540 AD to 1900 AD).
Timber framed building (Post Medieval—1540 AD to 1900 AD).

Similar successions of use can be found for other sites in Norwich—particularly those that are adjacent to the river such as Fishergate (also a site of landing of shellfish) and near to King Street.

The nature of the access from Norwich to the sea at Great Yarmouth and the tidal nature of that access were important factors in the building of boats suited to local conditions. One of the most successful of these was the Norfolk keel—a clinker-built boat that was widely used in the region for transport of goods and people for up to 1000 years. Clinker (or clincher) building is a method of constructing hulls of boats by fixing wooden planks to each other so that the planks overlap along their edges [43]. This method was developed in northern Europe and extensively used by the Vikings and by other communities adjoining the North Sea [44]. Its use in England is considered to be evidence of the influence of Scandinavian boat building technology. While other forms of boat building (for example, Carvel construction) came to be favoured for boats used for sea fishing due to improved stability, the persistence of the clinker technology is evidence of its suitability for use in Norfolk over a long period of time. Although the type of boats referred to as 'wherries' are often associated with Norfolk and were extensively used to transport goods, the boats that were used to transport goods and passengers across the Thames in London prior to the building of bridges in the eighteenth century were also referred to as wherries and evolved into the Thames skiff, a common sculling/rowing boat that was used for leisure and racing. Strictly speaking, sculling is the act of propelling the boat with a pair of oars (or blades), as opposed to rowing which requires both hands on a single oar. In the context of Norfolk, wherries were decked sailing vessel of very shallow draught used for the transport of freight. They are broad, shallow draught craft and are fitted with a single mast (had a distinctive 'gaff rig' of a single, high-peaked sail and the mast stepped well forward) that can be lowered to the deck when passing under bridges, etc. Wherries were used to transport goods from the coast (Great Yarmouth) to inland sites on the Norfolk Broads and were also more widely used for transport of goods along canals in England. While they were primarily used for the transport of goods, they were also used, in summer, for moving passengers. As with the Norfolk keel, wherries were largely clinker-built and, as a consequence of established trade routes, Norwegian or Baltic timber was readily available [45].

Evidence for pre-Norman leather working has been noted from archaeological finds (see above). The remains of a thirteenth century tannery were discovered in Norwich and this included freshwater pits and other residues of materials required for the production of leather and vellum (probably for ecclesiastical writing purposes) [46]. Cordwainers had been making shoes in Norwich for many centuries (the word Cordwainer derives from the Spanish "Cordovaner", a leather worker from Cordoba in Andulucia. The word later became incorporated into the English language in its present context. The Worshipful Company of Cordwainers was established in the City of London in 1272 and were established to make the finest shoes from Spanish goatskin, unlike cobblers who worked in lower grade leather. This is reflected in the badge of the Cordwainer's Company which consists of a gold chevron between three goats' heads and is its coat of arms.) however this was, until about 1800, a relatively minor trade compared to cloth making and weaving. The arrival of the railway network to Norwich in 1846 (the last major city to be connected to the rest of England in this way) likely aided the development of the shoe trade and enabled trade to be developed. In addition to making final products, the wool and related industries ensured a supply of leather from animal skins. Unfortunately, leather does not survive in archaeological samples particularly well (particularly if waterlogged) [47] so finding conclusive evidence can be challenging. The process of 'tanning' is a way of preserving animal skins by steeping them in tannin obtained from tree (oak) bark [48]. In 1800, Sir Humphrey Davy discovered that materials from other trees such as chestnut and ash had the same properties and could also be used for tanning [49]. Davy continued his studies on tanning and published more on this subject [50]. While the use of vegetable material as part of the tanning process continued for a considerable period, the process was slow and the demand for leather for belts for industrial machines in factories, as well as for domestic use, led to the adoption of other, newly prepared, chemicals as a way of carrying out tanning more quickly. In particular, the introduction of chrome tanning using chromium sulphate in 1858, greatly speeded up the tanning process even if the mechanism of the process of tanning in this case is unclear [51]. Chromium sulphate is toxic [52] and Cr (VI) is the most toxic (and bioaccumulated) form of the metal. In addition, the possibility of leaching of Cr causing an allergic effect has been suggested [53]. While this seems to be a relatively unlikely possibility, the toxic effect of Cr (VI) in particular has led to concerns about the use of such a toxic metal as part of an important industrial process [54] and alternative procedures have been suggested [55] including limiting the use of metals [56]. The leather economy is an example of how, as technology developed, small scale industry was adapted to an industrial level scale [57]. In addition, the history of shoemaking in Norwich has been documented [58]. As was the case of the dyeing industry, the development of tanning required access to water, and this was again provided by the Wensum. Unlike the case of the Flemish and Walloon weavers who came to Norfolk and settled bringing technology and trade opportunities [1], there is little evidence of any cross-fertilisation of trades associated with the leather industry between Norfolk and northern Europe although, prior to the introduction of chrome-based tanning methods, a range of vegetable tannins were used across Europe [59].

The relatively small-scale production of wire was a trade that was developed in the late nineteenth and early twentieth centuries in Norwich. The technological basis for making wire netting derives from the same type of thinking that underpins the production of cloth by weaving. The founder of the company of Barnard, Bishop and Barnard (Charles Barnard) invented a machine (loom) for making wire netting 40 foot wide in 1844. Charles Barnard began in business near to Norwich Market in

1826 as an ironmonger, oil and colourman. He later established a collaboration with John Bishop in 1846. They became known for the manufacture of household and garden equipment made of iron. Barnard did not patent the machine he invented for making wire netting. Indeed, a patent for making wire netting was issued much later in the US in the twentieth century [60]. Another Norwich company, Boulton and Paul (derived from an ironmongers founded in Norwich in 1797), became a major producer of wire netting—particularly for agricultural use in Australia and for wartime use. In addition, Boulton and Paul evolved into a construction engineering company and, among other products (particularly prefabricated buildings), were involved in aircraft manufacture. The manufacturing base in Norwich consisted of a number of interconnected companies. For example one of the Board members of Boulton and Paul was Richard Jewson of the Jewson timber merchants (see below). In 1859, Barnard's two sons joined the Barnard and Bishop business and it became Barnard, Bishop and Barnards and its symbol became 4 bees in a roundel. The ironworks remained a significant feature of the company and the city. The importance of wire netting to prevent crop damage by animals ensured the company did well from its core business. In addition to wire netting and cast-iron metal furniture, the firm made chairs, dressing-tables, lamps and other household equipment using wire and exhibited at the Great Exhibition (1851). The combination of weaving (metal rather than worsted wool) and the arable tradition in Norfolk that required the use of fencing to protect crops from animals, contributed to the success of Barnard, Bishop and Barnards. One of the designers at the company was Thomas Jeckyll who, through his involvement with London-based artists such as Whistler, became a key figure in the Anglo-Japanese Aesthetic Movement but who ended his days in a lunatic asylum in Norwich [61]. While there are few things more utilitarian than wire netting, the lasting reputation of Barnard, Bishop and Barnards lies with their later embrace of art and design rather than a means of preventing farm animals from straying into crops. An example of the design ethos of the company were the Norwich gates which were the wedding gift from the Gentlemen of Norwich to the Prince of Wales and his bride, Alexandra of Denmark and were installed at Sandringham. While the major aspects of the industrial revolution were centred around the great manufacturing cities of the north and the midlands, Norwich continued with a small but diverse culture of generation of economic benefits. These ventures built upon the individual, institutional and societal skills and knowledge and included a significant and important timber merchant (Jewson) and Laurence, Scott and Electromotors as well as banking (Gurney) and insurance (Norwich Union). The forerunner of Laurence, Scott and Electromotors (Scott and Paris) was established in 1884 after Scott and Paris were working for a company (Hammonds, which, about that time, amalgamated with Ferranti in 1884) who were engaged to supply a dynamo and install electric lighting for another Norwich business, Colman's ((Theme Four, Chap. 8). Scott and Paris with the support of Jeremiah Colman set up an independent electrical business at the Gothic Works in King Street, Norwich and later (1898) this enterprise moved to Hardy Road on a site originally owned by Colman, and it retained the name, Gothic Works. This move was made possible, in part as a consequence of investment by Reginald Edward

Laurence resulting eventually in the renaming of the company to Laurence, Paris and Scott and later to Laurence, Scott Ltd. The acquisition of Electromotors Ltd. (a Manchester-based company) in 1929 led to the formation of Laurence, Scott and Electromotors Ltd. The company was a supplier of dynamos, motors and electrical apparatus and later, other electrical parts including control panels for nuclear submarines. The intertwined nature of the businesses that were established in Norwich in the nineteenth century is not surprising, given the size of the city and the diminished value of the natural resources (particularly the lack of coal or mineral deposits) or significant rivers to supply water for factory machinery. In addition, the lack of efficient transport links with the rest of England (canals and railways) and the increased emphasis on trade with the new markets in the Americas rather than the traditional and mature (and unstable due to conflict) markets in Europe resulted in a loss of wealth and importance for the region. In addition, much of the existing wealth in Norfolk was associated with land ownership. It has been estimated that by 1796, two-thirds of the county of Norfolk was composed of arable farms and the majority of these were 'enclosed'. The vast majority of the population were by then employed in agriculture. According to the census of 1871 and the agricultural Returns of 1870, Norfolk had more farmers and agricultural holdings than any other county in England except Lincolnshire (which was much larger in size). In addition, the lack of competitiveness of the weaving industry in Norfolk compared to the factories of the North of England resulted in a loss of wealth generation. The loss of agriculture income encouraged the population of the county to move to Norwich and the city population within the medieval city walls tripled in the nineteenth century and this, combined with the depression in the textile trade, led to there being significant pressure on land for cheap housing within the city and less availability of industrial sites. Even into the twentieth century, Poor Relief was required for a significant proportion of the population whose major asset was only their labour [62].

7.4 The Post-Industrial Age: The Return of Agriculture

In a primarily agricultural county such as Norfolk, the developments of the early twentieth century in chemistry—particularly the Haber-Bosch process for the production of ammonia [63]—which enabled increases in yields from arable land [64], should have been particularly advantageous for Norfolk and other agricultural areas [65]. In some cases (particularly in the case of cereal crops and oil seed rape), the improvement has been as much due to the consequence of the development of varieties with improved characteristics [66] as the development of new agronomic practices. Others have emphasised the importance of human interventions through the use, for example, of pesticides and herbicides [67]. The development and application of a range of new technologies in the twentieth century including fertilisers, new varieties and better plant protection products led to a significant increase in the yields obtained from agriculture and were, together termed 'the Green Revolution'—an expression also used to describe advances in mechanisation and precision agriculture [68]. Precision is a descriptor used in the context of a number of fields including medicine [69], nutrition [70] and even chemistry [71]. Overall, research related to the Green Revolution has enabled significant increases in agricultural production without the necessity of increasing the amount of arable, farmed land [72]. In the case of Norfolk where a substantial proportion of the potential arable land was already under cultivation, the opportunity to create increased wealth and value from an existing, finite resource was attractive. While significant industry, even related to agriculture, was not established to any great extent in Norfolk, one of the first factories in England was the 'Vitriol and Manure Works' opened by James Fison and Sons (a local company established in 1808) in 1853 at a site near Thetford. A significant centre for plant breeding in the UK was established on the outskirts of Norwich after being translocated from London [73] alongside expertise in food and health research and, for example, in biorefining [74]. The development of expertise in the study of secondary metabolites from microbial species [75] or from plants [76] has been established in Norwich [77] even if there remains some uncertainty concerning the use of such compounds to treat non-communicable disease conditions [78]. The purpose of natural products and if that purpose is dislocated from their use by consuming species is an important topic from which valuable insights (and products) may be derived [79]. Multidisciplinary techniques for the investigation and manipulation of agricultural products and their optimised use remains a core set of skills in Norfolk from the standpoint of individuals, institutions and society as a whole. The development of agriculture and the use of agriculture-based technologies continued with processing of products and the creation of added value underpinned equally by the resources and the expertise available.

References

1. Joby C. Flemish and Walloon exile communities in sixteenth-century Norwich: a case study of local and national responses to large-scale migration from the Low Countries. Immigr Minorities. 2025;43(3):385–423. https://doi.org/10.1080/02619288.2024.2392831.
2. Carneiro A. Adolphe Wurtz and the atomism controversy. Ambix. 1993;40(2):75–95. https://doi.org/10.1179/amb.1993.40.2.75.
3. Rocke AJ. Chemical atomism and the evolution of chemical theory in the nineteenth century. In: Klein U, editor. Tools and modes of representation in the laboratory sciences. Boston studies in the philosophy and history of science, vol. 222. Dordrecht: Springer; 2001. https://doi.org/10.1007/978-94-015-9737-1_1.
4. Sakkopoulos SA, Vitoratos EG. Empirical foundations of atomism in ancient Greek philosophy. Sci Educ. 1996;5:293–303. https://doi.org/10.1007/BF00414318.
5. Smith J, Zadeh Haghighi H, Salahub D, et al. Radical pairs may play a role in xenon-induced general anesthesia. Sci Rep. 2021;11:6287. https://doi.org/10.1038/s41598-021-85673-w.
6. Li N, et al. Nuclear spin attenuates the anesthetic potency of xenon isotopes in mice implications for the mechanisms of anesthesia and consciousness. Anesthesiology. 2018;129:271–7.
7. Sadek B, Saad A, Schwed JS, Weizel L, Walter M, Stark H. Anticonvulsant effects of isomeric nonimidazole histamine H3 receptor antagonists. Drug Des Devel Ther. 2016;7(10):3633–51. https://doi.org/10.2147/DDDT.S114147.

8. Kitsikopoulos H. Steam power diffusion in the British cotton and woolen industries, 1774–1800: the role of firm size. Rev Evol Polit Econ. 2024;5:559–80. https://doi.org/10.1007/s43253-024-00129-x.
9. Bottomley S. Stationary steam power in the United Kingdom, 1800–70: an empirical reassessment. Econ Hist Rev. 2025;78:825–48. https://doi.org/10.1111/ehr.13375.
10. Jonell TN, Jones P, Lucas A, Naylor S. Limited waterpower contributed to rise of steam power in British “Cottonopolis”. PNAS Nexus. 2024;3(7):251. https://doi.org/10.1093/pnasnexus/pgae251.
11. Williams DM, Armstrong J. An appraisal of the Progress of the steamship in the nineteenth century. In: Harlaftis G, Tenold S, Valdaliso JM, editors. The world’s key industry. London: Palgrave Macmillan; 2012. https://doi.org/10.1057/9781137003751_4.
12. Sabbatani S. Epidemics on the sea: migrants journeys in the nineteenth century. Infez Med. 2015;23(2):195–206.
13. Ville S, editor. Shipbuilding in the United Kingdom in the nineteenth century: a regional approach. Liverpool: Liverpool University Press; 1993.
14. Wallace K, Crone A, Wood A, et al. New insights into the development of a Scottish port: the archaeological investigation of Leith’s nineteenth century docklands. J Mari Arch. 2022;17:219–44. https://doi.org/10.1007/s11457-021-09323-y.
15. Armstrong J. Coastal shipping: the neglected sector of nineteenth-century British transport history. In: The vital spark: the British coastal trade, 1700-1930. Liverpool: Liverpool Scholarship; 2009. https://doi.org/10.5949/liverpool/9780986497308.003.0006.
16. Fenton R. The evolution and significance of the powered bulk carrier: the black freighters. Liverpool: Liverpool University Press; 2023. https://doi.org/10.2307/j.ctv33b9wgc.
17. Martin J. British agricultural revolutions: the dissemination and assimilation of more productive methods. Icon. 2023;28(1):24–47.
18. Doherty SP, Alexander MM, Henderson S, Newton J, Finch J, Collins MJ. Tracking the British agricultural revolution through the isotopic analysis of dated parchment. Sci Rep. 2023;13(1):61. https://doi.org/10.1038/s41598-022-26013-4.
19. Fletcher TW. The great depression of English agriculture 1873-1896. In: Perry PJ, editor. British agriculture 1875–1914. London: Methuen; 1973. p. 31.
20. Rynne C. 'The linen and wool Industries in Britain and Ireland. In: Casella EC, Nevell M, Steyne H, editors. The Oxford handbook of industrial archaeology. Oxford: Oxford Handbooks; 2022.
21. Hedrick MC. The Australian woolen industry; British Investment in Colonial Australia: unraveling the threads of economic development 1788-1850. Doctoral Dissertations and Projects 5753. 2024. https://digitalcommons.liberty.edu/doctoral/5753.
22. Alegbe EO, Uthman TO. A review of history, properties, classification, applications and challenges of natural and synthetic dyes. Heliyon. 2024;10(13):e33646. https://doi.org/10.1016/j.heliyon.2024.e33646.
23. Cobbold C. Adulation or adulteration? Representing chemical dyes in the Victorian media. Ambix. 2019;66(1):23–50. https://doi.org/10.1080/00026980.2019.1574385.
24. Hagan E, Poulin J. Statistics of the early synthetic dye industry. Herit Sci. 2021;9:33. https://doi.org/10.1186/s40494-021-00493-5.
25. Sigurdson GT, Tang P, Giusti MM. Natural colorants: food colorants from natural sources. Annu Rev Food Sci Technol. 2017;28(8):261–80. https://doi.org/10.1146/annurev-food-030216-025923.
26. Ghigo T, Occhipinti M, Beeby A, et al. Concerns over colour durability in the nineteenth-century industrial revolution: insights from John Ruskin’s teaching collection. Herit Sci. 2023;11:168. https://doi.org/10.1186/s40494-023-01010-6.
27. Russell PL. World history and resources of oil shale. Energy Explor Exploit. 1986;4(4):301–20.
28. Kerker M. Science and the steam engine. Technol Cult. 1961;2(4):381–90.
29. Marriott J, Macalister T. Crude Britannia: how oil shaped a nation. London: Pluto Press; 2021. https://doi.org/10.2307/j.ctv1n9dkgb.

30. Volta A. On the electricity excited by the mere contact of conducting substances of different kinds. Philos Trans R Soc Lond. 1800;1:27–9.
31. Piccolino M. The bicentennial of the voltaic battery (1800-2000): the artificial electric organ. Trends Neurosci. 2000;23(4):147–51. https://doi.org/10.1016/s0166-2236(99)01544-1.
32. Smil V. 'Internal combustion engines', creating and transforming the twentieth century, revised and expanded: technical innovations and their lasting impact. New York: Oxford University Press; 2025.
33. Bryant L. The origin of the four-stroke cycle. Technol Cult. 1967;8(2):178–98. https://doi.org/10.2307/3101967.
34. Li W, et al. Lessons from internal-combustion engines should aid the development of electric vehicles. Cell Rep Phys Sci. 2025;6(8):102783.
35. Nemitallah MA, Habib MA, Abdelhafez A. Engines, fuels, and thermodynamics of combustion. In: Hydrogen for clean energy production: combustion fundamentals and applications. Singapore: Springer; 2024. https://doi.org/10.1007/978-981-97-7925-3_2.
36. Castaldi C, Nuvolari A. Technological revolution and economic growth: the "age of steam" reconsidered, LEM working paper series. Pisa: Laboratory of Economics and Management (LEM); 2004.
37. Fouquet R, Broadberry S. Seven centuries of European economic growth and decline. J Econ Perspect. 2015;29(4):227–44.
38. Broadberry S, Chadha JS, Lennard J, Thomas R. Dating business cycles in the United Kingdom, 1700–2010. Econ Hist Rev. 2023;76:1141–62. https://doi.org/10.1111/ehr.13238.
39. Bogart D. Turnpike trusts, infrastructure investment, and the road transportation revolution in eighteenth-century England. J Econ Hist. 2005;65(2):540–3. https://doi.org/10.1017/S0022050705260187.
40. Ayers BS, Leah M. Report of trial excavations at quayside, Norwich (report no 75). NAU Archaeology. 1993.
41. Ayers BS. The German Ocean: medieval Europe around the North Sea, studies in the archaeology of medieval Europe (equinox, 2016).
42. Gilchrist R, Cattermole P. Church Bells and bell-ringing. A norfolk profile, The Boydell Press, Woodbridge, 234pp. Rural Hist. 1991;2(2):256–7. https://doi.org/10.1017/S0956793300002880.
43. Hornell J. Evolution of the clinker-built fishing Lugger. Antiquity. 1936;10(39):341–5. https://doi.org/10.1017/S0003598X00011881.
44. Ounanian K, Howells M. Clinker, sailor, fisher, why? The necessity of sustained demand for safeguarding clinker craft intangible cultural heritage. Maritime Stud. 2022;21(4):411–23. https://doi.org/10.1007/s40152-022-00260-7.
45. Hutchison R. The Norwegian and Baltic timber trade to Britain 1780–1835 and its interconnections. Scand J Hist. 2012;37(5):578–99. https://doi.org/10.1080/03468755.2012.730057.
46. Atkin M, Evans DH. Excavations in Norwich 1971–1978 part III. East Anglian Archaeology 100. 2002.
47. Hallsdordottir HH, Williams R, Greene EM, Taylor G. Rapid deterioration in buried leather: archaeological implications. RSC Adv. 2024;14(6):3762–70. https://doi.org/10.1039/d3ra07020d.
48. Hathway DE. Oak-bark tannins. Biochem J. 1958;70(1):34–42. https://doi.org/10.1042/bj0700034.
49. Davy H. An account of some experiments and observations on the constituent parts of certain astringent vegetables; and on their operation in tanning. Abstr Pap Print Philos Trans R Soc Lond. 1800;1:114–8.
50. Davy H. Observations on the processes of tanning. Philos Mag. 1803;17(65):58–62.
51. Nashy EH, Osman O, Mahmoud AA, Ibrahim M. Molecular spectroscopic study for suggested mechanism of chrome tanned leather. Spectrochim Acta A Mol Biomol Spectrosc. 2012;88:171–6. https://doi.org/10.1016/j.saa.2011.12.024.

52. Hossini H, Shafie B, Niri AD, Nazari M, Esfahlan AJ, Ahmadpour M, Nazmara Z, Ahmadimanesh M, Makhdoumi P, Mirzaei N, Hoseinzadeh E. A comprehensive review on human health effects of chromium: insights on induced toxicity. Environ Sci Pollut Res Int. 2022;29(47):70686–705. https://doi.org/10.1007/s11356-022-22705-6.
53. Moretto A. Hexavalent and trivalent chromium in leather: what should be done? Regul Toxicol Pharmacol. 2015;73(2):681–6. https://doi.org/10.1016/j.yrtph.2015.09.007.
54. Salmi O, Laudisa G, Rossi F, Masi M. Evaluation of antioxidant performance in chromium oxidation prevention. Materials. 2025;18(8):1858. https://doi.org/10.3390/ma18081858.
55. China CR, Maguta MM, Nyandoro SS, Hilonga A, Kanth SV, Njau KN. Alternative tanning technologies and their suitability in curbing environmental pollution from the leather industry: a comprehensive review. Chemosphere. 2020;254:126804. https://doi.org/10.1016/j.chemosphere.2020.126804.
56. Ahmed MD, Maraz KM. Benefits and problems of chrome tanning in leather processing: approach a greener technology in leather industry. Mater Eng Res. 2021;3(1):156–64. https://doi.org/10.25082/MER.2021.01.004.
57. Henderson S. Life, art, and industry: a history of the English and welsh leather economy, 1700-1900 [Apollo - University of Cambridge Repository]. 2024. https://doi.org/10.17863/CAM.112821.
58. The story of shoemaking in Norwich from the earliest times to the present day (1949) By W. L. Sparks, Published by National Institution of the Boot and Shoe Industry, Northampton.
59. Falcão L, Araújo MEM. Vegetable tannins used in the manufacture of historic leathers. Molecules. 2018;23(5):1081. https://doi.org/10.3390/molecules23051081.
60. Mafera, F. Method of forming wire fence fabric. (U.S. patent no. 1811511A). U.S. patent and trademark office. 1930.
61. Soros SW, Arbuthnott C. Thomas Jeckyll: architect and designer, 1827–1881. The bard graduate Centre for Studies in the decorative arts, design and culture. New Haven: Yale University Press; 2003.
62. Grover C. Wage supplements and poor relief in the 1920s: Norfolk's agricultural labourers. In: Social security and wage poverty. London: Palgrave Macmillan; 2016. https://doi.org/10.1057/9781137293978_3.
63. Li J, Xiong Q, Mu X, Li L. Recent advances in ammonia synthesis: from Haber-Bosch process to external field driven strategies. ChemSusChem. 2024;17(15):e202301775. https://doi.org/10.1002/cssc.202301775.
64. Miflin B. Crop improvement in the 21st century. J Exp Bot. 2000;51(342):1–8.
65. Page A. "The greatest victory which the chemist has won in the fight (…) against nature": nitrogenous fertilizers in Great Britain and the British empire, 1910s-1950s. Hist Sci. 2016;54(4):383–98. https://doi.org/10.1177/0073275316681801.
66. Mackay I, Horwell A, Garner J, White J, McKee J, Philpott H. Reanalyses of the historical series of UK variety trials to quantify the contributions of genetic and environmental factors to trends and variability in yield over time. Theor Appl Genet. 2011;122(1):225–38. https://doi.org/10.1007/s00122-010-1438-y.
67. Tudi M, Daniel Ruan H, Wang L, Lyu J, Sadler R, Connell D, Chu C, Phung DT. Agriculture development, pesticide application and its impact on the environment. Int J Environ Res Public Health. 2021;18(3):1112. https://doi.org/10.3390/ijerph18031112.
68. Ullah Z, et al. In: Fahad S, Saud S, Nawaz T, Gu L, Ahmad M, Zhou R, editors. Environment, climate, plant and vegetation growth The green revolution: promoting environmental stewardship and plant growth. Cham: Springer; 2024. https://doi.org/10.1007/978-3-031-69417-2_15.
69. Evans W, Meslin EM, Kai J, Qureshi N. Precision medicine-are we there yet? A narrative review of precision medicine's applicability in primary care. J Pers Med. 2024;14(4):418. https://doi.org/10.3390/jpm14040418.
70. Kirk D, Catal C, Tekinerdogan B. Precision nutrition: a systematic literature review. Comput Biol Med. 2021;133:104365. https://doi.org/10.1016/j.compbiomed.2021.104365.

71. Zhong W, Liu Y, Fan R. Precision chemistry in precision diagnostics. Precis Chem. 2025;3(11):640–3. https://doi.org/10.1021/prechem.5c00208.
72. Stevenson JR, Villoria N, Byerlee D, Kelley T, Maredia M. Green revolution research saved an estimated 18 to 27 million hectares from being brought into agricultural production. Proc Natl Acad Sci U S A. 2013;110(21):8363–8. https://doi.org/10.1073/pnas.1208065110.
73. Palladino P. Science, technology, and the economy: plant breeding in Great Britain, 1920-1970. Econ Hist Rev. 1996;49(1):116–36.
74. Wood IP, Cook NM, Wilson DR, Ryden P, Robertson JA, Waldron KW. Ethanol from a biorefinery waste stream: saccharification of amylase, protease and xylanase treated wheat bran. Food Chem. 2016;198:125–31. https://doi.org/10.1016/j.foodchem.2015.09.108.
75. Hopwood DA, Sherman DH. Molecular genetics of polyketides and its comparison to fatty acid biosynthesis. Annu Rev Genet. 1990;24:37–66. https://doi.org/10.1146/annurev.ge.24.120190.000345.
76. Erb M, Kliebenstein DJ. Plant secondary metabolites as defenses, regulators, and primary metabolites: the blurred functional trichotomy. Plant Physiol. 2020;184(1):39–52. https://doi.org/10.1104/pp.20.00433.
77. Rhodes MJ. Physiological roles for secondary metabolites in plants: some progress, many outstanding problems. Plant Mol Biol. 1994;24(1):1–20. https://doi.org/10.1007/BF00040570.
78. Noubissi FK, Omolekan TO, Walker AL, Chamcheu JC. Bioactive natural and synthetic products in human health and diseases: basic, preclinical and clinical studies. Nutrients. 2024;16(19):3265. https://doi.org/10.3390/nu16193265.
79. Dixon RA, Dickinson AJ. A century of studying plant secondary metabolism-from “what?” to “where, how, and why?”. Plant Physiol. 2024;195(1):48–66. https://doi.org/10.1093/plphys/kiad596.

Part IV
Chemistry and Products from Nature. Food Processing for Value

Chapter 8
Chemistry and Products from Nature. Valuable Natural Products: Mustard and Mint

Abstract The predisposition of certain groups to disease—genetic polymorphisms. The history of domestication and breeding of plants demonstrated in herbariums, botanic gardens and paintings. The production of dyes from different natural sources. Mustard and mint—their exploitation and differences between them in the context of secondary metabolite production. Different varieties of mustard and mint and their organoleptically active principles. The use of mint and isothiocyanates as a preventative measure for halitosis. Mastication and mint in the prevention of cognitive decline. The Lossen Rearrangement and the formation of the components of mustard oil. Sir Thomas Browne in Norwich and Herman Boerhaave in Leiden—connection and correlation. The importance of extra-oral taste receptors—GPCRs. Natural and synthetic plant protection factors.

Keywords Nature · Association/correlation · Population genetics · Mustard · Mint · Dithiocarbamate · Glucosinolate · Terpene · Lossen rearrangement

Chemical products from nature include compounds that are both intrinsically valuable and also provide an insight into how natural systems operate and respond at a molecular level. While part of the attraction of the study of chemicals lies in their acute and measurable biological activity and this has led to the growth in the study of natural products as pharmaceuticals [1], the effects of natural bioactive principles on chronic disease outcomes has been more difficult to define and this has resulted in the investigation of products with a more general spectrum of properties rather than a targeted effect on a molecular level including, for example, anti-inflammatory [2] and putative anti-tumourigenic effects [3]. In many cases, details of the mechanism of action remains unclear in part because multiple pathways and effects may be implicated. This is particularly the case when studying the link between diet and chronic disease [4]. In many cases the correlation is associative (implied) rather than causal (proven) and mechanistic information is either unavailable or cannot be deconvoluted with any degree of certainty due to the cumulative and multiple effects of a number of factors and/or the presence of confounders. Nonetheless, the impact of 'Western' diets on chronic, non-communicable disease outcomes (and incidence)

B. Hanley, *The Chemistry of a County*,
https://doi.org/10.1007/978-3-032-21491-1_8

has been confidently stated by numerous authors [5], as has the impact of food processing [6]—although correlation remains more common than causality. Part of the confusion about the impact of specific dietary components on chronic disease outcomes arises from the clash between the traditional, reductionist approach of chemists and the complexity multifactorial nature of living systems. The other confounding factor is the impact of genetic predisposition which in turn has been affected by the mobility of the population with concomitant interbreeding of populations. The long-term history of mobility of populations has been investigated by studying ancient DNA [7]. In addition to heritable characteristics, the movement of people from an agricultural (village or small town) environment to an urban environment (the 'rural exodus') that can be documented by consideration of census information [8], together with epidemiological observations of chronic disease occurrence, provides information about the health consequences of changes in habits and diets. Genetic flow has been investigated on a population level and this has revealed the extent to which external factors may have influenced population dynamics [9]. In addition to creating conditions for the prevalence and spread of different types of diseases (exacerbated by the types of work being carried out in rural and urban communities), the fluidity of populations means that the spread of heterogenous infections by, for example *Mycobacterium tuberculosis*, which is transmitted between humans [10] may be surrogates for the measurement of the intermixing of population genomes. Fine scale genetic investigations [11] can also facilitate the study of genetic effects on disease susceptibility. Ultimately, such studies may provide some illumination regarding the complex interplay between the genetic ancestry of distinct populations, diet and health and morbidity/mortality [12]. The relative national isolation of the pre-industrial population of Norfolk (in the context of the rest of England) suggests that some health traits that have a genetic basis might be more likely to be shared with northern European or Flemish populations since large cohorts of immigrants from there moved to Norfolk. This may be observed particularly in the impact of the Flemish and Walloon populations which constituted 2000 individuals in the 1568 census of Norwich and by 1578 had risen to about 7000, or about 40% of the city's population [13]. This immigrant population will have had an impact upon health and disease in the city [14]. For example, the immigrant population seem to have suffered a particularly high level of casualties due to plague in the sixteenth century. It is unclear if there is any genetic predisposition or if it is a consequence of the spread of the disease due to crowded urban living conditions. In some cases, there are insufficient unambiguous genetic markers to determine the origin of individuals in populations. In some cases, the populations are more easily mapped by considering surnames [15]. Surnames can, in some cases, provide a guide to occupations or origin (van Brughen—of Brughes; de Bakker—baker). In addition, the spread of particular social habits can be indicative of population migration (or at least the establishment of trade routes). For example, tulips were introduced into a number of countries in the sixteenth century [16]. These flowers originated in Flanders and their cultivation together with, for example, the creation of a botanical garden at Leiden in the sixteenth century is an indicator of the importance of plants in Flanders at that period. Indeed, Leiden was an important botanical centre and was directed by Professor Herman Boerhaave—perhaps the most significant medical authority of his time [17, 18] who was known throughout Europe and the practices he introduced remain impactful [19]. Plants existed as an important resource that was just beginning to be appreciated for medicinal purposes [20]. An example of the gradual increase in popularity of plant-based remedies can be found from the assessment of Bryophytes (plants are classified either as Bryophytes which includes hornworts, liverworts and mosses or as vascular plants) in central Europe in the sixteenth century [21]. Another source of information about the domestication of plants are paintings from Flemish and Dutch masters [22]. Reviews of herbal practices are also a useful source of information about which plants were available and the types of conditions they were used for [23]. In addition to medicinal plants, a range of dyes were commonly used in Flanders and dye analysis of paintings may also reveal the source plants for some pigments [24]. Analysis of the waste from riverbeds suggests that certain plants were in active use by weavers and dyers during the period of emigration to Norfolk [25]. Indeed, by the early 1500s cultivation of specific plants for dye production was common in Europe. The importation of dye materials or partially processed plants (rather than growing the dye plants themselves) was an important aspect of trade in the sixteenth century and was facilitated by the spread of trade routes. The close marine links between Norfolk and Europe will have been convenient for the transport of high value goods such as pigments. In addition to dyes and pigments, some of the most significant later (nineteenth century) contributions to the industrial landscape in Norwich and Norfolk was achieved through exploitation of the chemistry of flavour compounds—particularly those found in mustard and mint.

8.1 Mustard and Mint: Overview

Both mustard and mint contain organoleptically active principles however the production, storage and properties of the groups of compounds are very different. These will be delineated in more detail in succeeding sections and will be preceded by an overview.

Mustard occurs as a number of distinct botanical species including brown, black, white and yellow. They contain a number of flavour precursors which are released upon crushing and wetting of the seed (Sect. 8.2). The principal flavour (and main bioactive) molecules in mustards are isothiocyanates although other compounds of varying degrees of volatility and toxicity are also released from the plant [26]. The production of volatile and pungent isothiocyanates by mustard seeds occurs when the seed is crushed in water and the precursor glucosinolate is hydrolysed by a co-occurring but separately compartmentalised enzyme, myrosinase. This is the reaction that produces the well-known pungency of culinary mustard. The different mustards mentioned previously belong to the commonly known and widespread family *Brassicaceae* of the order Brassicales. Other members of the class include cabbage, cauliflower and rapeseed. White mustard (*Sinapis alba*) is probably the

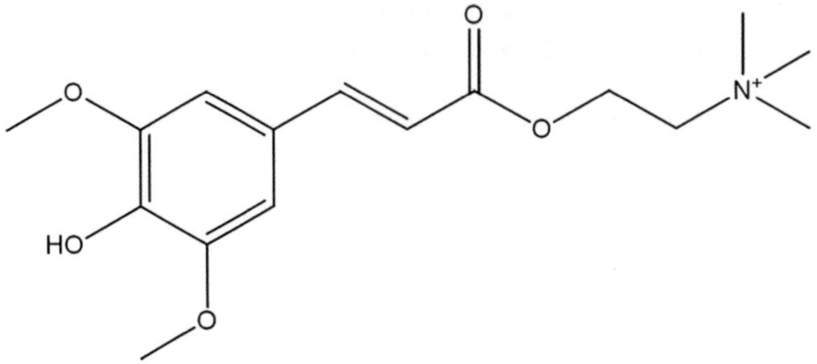

Fig. 8.1 The structure of sinapine

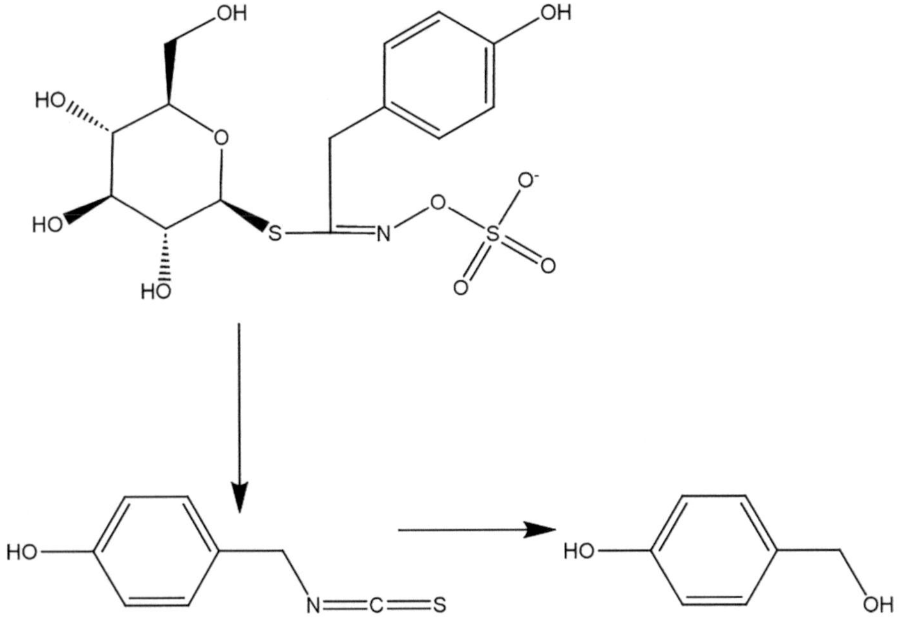

Fig. 8.2 Production and hydrolysis of p-hydroxybenzyl isothiocyanate

most widely cultivated mustard species and it has been extensively studied both as a food ingredient and as a source of secondary metabolites. In addition to a glucosinolate (sinalbin) plants of white mustard also contain an alkaloidal quaternary amine, sinapine (Fig. 8.1) which is the choline ester of sinapic acid.

Sinapine also occurs in other mustards and in related species and was discovered by Etienne Ossian Henry in 1825 [27]. The glucosinolate, sinalbin, produces the pungent compound p-hydroxybenzyl isothiocyanate upon crushing and wetting of the seed (see above). This compound is relatively unstable and slowly decomposes (hydrolysis) to give the corresponding alcohol and thiocyanate ion (Fig. 8.2) [28].

The cultivation of mustard and related species (including oilseed rape—*Brassica napus* subsp. *Napus*) has been carried out principally for the extraction of oil which is has variable amounts of polyunsaturated fatty acids (PUFAs) [29] and for the residual protein cake which is primarily used for animal feed. In addition to *Sinapis alba*, two other mustards are commonly used as condiments—black mustard (*Brassica nigra*) and brown mustard (*Brassica juncea*). All three contain structurally similar but slightly different glucosinolates and produce, therefore, a slightly different spectrum of isothiocyanates and these will be considered in more detail later. The mixture of different of these compounds is largely responsible for the organoleptic differences between different mustard preparations but it is also a consequence of the processing method. In addition to intact glucosinolates (and their enzymically-derived breakdown products) mustards contain a number of other chemical constituents and the presence of some of these have been correlated with plant part and maturity [30]. In addition, the nutritional characteristics of the three major mustards (black, brown and white) (particularly protein digestibility) have been studied [31] and some of the differences ascribed to the presence of allyl isothiocyanate. The reaction of isothiocyanates with SH groups to form thioureas and to then participate in further reactions including the release of hydrogen sulphide, is known [32].

In addition to having potential biological relevance as a consequence of the release of hydrogen sulphide [33], this reaction may inhibit the formation of disulphide linkages thereby affecting both protein stability and its physical properties as well as its nutritional characteristics. In addition, the reaction of organic isothiocyanates with lysine residues in peptides has been implicated in the realisation of the allergenic potential of isothiocyanates [34]. The reaction of isothiocyanates with a range of food components has been studied and this may have implications for the physiological effects of *Brassica*-derived isothiocyanates *in vivo*. The extent to which different isothiocyanates with varying reactivities are formed (they are electrophiles and will react with nucleophilic carbon atoms such as amines, thiols and alcohols—hence the reactions with cysteine and lysine residues in proteins, in particular) will affect observed biological properties. Common food-related isothiocyanates such as sulforaphane, phenethyl isothiocyanate, and benzyl isothiocyanate can be shown, through direct assay and proteomic analysis to react with a range of biological targets [35].

As noted previously, different types of commercial mustard preparations contain different phytochemicals (particularly isothiocyanates) and this affects characteristics such as pungency and the interaction with biological targets. The simplest correlation is between pungency and the levels of allyl isothiocyanate [36] which is derived from the parent glucosinolate sinigrin (Fig. 8.3).

The most prominent isothiocyanate in white mustard is that derived from sinalbin (4-hydroxybenzyl isothiocyanate), while in black mustard it is allyl isothiocyanate. For brown mustard, the main isothiocyanates formed are allyl and butenyl isothiocyanates [37]. As noted previously, the level of allyl isothiocyanate present is the simplest surrogate analytical measurement for pungency and a higher allyl

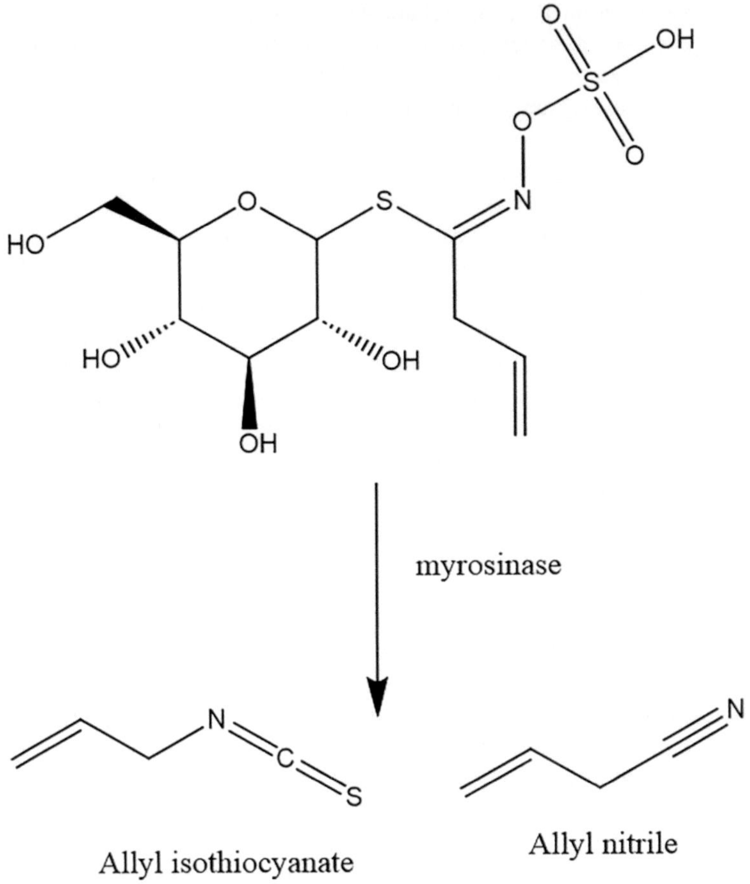

Fig. 8.3 The formation of allyl isothiocyanate from sinigrin

isothiocyanate content induces an intensified trigeminal pungency sensation in the consumer [38].

The discovery of the pungency precursor (sinigrin) by Bussy in 1840 [39] and the proposed structure which was suggested by Gadamer in 1897 [40] although later amended [41] demonstrated how mustard flour could be produced and stored as a stable pre-condiment before being mixed with water to release the pungent principle, allyl isothiocyanate. Blending of types of mustard can give a product with a range of characteristics. For example, Colman's mustard is a mixture of brown and while mustard seeds ground and mixed with wheat flour. The founder of the Colman's mustard business, Jeramiah J Colman came from a family of flour millers in Norfolk, so the use of wheat flour in mustard preparation was not unexpected. The milling process for mustard was not developed in Norfolk but in Durham and other places in England also developed the technology to process mustard seed to give the condiment however the arable land available in Norfolk and the tradition of growing crops (including crop rotation) favoured the area as an agricultural centre for the production of mustard. In addition to growing mustard for harvest, the crop is also thought to reduce the levels of some pathogens thereby improving subsequent crop yields [42]. The importance of crop rotation in East Anglian farming has already been mentioned. The potential of allyl isothiocyanate itself as a biofumigant has been investigated [43] and the presence of residues of this compound and its precursor glucosinolate (sinigrin) may contribute to the effectiveness of mustard as a rotation crop. The use of mustard predated the mechanisation of the process for its production by Colman and others. Indeed it was first introduced into England in the twelfth century however production on a large scale was driven by marketing and mass production in the nineteenth century. While mustard was first developed in Paris, the French home of mustard since the late Middle Ages is Dijon where brown mustard seeds (Brassica juncea) and a mixture of white wine, vinegar, water, and salt are used. The hydrolytic reaction that produces the hot taste of mustard (allyl isothiocyanate) can also give rise to other products including the corresponding nitrile [44] at low pH and in the presence of a nitrile specifier protein [45]. The use of vinegar in a Dijon mustard preparation might, therefore, lead to the formation of less allyl isothiocyanate and a less pungent condiment. In addition the Lossen Rearrangement which leads to formation of the isothiocyanate occurs optimally at pH 8.5 and in the presence of a base [46] so alternative hydrolysis products are likely to be formed at lower pH as a consequence of the blocking of the Lossen Rearrangement by protons [47]. The development of Colman's mustard (and its positioning as the market leader among English mustard makers) was a triumph of marketing and creation of brand awareness. In addition, the move from a factory outside Norwich to a site on the river Wensum next to the train station ensured that the product could be quickly and effectively transported. It was, however, a small volume, relatively expensive product with a particular niche market. Nonetheless the business thrived and expanded into other areas of processed food such as flavoured water (barley water) and household starch production. In 1856, Colman's employed 200 people, by 1862 this had risen to 600 and by 1898 it was closer to 2000. This substantial growth was a result of internationalisation of the mustard brand and the extension into other household products. It appears that Colman's (unlike many traditional Victorian employers who claimed that paying their workers more or improving their working conditions would wreck their businesses and ruin the national economy) was a benevolent employer who provided housing, medical and other services for workers and their families. Another reason for the expansion of the mustard business was the use of mustard for medicinal purposes (including as poultices). This application was known and appreciated in the nineteenth century and before [48]. More recently, the use of mustard preparations in the treatment of respiratory tract infections has been suggested [49]. Mustard and other 'natural oils' have been associated with improvements in hair health and growth [50] however the validity of such claims is often unproven and they should be treated with caution. Nonetheless, antimicrobial and anti-inflammatory properties have been associated with constituents derived from mustard [51] and the specific properties of sinigrin in Ayurvedic medicine have been reviewed [52]. While the growing and processing of mustard became particularly important in Norfolk in the nineteenth century, the traditional use of mustard in agriculture can be found in the arable farming and crop rotation practices introduced in the eighteenth and nineteenth centuries [53].

Mint is, like mustard, a relatively high value crop which is grown both for its flavour and its medicinal properties [54]. Unlike mustard which contains a limited number of sensorially and biologically active compounds primarily derived from precursor molecules (glucosinolates), mint contains a chemically diverse group of molecules which are biosynthesised *in situ* and stored as active compounds in the plant [55]. The chemistry and enzymology of both mint and mustard will be considered in more detail in Sect. 8.2. The main active compounds in dementholised mint are terpenes (Fig. 8.4) and their biosynthesis together with their occurrence in different cultivars of mint will be described in Sect. 8.4.

One of the most widely appreciated attributes of mint is its use to mask odours and, in particular, halitosis. Indeed, mint remains a significant component of many oral health products including toothpaste, mouthwash, chewing gum and, perhaps paradoxically, breath freshening confectionery (dragees). There are two alternatives to the use of masking agents for halitosis. The first of these relies upon a knowledge of the chemistry of the compounds that cause the condition. It has been found that, while halitosis may be caused by extra-oral conditions, particularly consumption of foods containing odour compounds (garlic, onion, spices etc.) about 90% of halitosis is a consequence of intra-oral effects [56]. The primary source of oral malodour is the production of volatile sulphur compounds (VSCs) which are produced generally as a consequence of microbial breakdown of proteins. The main odiferous compounds

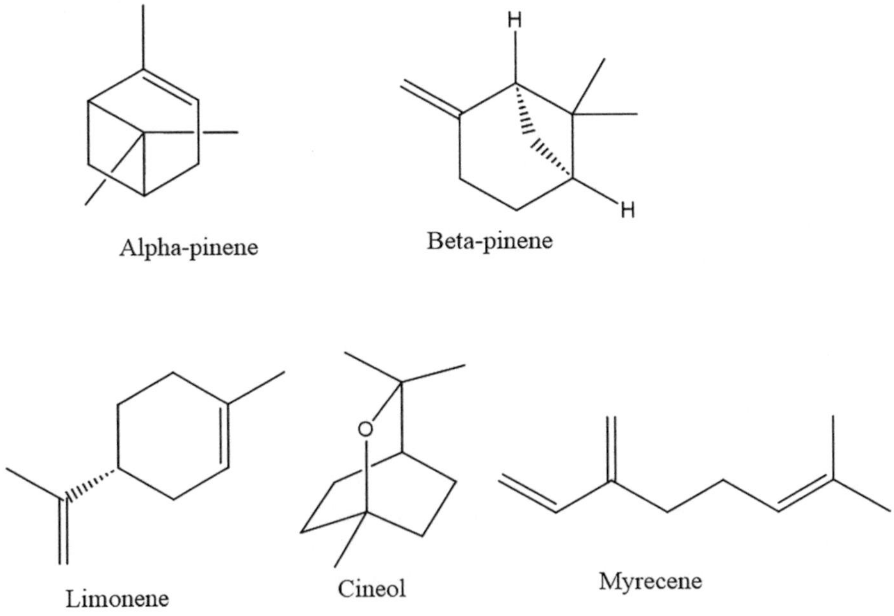

Fig. 8.4 Structure of terpenes derived from mint

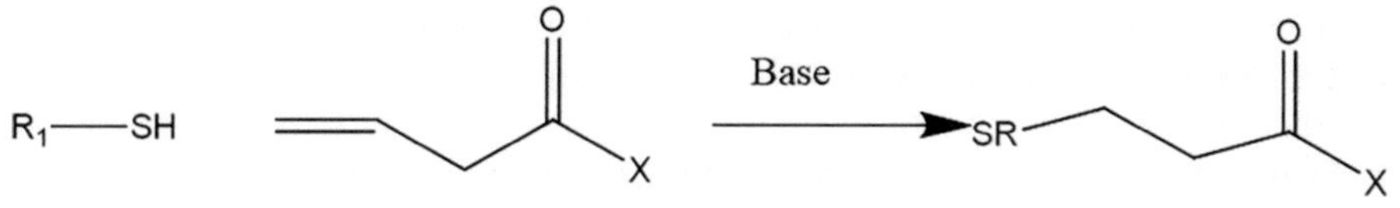

Fig. 8.5 The Michael addition reaction of thiol groups

are hydrogen sulphide (H_2S) and methyl mercaptan (CH_3SH) and arise from the degradation of cysteine and methionine [57]. The chemistry of thiol groups suggests one approach—the Michael reaction of the -SH moiety with a double bond (Fig. 8.5).

The reaction is relatively slow however kinetic studies (often using cysteine-containing proteins) have been carried out [58]. Hydrogen sulphide is synthesised *in vivo* by the action of three enzymes [59], cystathionine γ-lyase, cystathionine β-synthetase and 3-mercaptopyruvate sulfotransferase and inhibition of the transcription or activity of these is likely to reduce the levels of hydrogen sulphide produced. The major producer of methyl mercaptan in the oral cavity is *Porphyromonas gingivalis.* The enzyme L-methionine-α-deamino-γ--mercaptomethane lyase produces ammonia and methyl mercaptan from L-cysteine and the expression of the gene *mgl* codes for the enzyme [60]. He down regulation of expression of the gene has been suggested as a novel route for inhibition of halitosis [61] by plants of the *Magnoliaceae* family. A second alternative to the Michael reaction or down regulation of the enzymes mentioned above is the removal of hydrogen sulphide and methyl mercaptan (and other thiol-containing compounds) by reaction with an organic isothiocyanate to form an odourless dithiocarbamate (this is the reverse of a standard method for the synthesis of isothiocyanates [62]. As noted previously, mustard contains allyl isothiocyanate and incorporation of this molecule into chewing gum has been demonstrated to be an effective means of removing the compounds that cause halitosis [63]. A third possibility is the manipulation of the oral microbiome to prevent the growth of microbes associated with VSC formation [64] however this is unlikely to be a quick or persistent solution. The oral microbiome has been shown to be resilient [65] and regulation of oral microbial homeostasis has been proposed as a means of maintaining oral (and systemic) health [66]. From this discussion it is apparent that two of the crops grown in Norfolk (mint and mustard) may have effects on a specific chronic condition (halitosis). Both are also implicated in a wider sense in the maintenance of health however this will be considered in more detail (alongside some other natural products) in Sect. 8.3.

8.2 Chemistry and Enzymology of Mustard and Mint

While both mustard and mint contain volatile secondary products the biogenesis of them is different (Sect. 8.4) and their production is dissimilar. The main volatile compounds found in mustard are isothiocyanates which, as noted previously, are produced through the action of an enzyme (thioglucoside glucohydrolase) on a precursor glucoside (glucosinolate) with the concomitant elimination of glucose and sulphate (Fig. 8.6).

One of the main drivers that underpinned the studies of mustard oil and mustard oil glycosides (as glucosinolates were referred to) was the study of medicinal plants [67] in the nineteenth century. The isolation of active principles from plants was demonstrated by the work of a pharmacist, Friedrich Wilhelm Adam Serturner, who isolated and purified morphine in 1805 [68]. The isolation of alkaloids such as morphine was an important step for both chemical and biological reasons [69]. From a chemical standpoint, these 'alkali-like' compounds could all be isolated using similar techniques since they could all act as bases by neutralising acids to form salts. Prior to the discovery and isolation of alkaloids, several naturally occurring acids (including tartaric, benzoic, citric, oxalic, malic, glucuronic and gallic acids) were isolated by the end of the eighteenth century [70]. Much early natural product chemistry was carried out by those trained as pharmacists since they had developed the isolative and extractive skills required and had a requirement to purify

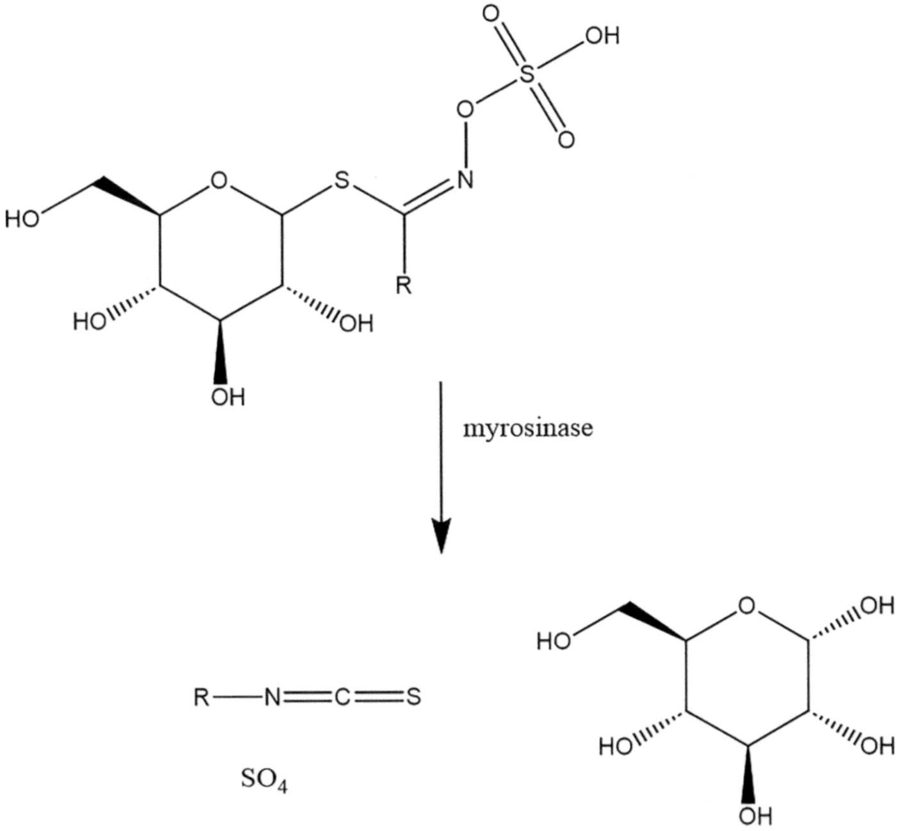

Fig. 8.6 The enzymic hydrolysis of glucosinolates

compounds to homogeneity (or at least to a constant and consistent formulation) in order to administer pharmacologically effective dosages of active compounds in their preparations. For example, Serturner (morphine) was a pharmacist as were Pierre Pelletier and Joseph Caventou who isolated quinine in 1820. In the case of mustard, the importance of the seeds was primarily as a spice or condiment rather than as a medicinally important product. Unlike alkaloids, the biologically active principles of mustard are not present in the plant material in the active form but are formed after the seed or other plant material is crushed in the presence of water. This is a consequence of the compartmentalisation of a hydrolytic enzyme (myrosinase) and the precursor glycoside (glucosinolate) in different parts of the plant cells. The myrosinase enzymes are located in myrosin cells (a specific type of idioblast) and are physically separated from the glucosinolate until tissue damage occurs—for example by herbivore grazing—as a result of which hydrolysis of the glucosinolate takes place [71] and volatile isothiocyanates (and other products) are formed. The localisation of parent glucosinolates (and their biosynthetic precursor) in myrosin cells has been studied [72]. Overall, the complex relationship between glucosinolates and myrosinases has been studied in detail. While glucosinolates are relatively biologically inert, the isothiocyanates produced from them have a range of primary (aversion to consumption due to pungency) and secondary biological properties [73] as well as anti-microbial and other properties [74]. In terms of plant protection, allyl isothiocyanate is toxic to some insects at different developmental stages [75] and to both specific insect predators [76] and other plant pests as a consequence of its effect on the plant defence mechanisms—closure of stomatal pores which prevent the entry of plant pathogens [77]. Key steps in the biosynthesis of different glucosinolates also takes place in specific cellular locations in plants [78] and this may suggest different sets of interrelationships with plant metabolic pathways.

Although the main marketing focus was on mustard as a condiment, unsurprisingly Colman's also built a secondary reputation based upon the medicinal uses of mustard preparations including for the relief of pain and to ease colds, rheumatism, chilblains, neuralgia, toothache, lumbago, insomnia, asthma, bronchitis, coughs and as a poultice. Mustard, in the form of plasters or poultices, was traditionally used for the relief of respiratory issues and it has been linked to alleviation of chronic obstructive pulmonary disease (COPD) [79]. While more recent evidence is not extensive and the mechanistic basis of the effect is unclear, a possible link between reduction in the severity of a respiratory tract infection and a mustard footbath has been suggested [49] as has a possible link between allyl isothiocyanate intake and minimisation of airway inflammation caused by allergens [80]. Traditional methods of extraction using water [81] are ineffective for the isolation of glucosinolates because of the action of myrosinase. In addition, glucosinolates, while they contain a N atom, have a sulphonated moiety and so carry a negative charge therefore extraction into acid or into an alkane will also be unsuccessful. Sinigrin was isolated by Bussy in 1839. Similarly to Serturner and to Pelletier and Caventou, Bussy was trained as a pharmacist but carried out significant work in inorganic chemistry and on acids and minerals and was, therefore, more broadly trained. He found that the active principle of the seeds of black mustard was a volatile oil which did not exist in the free state in the seeds themselves. Bussy isolated two fractions—a bitter solid which he called potassium myronate and another component which was similar to albumin and hence was named myrosyne [82]. Bussy carried out a number of investigations of the fractions from black mustard and their transformation into mustard oil, some of the principles he discovered may have helped to advise manufacturers on methods of preparation of mustard powder however the method developed by Colman that led to the establishment of his factory in 1814, significantly pre-dated Bussy's publication of his discovery of sinigrin. Colman sourced mustard seed from throughout England but also (significantly) from Holland. The technology used for the preparation of wheat flour has characteristics in common with the preparation of mustard flour and the Colman family were established farmers and flour producers. In both cases, an enzyme amylase (wheat) or myrosinase (mustard)) must be prevented from operating either by heating the material before grinding or by the careful exclusion of water or addition of vinegar to lower the pH of the ground material. A detailed understanding of the chemistry and enzymology of the mustard is not a requirement to enable a stable powder to be produced. A patent for the production of 'mustard flour' was issued to Samuel Hopkins in the US in 1813 and 1817. Prior to that, Colman was also granted a patent at the beginning of the nineteenth century (1804) for the production of 'double superfine' mustard powder.

In the case of mint, the flavour (and bioactive) ingredients are present in the plant so their extraction and isolation is, in principle, simpler. Mint is a member of the *Lamiaceae* family which also includes basil, rosemary, sage, savory, oregano, and thyme. Many of these are used for culinary purposes and have biological activity ascribed to them [83]. For the sake of simplicity, reference to mint will be restricted principally to those members of the family that are traditionally referred to as mint—plants of the genus *Metha.* Peppermint (*M. piperita*) is a natural hybrid of spearmint (*M. spicata*) and water mint (*M. aquatica*). It was first discovered in England in Hertfordshire in the seventeenth century. The plant is sterile and spreads through underground rhizomes. While much of the interest in mint is a consequence of the natural oils (with the major component being menthol), peppermint, for example, also contains flavonoids (such as hesperidin, eriodictyol, naringenin, quercetin, myricetin, and kaempferol), and nonflavonoid phenolcarboxylic acids. The principal volatile components of the *Mentha* genus include (particularly peppermint) menthols and terpenes—such as carvone (Fig. 8.7).

Carvone occurs as two different isomers. The R-(−) form (D-carvone) gives mint its particular flavour while the S-(+) form (L-carvone) is responsible for the odour of caraway. There has been considerable economic interest in optimising the growth and, in particular, the essential oil production of mint [84]. These essential oils are the most economically, organoleptically and biologically significant aspects of mint. The impact on health and the functional properties of members of the *Mentha* genus have been studied and will be described in more detail in Sect. 8.3. However, briefly, the chemical and biological (antimicrobial) properties of different types of mint have been studied [85] and a correlation between the spectrum of chemical constituents in the volatile oils and the antimicrobial activity noted. The evolution of the production of the chemically diverse characteristics of the *Lamiaceae* family

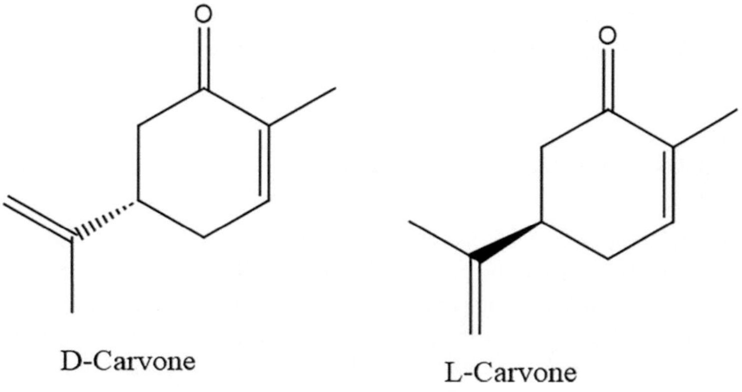

Fig. 8.7 Structures of isomers of carvone

as a whole has been found to implicate several phylogenetic mechanisms [55]. The effect of different agronomic conditions on a range of growing characteristics of spearmint, including the production of secondary metabolites, have been reviewed [86]. In addition, the effect of different fertiliser regimens and of stress induced by salinity has been studied [87]. Unlike mustard, which is milled and prepared as a dried seed powder which is activated by the addition of water, the commercial exploitation of mint makes uses the vegetative part of the plant. The development of a mint business by Colman's did not start until the early 1960s. The study of the properties of mint and other plant species was a consequence of the drawing of cultivated plants and the 'binomial nomenclature' introduced by Carl Linnaeus [88] much earlier (1707–1778). The prominent Norwich polymath, Sir Thomas Browne founded a 'physic garden' in Norwich (1667) [89] following a tradition of plant identification and characterisation (for medicinal purposes) that included the founding of the Oxford (1621) and Chelsea (1674) Physic gardens. It is likely that mint was one of the species cultivated by Browne in his garden [90]. Interestingly, given the link between Boerhaave (a prominent Calvinist and botanist who also created and cultivated a garden in Leiden) and Norwich (with Browne's reputation in medicine (Religio Medici) and his garden and its population of Flemish, protestant artisans), Browne also studied at Leiden and obtained an MD in December 1633. This predates the time Boerhaave spent at Leiden (1701–1729); however, it acknowledges the reputation of Leiden that both Browne and Boerhaave (1690–1693) studied there. In addition, some of Boerhaave's work was translated for the Sir Thomas Browne Institute in Leiden [91] and this may suggest a further link between them. Browne's Religio Medici (1643) was circulated widely in Europe and was translated into Dutch and then into Latin by a Dutchman as well as into the major European languages and would have been known to scholars at Leiden, including Boerhaave. The decision by Browne to reside in Norwich from 1637 onwards [92] rather than the university towns of Oxford or Cambridge may be an acknowledgement of the importance of a connection to Europe for Browne although he had previously studied at Oxford as well as in Europe and he graduated from Pembroke

College [93]. The links between Norwich and Norfolk and northern Europe may nonetheless have been a factor in his choice of residence.

8.3 Natural Products and Health

While a range of plant-derived natural products have been described and their health promoting properties noted, the present discussion will be primarily concerned with mustard and mint. Plants of the genus *Mentha* have a significant history of medicinal use [54]. In addition to antimicrobial and antioxidative activity (noted above) mint oil also has anti-inflammatory activity [94] and to enhance intestinal barrier function. Both mint and mustard demonstrate a link between sensory properties and consequential biological effects. In the case of mint, the interaction with taste and pain receptors has enabled a health profile to be built up that includes, in particular, digestive and respiratory effects [95], anti-inflammatory activity [94] as well as an association with freshness and cleanliness. This is most commonly underpinned by studies of the effect of mint oils on the oral microbiome [96] however there are a number of potential confounders that were identified. The effect of menthol on cooling and freshness perception has also been noted [97] as well as an impact on human performance outcomes. As noted previously, mint contains different classes of phytochemicals and this (together with synergisms) may account for some of the diverse clinical effects of crude extracts. In the case of modulation of microbial activity, components present in mint may impact upon bacterial quorum sensing or quorum quenching either directly [98] or as a consequence of induction of lactonase expression or other quorum quenching enzymes [99]. In addition, chemical quorum sensing inhibitors such as halogenated furanones have also been found to have quorum quenching activity [100] and this may be a useful area of future phytochemical investigation—particularly given the antimicrobial activity of essential oil from *Mentha* species [101]. The range of biological activities exhibited by plants of the genus *Mentha* have been reviewed [102].

In the case of mustard, the major source of biologically active compounds are glucosinolates and their breakdown products however a range of other bioactive components—in particular fatty acids and phenolic compounds have also been identified [103] and other types of compounds have also been identified [104] including increased amounts of bioactive phenolic compounds at different stages of mustard plant maturity [105]. The majority of studies have been carried out on mustard seeds since the concentration of biologically active principles is greatest in this part of the plant however, as noted above, other vegetative parts of the plant have also been analysed and assessed [30]. Because of the nature of the most potent bioactive compounds in mustard (isothiocyanates and nitriles) and because the proportions of each that are present depends on extraction conditions, the use of a range of systems including supercritical fluids [106] and deep eutectic solvents [107] have been investigated. An interesting potential physiological application for both mustard and mint is in the area of cognition. In the case of spearmint, for example, an effect on working memory has been hypothesised [108] and mint extracts more generally have been proposed to have neuroprotective and neurochemical properties [109]. Many studies have been relatively poorly designed and lack mechanistic rigour and therefore a critical evaluation is necessary. For example, potential synergistic effects between chewing gum and mint flavour have been suggested but the robustness of effects have been questioned [110]. In addition, a review of the effects of mastication in a different psychological area (self-reported hunger) would suggest that mechanistic information is lacking [111] and it will, therefore be challenging to deconvolute the psychological or cognitive effects of mint or some of its components in isolation [112] and to ensure that the results are consistent between different model systems. For example, in the case of brown mustard (*Brassica juncea*) an effect on cognition has been reported [113] however this was a study carried out on rats rather than humans and the effect was observed in a constrained model system comprising rats challenged with scopolamine or with induced diabetes. The effects of specific active components from mustard such as allyl isothiocyanate have been attributed to a specific effect [114] on glycerophospholipids in ovariectomized mice (again a highly specialised model with questionable relevance to a human population). Erucic acid, a known component of mustard oil, has been proposed to have an effect on the pathogenesis of Huntington's disease [115] but much of the work has been validated in animal models. An alternative mechanism may be a consequence of the pro-inflammatory pathology of the brains of individuals suffering from Alzheimer's disease [116] and the anti-inflammatory Activity ascribed to components of both mint and mustard. While anti-inflammatory activity has been suggested for mustard in the context of oral pathogens [117] and the transmission of such pathogens from the oral cavity to the brain has been proposed [118] the causal, mechanistic link (rather than association) remains unproven. Nonetheless, the possible implication of mustard in the prevention of cognitive decline together with the proposed effects of mint and chewing on the development of cognition, suggest that chewing gum containing both menthol and allyl isothiocyanate may have some efficacy in optimising cognition by different, complementary mechanisms.

8.4 Biosynthesis: Glucosinolates and Terpenes

Much of the interest in the biosynthesis of glucosinolates and terpenes derives from fairly recent attempts to understand the role and to then develop methods for the manipulation of secondary metabolites. Previously, plant breeding was oriented towards optimisation of yields, improvements in agronomy and ease of processing. Secondary metabolites were comparatively little considered because their manipulation did not generally increase the utility or value of a crop—except for a few specific characteristics such as the production of colour compounds. The exceptions to this general position arose from the realisation that some secondary metabolites had pharmacological activity [119], and others had desirable taste profiles [120]. In the case of most glucosinolates, the biosynthetic precursor is an amino acid which is modified prior to the incorporation of the thioglucose and sulphonated oxime moieties (Fig. 8.8).

The immediate precursor of the glucosinolate is the corresponding desulpho compound which was originally thought to simply not be a substrate for the co-occurring thioglucosidase enzyme (myrosinase) but was later found to be an inhibitor of the hydrolytic enzyme [121]. The production of (particularly) isothiocyanates by the enzymatic hydrolysis of glucosinolates (Fig. 8.9) gives glucosinolates a toxicity profile that may deter consumption by animal and insect herbivores—both of which may contain myrosinase (either constitutively or as a consequence of colonisation by gut bacteria [122]) which may catalyse the *in situ* formation of toxic volatile compounds.

Some microbial species contain sulphatases [123] and the presence of these microbes in the gut of consuming species could result in the formation of desulpho-glucosinolates and the inhibition of plant-derived or endogenous myrosinases. In addition, some insects also secrete a sulphatase [124] and these detoxifying enzymes are also found in pests such as flea beetle [125]. The snail, *Helix pomatia*, contains a glucosinolate sulphatase in such abundance and activity that it is isolated for commercial use [126, 127]. One of the starting points of any programme that relies on manipulation of a biosynthetic pathway is an assessment of natural cultivars [128], their properties and the development of hybrids with particular characteristics [129]. It may also be important to modify the toxicological profile of the part of the plant that is a target [26]. In addition, in the case of glucosinolates, their impact upon predation and growth of a plant may mean that there are unexpected consequences of modulation of secondary metabolite production and this includes the complete

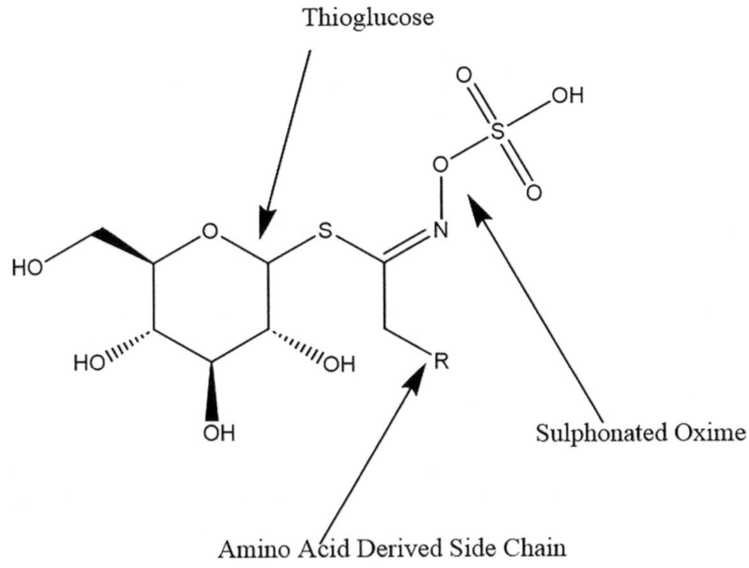

Fig. 8.8 Structural characteristics of glucosinolates

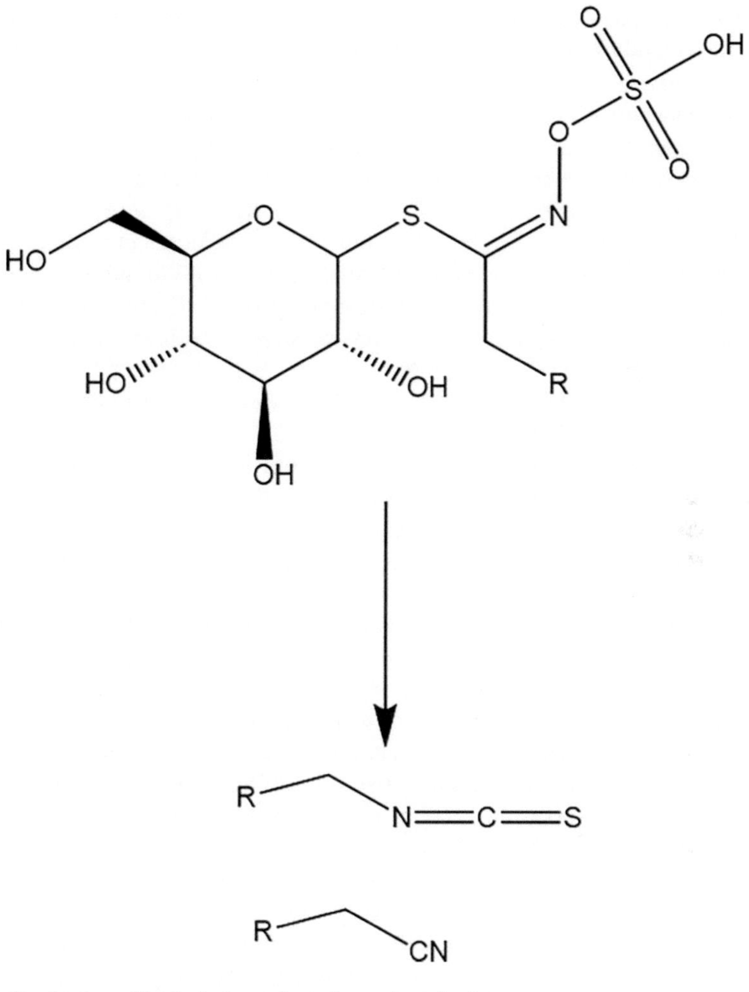

Fig. 8.9 Production of hydrolysis products from glucosinolates

glucosinolate/myrosinase system [130]. The use of glucosinolates and their breakdown products as natural plant protective factors may drive the optimisation of production for endogenous or exogenous use [131]. In the case of mustard seed, alternative processing methods may also affect the secondary metabolite profile of the consumed product [132] so this should (together with the potential creation of novel hybrids with unanticipated properties) be considered before embarking on a biosynthesis modification programme. Ultimately, however, a combination of metabolite and transcriptome analysis will provide a more complete picture of glucosinolates, their genetic manipulation and the consequences of that action [133]. While cooking and processing as well as biosynthesis can have a significant effect on glucosinolate levels, environmental and growing conditions can also be impactful [134] as can abiotic stress [135].

The biosynthesis of the volatile active components (terpenoids/isoprenoids) in mint is, in principle simpler since it does not include a post-biosynthetic reaction as does the synthesis of mustard isothiocyanates. Isoprenoids are derived from a C5-isoprene subunit (2-methyl-1,3-butadiene). As noted previously, mint is a member of the genus *Mentha* and of the family *Lamiaceae*. Much of the interest in the manipulation of the terpenes in the plants is a consequence of their biological activity [136] and flavour properties [137]. As with mustard, before embarking on the manipulation of biosynthetic pathways, the natural variation in cultivars and the possibility of creating hybrids should be investigated [138] along with analysis of gene flow as a consequence of the hybridisation process. If the route of changing biosynthetic pathways is chosen then care must be exercised since changing or increasing the metabolic flow towards secondary metabolites may also impact on primary metabolism. Some of the essential oil components of plants of the genus *Mentha* are illustrated in Fig. 8.10.

The economically predominant varieties within the genus are *Mentha spicata* (spearmint or common mint) which typically contains carvone, peppermint (a natural cross between *M. spicata* and *M. aquatica*) and *M. arvensis* (corn mint) which contains high levels of menthol. While a detailed consideration of the complex genetics of mint lies outwith this chapter, variations in the types and amounts of monoterpenes present have led to the isolation and characterisation of various chemotypes—for example the linalool-rich chemotype, which has been reported for *M. spicata* [139], the different profiles of *M. arvensis* [140] and *Mentha longifolia* [141] among others. This complexity in natural product formation can complicate the design of hybrid or transgenic experiment. Another example is the discovery of three chemotypes of *M. spicata* which had oil profiles dominated by linalool, carvone or piperitone oxide or piperitenone oxide. Similar essential oil chemical diversity is observed in other *Mentha* species. As with mustard, agronomic conditions (particularly stress) can affect secondary metabolite production in mint [142], however homologous and heterologous manipulation of pathway enzymes has been reviewed and potential directions for future studies noted [143]. Such reviews are fairly common and are a useful source of information [144]. Much of the work that led to the identification of specific pathways resulted from the analysis of the results of extensive natural crossing experiments to produce hybrids [145] which is analogous to the work on peas carried out by Mendel [146] and later the elucidation of functional characteristics leading to plants with improved agronomic characteristics [147] and the genetic basis of this and other characteristics [148]. The development of chemical (in the case of mint) or phenotypic (in the case of pea) characterisation can be seen to develop from the application of observational and inductive science regarding natural product biosynthesis to the development of theories concerning the molecular basis of particular aspects of the growth of living systems. This then enables the selective choice of the properties of systems based upon molecular manipulation through genetic engineering or gene editing. Such studies have been developed in Norwich at the John Innes Centre [149] and these follow on from the tradition of the study of agronomic practice that led to the development of, for example, the Norfolk crop rotation system [150].

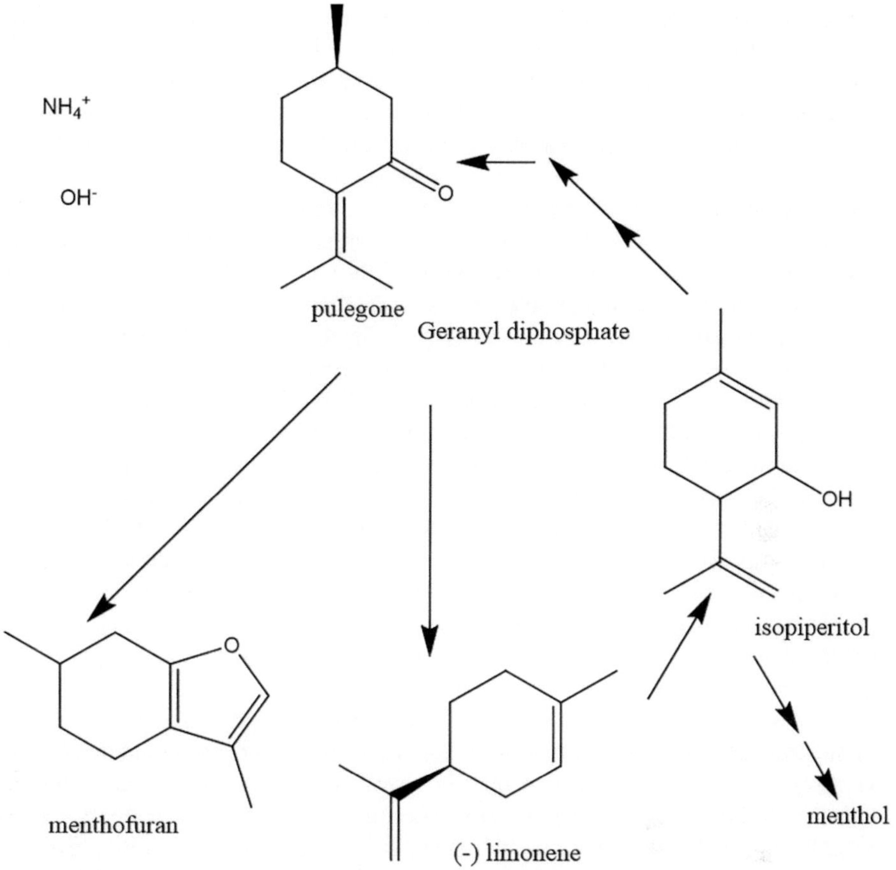

Fig. 8.10 Essential oils of mentha species and their relationships

8.5 Taste Receptors: Oral and Extra-Oral Significance

Taste receptors can, in some ways, be considered to be the gatekeepers of consumption of foods. The physiological importance of the sense of taste has been noted and extensively reviewed [151]. One key insight was that receptors can be found in places other than the oral cavity (extra-oral or non-gustatory taste receptors) and have functions beyond that of taste perception [152] and this realisation is crucial to the establishment of an understanding of the physiological role of taste receptors. Since many of the compounds that are ligands for taste receptors are secondary metabolites, this raises the question of how the distribution of taste receptors may affect the biological role of secondary metabolites and whether this role may extend beyond the aversive properties that mediate consumption. Even within the physiological boundaries of the sense of taste, there are interactions and crosstalk between different taste receptors and this can have physiological consequences [153].

Distinct types of taste (sweet, salty, umami, bitter, and sour) are mediated by different classes of receptor which are located, orally, on different types of cells in taste buds [154]. While salty and sour receptors represent ion channels, the receptors for sweet, umami and bitter belong to the G protein-coupled receptor superfamily. Perhaps, therefore of greatest interest in terms of taste modulation are Type II cells which detect and transduce the more complex (in terms of receptor design) sweet, bitter, and umami stimuli. These receptors are G protein-coupled receptors (GPCRs) and, apart from their differential ligand binding characteristics, the downstream signalling cascades are the same [155]. The GPCRs are membrane bound and have trans membrane regions. The outer membrane sections contain the ligand binding domains which are activated by the binding of a ligand and linked intracellular heterotrimeric G proteins initiate a cascade of downstream second messenger pathways. In the case of the sweet receptor (see below) knowledge of the binding characteristics of different sweet molecules (including artificial sweeteners) may be of value in the design of novel sweet agents [156]. Depending on the location of the receptor, the second messengers can produce a number of outcomes so some designer ligands may have extra-oral activity and hence novel properties. The GPCR taste receptors are classified according to structure and function and some awareness of this is important in the design of novel ligands. There are three TAS1 receptors and these form heterodimers to give the umami taste receptor, TAS1R1/TAS1R3 and the sweet taste receptor, TAS1R2/TAS1R3. These have long extracellular domains that are typical for class C GPCRs. There are a larger number of bitter receptors which are defined as Type 2 GPCRs (TAS2Rs) and all of these have short amino terminal sequences. Sweet, bitter and umami receptors are found throughout the gastrointestinal (GI) tract and may be involved in a number of physiological effects including nutrient sensing. They are also found in enteroendocrine cells where ligand binding results in the release of important peptide hormones involved in metabolic regulation and satiety, such as GLP-1 and ghrelin. Because of the number of different members of the class and the concomitant number of potential ligands that may bind, of the different types of taste receptors, bitter receptors (T2Rs) have probably attracted most interest [157]. In addition to being located throughout the GI tract, bitter receptors have particularly been found in the respiratory tract where they have been described as "gatekeepers of the airway epithelium" [158]. One role of bitter taste receptors in airways appears to be modulation of response to bacterial and other pathogens and infection. Taste receptors are also activated by quorum sensing molecules such as homoserine lactones [159, 160] and may thereby stimulate an immune response. One other role of TAS2Rs is in the excretion of bitter (potentially toxic) molecules from cells through the ABCB1 ligand transporter system [161] through a mechanism involving CCK [162]. Modulation of the ABCB1 transporter may be associated with multi-drug resistance—particularly in the stimulation of the efflux from cells of anti-cancer drugs [163]. The binding of the bitter component of *Brassica* vegetables (goitrin) to bitter receptors (particularly TAS2R38) [164] and the discovery of this receptor being associated with pancreatic cancer cells and its activation being associated with increased expression of ABCB1 [165] suggests that intake of *Brassica* vegetables might be associated with a diminished chance of recovery from cancer due to enhanced drug efflux from cells.

8.6 Herbicides and Pesticides. May and Baker in Norwich

Among the most important science-based series of events of the twentieth century were those that led to the Green Revolution. Much of the impact of the Green Revolution in the middle of the twentieth century was, in part, a consequence of the invention of ways to make fertilisers and plant protection products and the development of higher yielding varieties of crops [166] however a more sustainable agriculturally based green revolution that makes use of plant breeding and genetic manipulation, such as that developed at the JIC in Norwich, is possible and may circumvent some of the sustainability and other issues associated with the Green Revolution adopted in the mid-twentieth century [167]. A number of these approaches will be discussed further in Sect. 8.7. In addition, the use of biological agents to control the growth and performance of undesirable crop predators (including insects, microbial species and competing weeds) has been reviewed [168]. May and Baker was a speciality chemical company registered in London as a limited liability company in December 1890. It became a subsidiary of Poulenc Freres in 1922 but continued to trade under the name May and Baker. The company became active in pharmaceuticals and the sulphonamide, sulphapyridine (M&B 693), was developed in 1937 and was used extensively as an anti-infective agent. After the war, the company expanded into three divisions (pharmaceuticals, fine chemicals and agrochemicals) and to support the last of these a new factory was opened in 1955. May and Baker initially specialised in the synthesis of chemicals that contained the heavy metals, mercury and bismuth. One of the most common preparations containing bismuth is the antacid. Pepto-Bismol (bismuth subsalicylate) whose structure was only solved some hundred years after its introduction [169]. Pepto-bismol was first introduced as a cure for infant diarrhea caused by cholera in the early twentieth century [170] by the Norwich Pharmacal Company of New York. The institutional expertise of the company in the use of metals continued across other parts of the organisation—including agrochemicals [171]. In 1986 analysis of riverbed sediments near Norwich showed the presence of contamination by copper and mercury derived from the May and Baker plant and this was raised in the UK Parliament. This type of contamination is of particular concern in ecologically rare environments such as the rivers in Norfolk, many of which are derived from chalk springs. Indeed, the river Wensum has Site of Special Scientific Interest and Special Area of Conservation status for its entire length. Discharges either into the river or which may find their way to the river must be carefully monitored and controlled. May and Baker was also a significant producer of herbicides including the first to be manufactured in Norwich, tropotox and others such as bromoxynil [172] and ioxynil (Fig. 8.11).

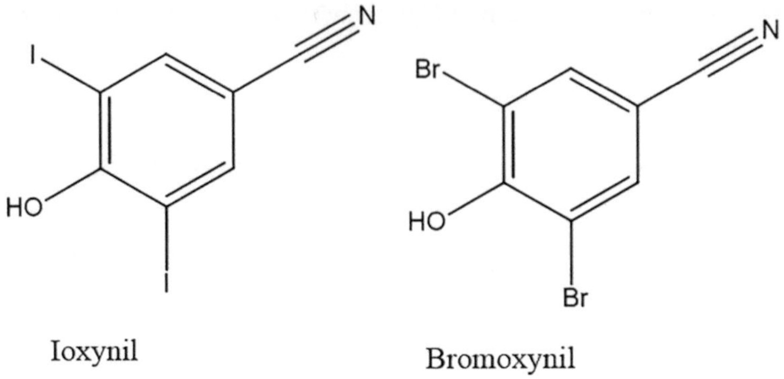

Fig. 8.11 Structures of bromoxynil and ioxynil

8.7 Plant Protective Factors: Natural and Synthetic

The development of initiatives such as crop rotation to improve crop yield and reduce the prevalence of pests and disease has already been mentioned (Chap. 2) and will not, therefore be repeated. Some initiatives such as the use of seaweed and fishwaste as fertilisers are of particular relevance to Norfolk since it is an agricultural/marine region. Seaweed and seaweed-based functional metabolites have been used to supplement feed for fish [173]. In particular, some components have been hypothesised to improve serum immune and antioxidant status and disease resistance in fish when incorporated in feed as a supplement. In addition, seaweed-based liquid fertilisers have been used to improve the growth and yield of terrestrial crops [174] and fish waste-based fertilisers have been used in the production of organic crops [175]. Fish by-products (particularly protein hydrolysates) have been proposed as biostimulants to promote plant growth [176] and this, together with nutrient retention by recycling of fish waste [177] is seen as a viable agronomic option. In addition to the potential use of fish waste products as plant growth stimulants, the microbial deterioration of fish is a problem with, globally, up to 30–40% of fish that is caught being discarded because of microbial contamination [178] and this presents a further sustainability challenge. Attempts to control this issue through the use of microbial quorum sensing inhibitors to prevent the growth of spoilage organisms has been proposed [179]. Acyl homoserine lactones (AHLs, autoinducer-1/AI-1), alkylquinolones (AQs), and α-hydroxyketones (cholera autoinducer-1/CAI-1) constitute the main classes of QS compounds in marine ecosystems. The marine environment is complex and the interplay between quorum sensing and inhibition is an important aspect of the maintenance of the marine environment and microenvironment [180]. For terrestrial plants, the development of agents that favour the growth of particular plants (crops) over others (weeds) is an important aspect of agronomy. As is the case with antibiotic resistance, the development of resistance to selective herbicides is a major challenge [181]. Omics technology has been used to investigate herbicide resistance mechanisms [182] however multiple resistance pathways may operate and this can provide a significant challenge. The essence of the action of herbicides is the identification of vulnerable aspects in the biochemical pathways of weeds. In many cases, however, the precise nature of the mechanism of action is unclear [183]. A mechanistic description of all the possible modes of action lies outwith the scope of this chapter however, unsurprisingly, photosynthesis is an important target. The development of pharmaceutical-type tools including the use of artificial intelligence and modelling studies has been advocated [184] and progress is likely to be achieved through the use of such tools. The identification of new natural phytotoxins has also been proposed [185] however the identification and isolation of natural chemicals with novel modes of action has met with very limited success. Nonetheless, attempts have continued to try to find ways of using non-synthetic (natural) compounds as herbicides [186]. The mechanistic similarity between the development of resistance to herbicides and responses to abiotic stress (particularly in terms of multiple crossover, linked resistance mechanisms) suggests that climate change and other stress events may also increase the resistance of weeds to herbicide treatment [187]. One of the most significant (and impactful) areas of the use of herbicides has been the development of strategies with a basis in genetic manipulation to confer resistance to herbicides on crop plants. This can almost be considered as a way of creating resistance in a controlled and specific manner. The concept of purposefully genetically manipulated herbicide-tolerant crops used in conjunction with specific herbicides has been the subject of intense debate [188] and an effect on other health factors—particularly in vulnerable populations such as peri-natal health has been suggested though causality has not been demonstrated [189]. Nonetheless the increase in herbicide use as a consequence of the introduction of genetically modified, herbicide resistant crops has fuelled some concerns [190]. The two major types of plant protection used are the herbicide glyphosate and an insecticidal toxin (Bt) derived from *Bacillus thuringiensis.* Glyphosate disrupts the shikimate pathway in plants and microorganisms thereby preventing the biosynthesis of aromatic amino acids [191]. There is no analogous pathway in higher animals such as humans and so it ought to be benign and a number of studies have suggested no chronic or acute effects using a rodent model [192]. The activity of Bt toxin is insecticidal and so will be considered in more detail below. Glyphosate resistance can develop as a consequence of the occurrence of a mutation in the target enzyme: 5-enolypyruvylshikimate-3-phosphate synthase (EPSPS) or by other mechanisms [193] and some of these may impact upon the effectiveness of glyphosate in controlling weeds. In addition to transgenic technology, gene editing has been proposed as a way of creating herbicide-resistant crops [194]. As herbicide resistance has increased in prevalence, multi-resistant crops (with so-called ‘stacked genes’) have been developed however the development of new ways to combat the spread of natural resistance will require that herbicides with novel modes of action be developed. Natural phytotoxins from plant and microbial species are considered to be one progressive way forward [195] but it will be important to develop deeper knowledge of plant (particularly weed) physiology and biochemistry if such approaches are to be identified and developed. Through a study of the mechanisms that have developed through evolution to ensure the primacy of certain plant species, it may be possible to design novel approaches. In the case of other plant protection products, the development of mechanistically based treatments has a long and varied history. In particular, insecticidal products such as dichlorodiphenyltrichloroethane (DDT), which was first synthesised in 1874 by Zeidler [196], although its insecticidal activity was not discovered until 1939, were developed. It was the first of the synthetic insecticides and was followed by the organophosphates, methylcarbamates, and pyrethroids—all of which are neuroactive chemicals. Previously a range of toxic entities including sulphur, arsenates and flowers from the pyrethrum plant which contain the natural insecticide, pyrethrin (Fig. 8.12) were used.

While a range of chemicals have been developed, resistance can build quickly in target species and novel approaches including modified and specific baculoviruses [197], entomopathogenic bacteria, neonicotinoids, DNA insecticides [198] and insecticides based on the toxin produced by *Bacillus thuringiensis.* In the case of this last group, the bacterium synthesizes parasporal crystalline inclusions containing Crystal (Cry) and Cytotoxic (Cyt) proteins, some of which are toxic against a wide range of insect orders and nematodes. The mechanism of action is known and has been studied [199]. Transgenic crops that express the Bt toxin have been created [200] and, while they are generally regarded as safe, concerns have been raised about potential problems—particularly with stacked gene systems [201]. The use of antifeedant strategies to control insect pests has been suggested [202] and the possibility of using plant extracts as nematicides has been proposed [203]. As with herbicides and developing optimal conditions for the growth of crop plants, investigation of the physiology of insect pests will be required to optimise approaches.

The continued and future success of arable farming in Norfolk (and elsewhere) will increasingly depend on the development of a more complete understanding of whole agronomic systems (including interactions between different components) and not simply the development of strategies to maximise yield and improve the harvesting and processability of a few main crops. The development of techniques

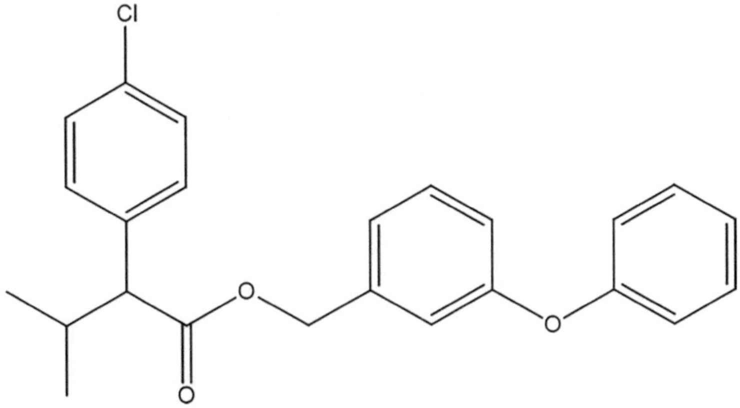

Fig. 8.12 Structure of pyrethrin

such as precision fermentation [204] will change the food production landscape and provide new opportunities for the creation of foods with specific and programmable qualities beyond the most effective production of macronutrients.

References

1. Bharate SW, Lindsley CW. Natural products driven medicinal chemistry. J Med Chem. 2024;67(23):20723–30.
2. Kim JH, Kismali G, Gupta SC. Natural products for the prevention and treatment of chronic inflammatory diseases: integrating traditional medicine into modern chronic diseases care. Evid Based Complement Alternat Med. 2018;2018:9837863. https://doi.org/10.1155/2018/9837863.
3. Matsuo Y. Natural products and cancer. Nutrients. 2023;15(24):5064. https://doi.org/10.3390/nu15245064.
4. Gropper SS. The role of nutrition in chronic disease. Nutrients. 2023;15(3):664. https://doi.org/10.3390/nu15030664.
5. Adolph TE, Tilg H. Western diets and chronic diseases. Nat Med. 2024;30:2133–47. https://doi.org/10.1038/s41591-024-03165-6.
6. Dai S, Wellens J, Yang N, Li D, Wang J, Wang L, Yuan S, He Y, Song P, Munger R, Kent MP, MacFarlane AJ, Mullie P, Duthie S, Little J, Theodoratou E, Li X. Ultra-processed foods and human health: an umbrella review and updated meta-analyses of observational evidence. Clin Nutr. 2024;43(6):1386–94. https://doi.org/10.1016/j.clnu.2024.04.016.
7. Williams MP, Huber CD. The genomic footprints of migration: how ancient DNA reveals our history of mobility. Genome Biol. 2025;26:206. https://doi.org/10.1186/s13059-025-03664-w.
8. Law CM. The growth of urban population in England and Wales, 1801-1911. Trans Inst Br Geogr. 1967;41:125–43.
9. Maca-Meyer N, Cabrera VM, Arnay M, Flores C, Fregel R, González AM, Larruga JM. Mitochondrial DNA diversity in 17th-18th century remains from Tenerife (Canary Islands). Am J Phys Anthropol. 2005;127(4):418–26. https://doi.org/10.1002/ajpa.20148.
10. Kay GL, Sergeant MJ, Zhou Z, Chan JZ, Millard A, Quick J, Szikossy I, Pap I, Spigelman M, Loman NJ, Achtman M, Donoghue HD, Pallen MJ. Eighteenth-century genomes show that mixed infections were common at time of peak tuberculosis in Europe. Nat Commun. 2015 Apr;7(6):6717. https://doi.org/10.1038/ncomms7717.
11. Leslie S, Winney B, Hellenthal G, Davison D, Boumertit A, Day T, Hutnik K, Royrvik EC, Cunliffe B, Lawson DJ, Falush D, Freeman C, Pirinen M, Myers S, Robinson M, Donnelly P, Bodmer W. The fine-scale genetic structure of the British population. Nature. 2015;519(7543):309–14. https://doi.org/10.1038/nature14230.
12. Karczewski KJ, Gupta R, Kanai M, et al. Pan-UK biobank genome-wide association analyses enhance discovery and resolution of ancestry-enriched effects. Nat Genet. 2025; https://doi.org/10.1038/s41588-025-02335-7.
13. Joby C. Law and order in exile communities in early modern Norfolk. History. 2025;110:227–43. https://doi.org/10.1111/1468-229X.13437.
14. Fay I. A note about the population of Norwich, 1100–1600. In: Health and the City: disease, environment and government in Norwich, 1200–1575. Rochester: Boydell & Brewer; 2015. p. 205.
15. Larmuseau M, Vanoverbeke J, Gielis G, et al. In the name of the migrant father—analysis of surname origins identifies genetic admixture events undetectable from genealogical records. Heredity. 2012;109:90–5. https://doi.org/10.1038/hdy.2012.17.

16. Stefanaki A, Walter T, van Andel T. Tracing the introduction history of the tulip that went wild (Tulipa sylvestris) in sixteenth-century Europe. Sci Rep. 2022;12:9786. https://doi.org/10.1038/s41598-022-13378-9.
17. Heniger J. Some botanical activities of Herman Boerhaave, professor of botany and director of the botanic garden at Leiden. Janus. 1971;58:1–78.
18. Offerhaus A, van Andel TR, Stefanaki A. What does the "true" Boerhaave herbarium tell us about the practice of collecting plant specimens in the botanical garden Leiden? Bauhinia Z Basler Bothanisch Gesellsch. 2023;29:129–30. https://doi.org/10.12685/bauhinia.1366.
19. Fiddes PJ, Komesaroff PA. Herman Boerhaave's clinical teaching: a story of partial historiography. J Bioeth Inq. 2023;20(2):295–313. https://doi.org/10.1007/s11673-023-10244-9.
20. Baldassarri F. Plants in 16th and 17th century: botany between medicine and science. Berlin: De Gruyter; 2023. https://doi.org/10.1515/9783110739930.
21. Drobnik J, Stebel A. Central European medicinal bryophytes in the 16th-century work by Caspar Schwenckfeld, and their ethnopharmacological origin. J Ethnopharmacol. 2015;175:407–11. https://doi.org/10.1016/j.jep.2015.09.033.
22. Zeven AC, Brandenburg WA. Use of paintings from the 16th to 19th centuries to study the history of domesticated plants. Econ Bot. 1986;40(4):397–4080.
23. Adams M, Berset C, Kessler M, Hamburger M. Medicinal herbs for the treatment of rheumatic disorders–a survey of European herbals from the 16th and 17th century. J Ethnopharmacol. 2009;121(3):343–59. https://doi.org/10.1016/j.jep.2008.11.010.
24. Chen VJ, Smith GD, Holden A, Arft S. Dye analysis of a 17th-century Mattens workshop tapestry based on Raphael's drawing. Miracul Draught Fish Heritage. 2024;7:1221–36. https://doi.org/10.3390/heritage7030059.
25. Speleers L, Vanden Berghe I, Ghesquière V, et al. Colourful rivers: archaeobotanical remains of dye plants from urban fluvial deposits in the southern low countries (Belgium). Veget Hist Archaeobot. 2025;34:291–306. https://doi.org/10.1007/s00334-024-01009-4.
26. Lietzow J. Biologically active compounds in mustard seeds: a toxicological perspective. Foods. 2021;10(9):2089. https://doi.org/10.3390/foods10092089.
27. Henry ÉO, Garot. Sur la Semence de Moutarde Blanche (Sinapis alba), et sur la Matière Cristallisable qui s'y Trouve (Sulfosinapisine) Désignée d'Abord sur le Nomd'Acide Sulfosinapique. J Pharm. 1831;17:1–16.
28. Borek V, Morra MJ. Ionic thiocyanate (SCN−) production from 4-hydroxybenzyl glucosinolate contained in Sinapis alba seed meal. J Agric Food Chem. 2005;53:8650–4.
29. Sircar S, Kansra U. Choice of cooking oils–myths and realities. J Indian Med Assoc. 1998;96(10):304–7.
30. Frazie MD, Kim MJ, Ku KM. Health-promoting phytochemicals from 11 mustard cultivars at baby leaf and mature stages. Molecules. 2017;22(10):1749. https://doi.org/10.3390/molecules22101749.
31. Garg S, Gairola K, Punetha H, Gangola S. An exploration of the biochemistry of mustard seed meals: a phytochemical and in silico perspective. Foods. 2024;13(24):4130. https://doi.org/10.3390/foods13244130.
32. Lin Y, Yang X, Lu Y, Liang D, Huang D. Isothiocyanates as H2S donors triggered by cysteine: reaction mechanism and structure and activity relationship. Org Lett. 2019;21(15):5977–80. https://doi.org/10.1021/acs.orglett.9b02117.
33. Xiao Q, Ying J, Xiang L, Zhang C. The biologic effect of hydrogen sulfide and its function in various diseases. Medicine. 2018;97(44):e13065. https://doi.org/10.1097/MD.0000000000013065.
34. Karlsson I, Samuelsson K, Ponting D, et al. Peptide reactivity of Isothiocyanates—implications for skin allergy. Sci Rep. 2016;6:21203. https://doi.org/10.1038/srep21203.
35. Mi L, Xiao Z, Veenstra TD, Chung FL. Proteomic identification of binding targets of isothiocyanates: a perspective on techniques. J Proteome. 2011;74(7):1036–44. https://doi.org/10.1016/j.jprot.2011.04.015.

36. Eib S, Schneider DJ, Hensel O, Seuß-Baum I. Relationship between mustard pungency and allyl-isothiocyanate content: a comparison of sensory and chemical evaluations. J Food Sci. 2020;85(9):2728–36. https://doi.org/10.1111/1750-3841.15383.
37. Daxenbichler ME, Spencer GF, Carlson DG, Rose GB, Brinker AM, Powell RG. Glucosinolate composition of seeds from 297 species of wild plants. Phytochemistry. 1991;30:2623–38. https://doi.org/10.1016/0031-9422(91)85112-D.
38. Eib S, Schneider DJ, Hensel O, Seuß-Baum I. Evaluation of trigeminal pungency perception of allyl isothiocyanate—a time intensity (TI) study. Food Qual Prefer. 2021;87:104039. https://doi.org/10.1016/j.foodqual.2020.104039.
39. Bussy A. Untersuchungen über die Bildung des ätherischen Senföls. Justus Liebigs Ann Chem. 1840;34:223–30. https://doi.org/10.1002/jlac.18400340211.
40. Gadamer J. Ueber das Sinigrin. Ber Dtsch Chem Ges. 1897;30:2322–7. https://doi.org/10.1002/cber.189703002220.
41. Waser J, Watson W. Crystal structure of Sinigrin. Nature. 1963;198:1297–8. https://doi.org/10.1038/1981297b0.
42. Jin X, Wang J, Li D, Wu F, Zhou X. Rotations with Indian mustard and wild rocket suppressed cucumber fusarium wilt disease and changed rhizosphere bacterial communities. Microorganisms. 2019;7(2):57. https://doi.org/10.3390/microorganisms7020057.
43. Zhu J, Ren Z, Huang B, Cao A, Wang Q, Yan D, Ouyang C, Wu J, Li Y. Effects of fumigation with allyl Isothiocyanate on soil microbial diversity and community structure of tomato. J Agric Food Chem. 2020;68(5):1226–36. https://doi.org/10.1021/acs.jafc.9b07292.
44. Halkier BA, Gershenzon J. Biology and biochemistry of glucosinolates. Annu Rev Plant Biol. 2006;57(1):303–33.
45. Lambrix V, Reichelt M, Mitchell-Olds T, Kliebenstein DJ, Gershenzon J. The Arabidopsis epithiospecifier protein promotes the hydrolysis of glucosinolates to nitriles and influences Trichoplusia in herbivory. Plant Cell. 2001;13(12):2793–807.
46. Li JJ. Lossen rearrangement. In: Name reactions. Cham: Springer; 2021. https://doi.org/10.1007/978-3-030-50865-4_85.
47. Uda Y, Kurata T, Arakawa N. Effects of pH and ferrous ion on the degradation of glucosinolates by myrosinase. Agric Biol Chem. 1986;50:2735–40.
48. Baldwin WO. Remarks on mustard poultices, applied extensively to the surface. West J Med Surg. 1845;3(1):1–11.
49. Goetz K, Hinz A, Steinhäuser J, von Rath U. Use of mustard seed footbaths for respiratory tract infections: a pilot study. Evid Based Complement Alternat Med. 2020;2020:5648560. https://doi.org/10.1155/2020/5648560.
50. Mysore V, Arghya A. Hair oils: indigenous knowledge revisited. Int J Trichol. 2022;14(3):84–90. https://doi.org/10.4103/ijt.ijt_189_20.
51. Al-Rajhi AMH, Alsalamah SA, Mashlawi AM, et al. Assessment of mustard oil and its ozonated form for their antimicrobial potency and their antioxidant and anti-inflammatory potential with molecular docking study. Appl Biol Chem. 2025;68:84. https://doi.org/10.1186/s13765-025-01058-x.
52. Mazumder A, Dwivedi A, du Plessis J. Sinigrin and its therapeutic benefits. Molecules. 2016;21(4):416. https://doi.org/10.3390/molecules21040416.
53. Turner ME, Beckett JV, Afton B. Farming practice and techniques. In: Farm production in England 1700–1914. Oxford: Oxford Academic; 2001.
54. Silva H. A descriptive overview of the medical uses given to Mentha aromatic herbs throughout history. Biology. 2020;9(12):484. https://doi.org/10.3390/biology9120484.
55. Mint Evolutionary Genomics Consortium. Phylogenomic mining of the mints reveals multiple mechanisms contributing to the evolution of chemical diversity in lamiaceae. Mol Plant. 2018;11(8):1084–96. https://doi.org/10.1016/j.molp.2018.06.002.
56. Memon MA, Memon HA, Muhammad FE, Fahad S, Siddiqui A, Lee KY, Tahir MJ, Yousaf Z. Aetiology and associations of halitosis: a systematic review. Oral Dis. 2023;29:1432–8. https://doi.org/10.1111/odi.14172.

57. Tangerman A, Winkel EG. Volatile sulfur compounds as the cause of bad breath: a review. Phosphorus Sulfur Silicon Relat Elem. 2013;188(4):396–402. https://doi.org/10.1080/10426507.2012.736894.
58. Sauerland M, Mertes R, Morozzi C, Eggler AL, Gamon LF, Davies MJ. Kinetic assessment of Michael addition reactions of alpha, beta-unsaturated carbonyl compounds to amino acid and protein thiols. Free Radic Biol Med. 2021;169:1–11. https://doi.org/10.1016/j.freeradbiomed.2021.03.040.
59. Huang CW, Moore PK. H2S synthesizing enzymes: biochemistry and molecular aspects. Handb Exp Pharmacol. 2015;230:3–25. https://doi.org/10.1007/978-3-319-18144-8_1.
60. Yoshimura M, Nakano Y, Yamashita Y, Oho T, Saito T, Koga T. Formation of methyl mercaptan from L-methionine by Porphyromonas gingivalis. Infect Immun. 2000;68(12):6912–6. https://doi.org/10.1128/IAI.68.12.6912-6916.2000.
61. Ouhara K, Iwasaki Y, Kajiya M, Savitri IJ, Kitagawa M, Tokunaga N, Shintani T, Ogawa I, Hino T, Fujita T, Shiba H, Kurihara H. The differential expression of mgl mRNA by Porphyromonas gingivalis affects the production of methyl mercaptan. Oral Dis. 2015;21(5):626–33. https://doi.org/10.1111/odi.12326.
62. Eschliman K, Bossmann SH. Synthesis of isothiocyanates: an update. Synthesis. 2019;51(8):1746–52. https://doi.org/10.1055/s-0037-1612303.
63. Tian M, Hanley AB, Dodds MW, Yaegaki K. Chewing gum containing allyl isothiocyanate from mustard seed extract is effective in reducing volatile sulfur compounds responsible for oral malodor. Am J Dent. 2013;26(4):180–4.
64. Foo LH, Balan P, Pang LM, Laine ML, Seneviratne CJ. Role of the oral microbiome, metabolic pathways, and novel diagnostic tools in intra-oral halitosis: a comprehensive update. Crit Rev Microbiol. 2021;47(3):359–75. https://doi.org/10.1080/1040841X.2021.1888867.
65. Wade WG. Resilience of the oral microbiome. Periodontol 2000. 2021;86(1):113–22. https://doi.org/10.1111/prd.12365.
66. Li ZY, Zhou XD, Peng X. Regulation of oral microbiota homeostasis and oral diseases. Zhonghua Kou Qiang Yi Xue Za Zhi. 2024;59(2):124–9. https://doi.org/10.3760/cma.j.cn112144-20231012-00189.
67. Petrovska BB. Historical review of medicinal plants' usage. Pharmacogn Rev. 2012;6(11):1–5. https://doi.org/10.4103/0973-7847.95849.
68. Krishnamurti C, Rao SC. The isolation of morphine by Serturner. Indian J Anaesth. 2016;60(11):861–2. https://doi.org/10.4103/0019-5049.193696.
69. Hosztafi S. The discovery of alkaloids. Pharmazie. 1997;52(7):546–50.
70. Drobnik J, Drobnik E. Timeline and bibliography of early isolations of plant metabolites (1770-1820) and their impact to pharmacy: a critical study. Fitoterapia. 2016;115:155–64. https://doi.org/10.1016/j.fitote.2016.10.009.
71. Shirakawa M, Ueda H, Shimada T, Hara-Nishimura I. Myrosin cells are differentiated directly from ground meristem cells and are developmentally independent of the vasculature in Arabidopsis leaves. Plant Signal Behav. 2016;11(4):e1150403. https://doi.org/10.1080/15592324.2016.1150403.
72. Thangstad OP, Bones AM, Holtan S, Moen L, Rossiter JT. Microautoradiographic localisation of a glucosinolate precursor to specific cells in Brassica napus L. embryos indicates a separate transport pathway into myrosin cells. Planta. 2001;213(2):207–13. https://doi.org/10.1007/s004250000491.
73. Rajakumar T, Pugalendhi P. Allyl isothiocyanate regulates oxidative stress, inflammation, cell proliferation, cell cycle arrest, apoptosis, angiogenesis, invasion and metastasis via interaction with multiple cell signaling pathways. Histochem Cell Biol. 2024;161(3):211–21. https://doi.org/10.1007/s00418-023-02255-9.
74. Alibrahem W, Nguyen DHH, Kharrat Helu N, Tóth F, Nagy PT, Posta J, Prokisch J, Oláh C. Health benefits, applications, and analytical methods of freshly produced allyl isothiocyanate. Foods. 2025;14(4):579. https://doi.org/10.3390/foods14040579.

75. Singh S, Diksha M, E., et al. Appraisal of growth inhibitory, biochemical and genotoxic effects of allyl isothiocyanate on different developmental stages of Zeugodacus cucurbitae (Coquillett) (Diptera: Tephritidae). Sci Rep. 2022;12:10363. https://doi.org/10.1038/s41598-022-14593-0.
76. Agrawal AA, Kurashige NS. A role for isothiocyanates in plant resistance against the specialist herbivore Pieris rapae. J Chem Ecol. 2003;29(6):1403–15. https://doi.org/10.1023/a:1024265420375.
77. Khokon MA, Jahan MS, Rahman T, Hossain MA, Muroyama D, Minami I, Munemasa S, Mori IC, Nakamura Y, Murata Y. Allyl isothiocyanate (AITC) induces stomatal closure in Arabidopsis. Plant Cell Environ. 2011;34(11):1900–6. https://doi.org/10.1111/j.1365-3040.2011.02385.x.
78. Nintemann SJ, Hunziker P, Andersen TG, Schulz A, Burow M, Halkier BA. Localization of the glucosinolate biosynthetic enzymes reveals distinct spatial patterns for the biosynthesis of indole and aliphatic glucosinolates. Physiol Plant. 2018;163(2):138–54. https://doi.org/10.1111/ppl.12672.
79. Zhu WT, Li CH, Dai TT, Tao FL, Wang MW, Wang CY, Han ZL, Sun NX, Zhao YN, Wang DL. Effects of allyl isothiocyanate on the expression, function, and its mechanism of ABCA1 and ABCG1 in pulmonary of COPD rats. Int Immunopharmacol. 2021;101:108373. https://doi.org/10.1016/j.intimp.2021.108373.
80. Lai YH, Chiang YF, Huang KC, Chen HY, Ali M, Hsia SM. Allyl isothiocyanate mitigates airway inflammation and constriction in a house dust mite-induced allergic asthma model via upregulation of tight junction proteins and the TRPA1 modulation. Biomed Pharmacother. 2023;166:115334. https://doi.org/10.1016/j.biopha.2023.115334.
81. Popova M, Bankova V. Contemporary methods for the extraction and isolation of natural products. BMC Chem. 2023;17(1):68. https://doi.org/10.1186/s13065-023-00960-z.
82. Bussy A. Note sur la formation de l'Huile Essentielle de Moutarde. Compt Rendus. 1839;9:815–7.
83. Avasiloaiei DI, Calara M, Brezeanu PM, Murariu OC, Brezeanu C. On the future perspectives of some medicinal plants within Lamiaceae botanic family regarding their comprehensive properties and resistance against biotic and abiotic stresses. Genes. 2023;14(5):955. https://doi.org/10.3390/genes14050955.
84. Parrey ZA, Islam S, Shah SH, et al. Agronomical strategies to improve growth, physio-biochemistry, yield and quality attributes of mint plants under the varied environmental conditions: a review. J Soil Sci Plant Nutr. 2023;23:1489–514. https://doi.org/10.1007/s42729-023-01194-7.
85. Sfaxi A, Tavaszi-Sárosi S, Flórián K, Patonay K, Radácsi P, Juhász Á. Comparative evaluation of different mint species based on their in vitro antioxidant and antibacterial effect. Plants. 2025;14(1):105. https://doi.org/10.3390/plants14010105.
86. Rhioui W, Al Figuigui J, Nekhla H, et al. A comparative study on phytochemical composition, yield, morphological, and physiological characteristics of Mentha spicata (L.) across conventional, organic, and agroecological farming systems. Org Agr. 2025;15:259–71. https://doi.org/10.1007/s13165-025-00496-7.
87. Hosseini SJ, Tahmasebi-Sarvestani Z, Mokhtassi-Bidgoli A, Keshavarz H, Kazemi S, Khalvandi M, Pirdashti H, Hamidreza Hashemi-Petroudi S, Nicola S. Functional quality, antioxidant capacity and essential oil percentage in different mint species affected by salinity stress. Chem Biodivers. 2023;20(4):e202200247. https://doi.org/10.1002/cbdv.202200247.
88. Charmantier I. Carl Linnaeus and the visual representation of nature. Hist Stud Nat Sci. 2011;41(4):365–404. https://doi.org/10.1525/hsns.2011.41.4.365.
89. Merton ES. The botany of sir Thomas Browne. Isis. 1956;47(2):161–71.
90. Jones PM. Herbs and the medieval surgeon. In: Dendle P, Touwaide A, editors. Health and healing from the medieval garden. Rochester: Boydell & Brewer; 2015. p. 162–79.
91. Cunningham A. Boerhaave's orations, translated with introductions and notes by E. Kegel-Brinkgreve and A. M. Luyendijk-Elshout, Leiden, E. J. Brill for the sir Thomas Browne

institute, 1983, 8vo, pp. vii, 374. Med Hist. 1985;29(2):222. https://doi.org/10.1017/S0025727300044100.

92. Breathnach CS. Sir Thomas Browne (1605-1682). J R Soc Med. 2005;98(1):33–6. https://doi.org/10.1177/014107680509800115.
93. Hughes T. The medical education of sir Thomas Browne, a seventeenth-century student at Montpellier, Padua, and Leiden. J Med Biogr. 2001;9:70–6.
94. Hejna M, Kovanda L, Rossi L, Liu Y. Mint oils: in vitro ability to perform anti-inflammatory, antioxidant, and antimicrobial activities and to enhance intestinal barrier integrity. Antioxidants. 2021;10(7):1004. https://doi.org/10.3390/antiox10071004.
95. Best R. Mint and menthol: a review of potential health benefits and wider human effects. Preprintsorg. 2022; https://doi.org/10.20944/preprints202212.0273.v1.
96. Abdelrahman SM, El Samak M, El-Baz LMF, Hanora AMS, Satyal P, Dosoky NS. Effects of mint oils on the human Oral microbiome: a pilot study. Microorganisms. 2024;12(8):1538. https://doi.org/10.3390/microorganisms12081538.
97. Best R. A review of the digestive, respiratory and nocioceptive benefits, associated performance outcomes and clinical considerations following mint and menthol application. Med Res Arch. 2023;11:3321.
98. Klein K, Garkov D, Rütschlin S, Böttcher T, Schreiber F. QSDB—a graphical quorum sensing database. Database. 2021;2021:baab058. https://doi.org/10.1093/database/baab058.
99. Sikdar R, Elias M. Quorum quenching enzymes and their effects on virulence, biofilm, and microbiomes: a review of recent advances. Expert Rev Anti-Infect Ther. 2020;18(12):1221–33. https://doi.org/10.1080/14787210.2020.1794815.
100. Garcia-Contreras R, Perez-Eretza B, Jasso-Chavez R, et al. High variability in quorum quenching and growth inhibition by furanone C-30 in Pseudomonas aeruginosa clinical isolates from cystic fibrosis patients. Pathog Dis. 2015;73(6):40.
101. Escobar-Muciño E, Arenas-Hernández MMP, Luna-Guevara ML. Mechanisms of inhibition of quorum sensing as an alternative for the control of E. coli and salmonella. Microorganisms. 2022;10(5):884. https://doi.org/10.3390/microorganisms10050884.
102. Tafrihi M, Imran M, Tufail T, Gondal TA, Caruso G, Sharma S, Sharma R, Atanassova M, Atanassov L, Valere Tsouh Fokou P, Pezzani R. The wonderful activities of the genus Mentha: not only antioxidant properties. Molecules. 2021;26(4):1118. https://doi.org/10.3390/molecules26041118.
103. Hu Y, Yan Z. Comprehensive advances in phytochemical components, bioactive functionality, and processing applications of mustard (Brassica juncea (L.) Czern.): a review. Front Nutr. 2025;12:1626333. https://doi.org/10.3389/fnut.2025.1626333.
104. Balkrishna A, Verma S, Priya Rani M, Nain P, Varshney A. Exploring the potential of mustard (brassica spp.) seeds through 'Kolhu' traditional method of extraction and novel identification of an anti-cancer dipeptide, Aurantiamide acetate (Asperglaucide) on ultra-performance liquid chromatography-mass spectrometry coupled with quadrupole time-of-flight (UPLC/MS-QToF) analytical platform. Food Chem. 2024;446:138870. https://doi.org/10.1016/j.foodchem.2024.138870.
105. Salah HA, Elsayed AM, Bassuiny RI, et al. Improvement of phenolic profile and biological activities of wild mustard sprouts. Sci Rep. 2024;14:10528. https://doi.org/10.1038/s41598-024-60452-5.
106. Chakraborty S, Paul K, Mallick P, Pradhan S, Das K, Chakrabarti S, Nandi DK, Bhattacharjee P. Consortia of bioactives in supercritical carbon dioxide extracts of mustard and small cardamom seeds lower serum cholesterol levels in rats: new leads for hypocholesterolaemic supplements from spices. J Nutr Sci. 2019;8:e32. https://doi.org/10.1017/jns.2019.28.
107. Đorđević BS, Todorović ZB, Troter DZ, et al. Extraction of phenolic compounds from black mustard (Brassica nigra L.) seed by deep eutectic solvents. Food Measure. 2021;15:1931–8. https://doi.org/10.1007/s11694-020-00772-y.
108. Herrlinger KA, Nieman KM, Sanoshy KD, Fonseca BA, Lasrado JA, Schild AL, Maki KC, Wesnes KA, Ceddia MA. Spearmint extract improves working memory in men and women

with age-associated memory impairment. J Altern Complement Med. 2018;24(1):37–47. https://doi.org/10.1089/acm.2016.0379.

109. López V, Martín S, Gómez-Serranillos MP, Carretero ME, Jäger AK, Calvo MI. Neuroprotective and neurochemical properties of mint extracts. Phytother Res. 2010;24(6):869–74. https://doi.org/10.1002/ptr.3037.
110. Johnson AJ, Miles C. Chewing gum and context-dependent memory: the independent roles of chewing gum and mint flavour. Br J Psychol. 2008;99(Pt 2):293–306. https://doi.org/10.1348/000712607X228474.
111. Miquel-Kergoat S, Azais-Braesco V, Burton-Freeman B, Hetherington MM. Effects of chewing on appetite, food intake and gut hormones: a systematic review and meta-analysis. Physiol Behav. 2015;151:88–96. https://doi.org/10.1016/j.physbeh.2015.07.017.
112. Kennedy D, Okello E, Chazot P, Howes MJ, Ohiomokhare S, Jackson P, Haskell-Ramsay C, Khan J, Forster J, Wightman E. Volatile terpenes and brain function: investigation of the cognitive and mood effects of mentha × Piperita L. essential oil with in vitro properties relevant to central nervous system function. Nutrients. 2018;10(8):1029. https://doi.org/10.3390/nu10081029.
113. Thakur AK, Chatterjee SS, Kumar V. Beneficial effects of Brassica juncea on cognitive functions in rats. Pharm Biol. 2013;51(10):1304–10. https://doi.org/10.3109/13880209.2013.789917.
114. Nagata A, Oishi S, Kirishita N, Onoda K, Kobayashi T, Terada Y, Minami A, Senoo N, Yoshioka Y, Uchida K, Ito K, Miura S, Miyoshi N. Allyl Isothiocyanate maintains DHA-containing Glycerophospholipids and ameliorates the cognitive function decline in OVX mice. ACS Omega. 2023;8(45):43118–29. https://doi.org/10.1021/acsomega.3c06622.
115. Altinoz MA, Ozpinar A, Ozpinar A, et al. Erucic acid, a nutritional PPARδ-ligand may influence Huntington's disease pathogenesis. Metab Brain Dis. 2020;35:1–9. https://doi.org/10.1007/s11011-019-00500-6.
116. Kinney JW, Bemiller SM, Murtishaw AS, Leisgang AM, Salazar AM, Lamb BT. Inflammation as a central mechanism in Alzheimer's disease. Alzheimer's Dement. 2018;4:575–90. https://doi.org/10.1016/j.trci.2018.06.014.
117. Bajpai D, Malaiappan S. Evaluation of anti-inflammatory and antimicrobial properties of mustard seed extract-based hydrogel: an in vitro study. Cureus. 2023;15(9):e45146. https://doi.org/10.7759/cureus.45146.
118. Wan J, Fan H. Oral microbiome and Alzheimer's disease. Microorganisms. 2023;11(10):2550. https://doi.org/10.3390/microorganisms11102550.
119. Seca AML, Pinto DCGA. Biological potential and medical use of secondary metabolites. Medicines. 2019;6(2):66. https://doi.org/10.3390/medicines6020066.
120. Pavagadhi S, Swarup S. Metabolomics for evaluating flavor-associated metabolites in plant-based products. Meta. 2020;10(5):197. https://doi.org/10.3390/metabo10050197.
121. Hanley AB, Kwiatkowski CA, Fenwick GR. Enzymic hydrolysis of glucosinolates in a low water system. J Sci Food Agric. 1990;51:417.
122. Abdel-Massih RM, Debs E, Othman L, Attieh J, Cabrerizo FM. Glucosinolates, a natural chemical arsenal: more to tell than the myrosinase story. Front Microbiol. 2023;14:1130208. https://doi.org/10.3389/fmicb.2023.1130208.
123. Toesch M, Schober M, Faber K. Microbial alkyl- and aryl-sulfatases: mechanism, occurrence, screening and stereoselectivities. Appl Microbiol Biotechnol. 2014;98(4):1485–96. https://doi.org/10.1007/s00253-013-5438-0.
124. Manivannan A, Israni B, Luck K, Götz M, Seibel E, Easson MLAE, Kirsch R, Reichelt M, Stein B, Winter S, Gershenzon J, Vassão DG. Identification of a sulfatase that detoxifies glucosinolates in the phloem-feeding insect Bemisia tabaci and prefers indolic glucosinolates. Front Plant Sci. 2021;12:671286. https://doi.org/10.3389/fpls.2021.671286.
125. Ahn SJ, Betzin F, Gikonyo MW, Yang ZL, Köllner TG, Beran F. Identification and evolution of glucosinolate sulfatases in a specialist flea beetle. Sci Rep. 2019;9(1):15725. https://doi.org/10.1038/s41598-019-51749-x.

126. Correia MSP, Ballet C, Meistermann H, Conway LP, Globisch D. Comprehensive kinetic and substrate specificity analysis of an arylsulfatase from Helix pomatia using mass spectrometry. Bioorg Med Chem. 2019;27(6):955–62. https://doi.org/10.1016/j.bmc.2019.01.031.
127. Boppana VK, Lynn RK, Ziemniak JA. Immobilized sulfatase:beta-glucuronidase enzymes for the qualitative and quantitative analysis of drug conjugates. J Pharm Sci. 1989;78(2):127–31. https://doi.org/10.1002/jps.2600780211.
128. Cheng W, Luo A, Li H, Chen Z, Guan Z, Escalona VH, Liu D, Zhou Y, Zhang F, Huang Z, et al. Variation in nutritional qualities and antioxidant capacity of different leaf mustard cultivars. Horticulturae. 2025;11:59. https://doi.org/10.3390/horticulturae11010059.
129. Ahmad M, Manisha, Arya S, Pakpu T, Chauhan A. Comparative analysis of different mustard hybrids: analyzing growth and yield perspectives. J Environ Bio-Sci. 2024;38:111–5. https://doi.org/10.59467/JEBS.2024.38.111.
130. Loitongbam A, Samal NK, Kumar NR, Kumar S, Annamalai M, Kundu A, Subramanian S, Bhattacharya R. Differential regulation of glucosinolate-myrosinase mediated defense determines host-aphid interaction in Indian mustard Brassica juncea L. Mol Biol Rep. 2024;51(1):1079. https://doi.org/10.1007/s11033-024-10002-z.
131. Sarwar M, Kirkegaard JA, Wong PTW, Desmarchelier JM. Biofumigation potential of brassicas. Plant Soil. 1998;201:103–12.
132. Cools K, Terry LA. The effect of processing on the glucosinolate profile in mustard seed. Food Chem. 2018;252:343–8. https://doi.org/10.1016/j.foodchem.2018.01.096.
133. Wang J, Yu S, Ren X, Wang Y, Wang Z, Meng Q, Huang Y, Wang Y. Integrative analyses of metabolites and transcriptome reveal the metabolic pattern of Glucosinolates in potherb mustard (Brassica juncea var. multiceps). Plants. 2024;13:2481. https://doi.org/10.3390/plants13172481.
134. Wu X, Huang H, Childs H, Wu Y, Yu L, Pehrsson PR. Glucosinolates in brassica vegetables: characterization and factors that influence distribution, content, and intake. Annu Rev Food Sci Technol. 2021;12:485–511. https://doi.org/10.1146/annurev-food-070620-025744.
135. Del Carmen M-BM, Moreno DA, Carvajal M. The physiological importance of glucosinolates on plant response to abiotic stress in brassica. Int J Mol Sci. 2013;14(6):11607–25. https://doi.org/10.3390/ijms140611607.
136. Masyita A, Mustika Sari R, Dwi Astuti A, Yasir B, Rahma Rumata N, Emran TB, Nainu F, Simal-Gandara J. Terpenes and terpenoids as main bioactive compounds of essential oils, their roles in human health and potential application as natural food preservatives. Food Chem X. 2022;13:100217. https://doi.org/10.1016/j.fochx.2022.100217.
137. Peters VCT, Dunkel A, Frank O, McCormack B, Dowd E, Didzbalis J, Dawid C, Hofmann T. A high throughput toolbox for comprehensive flavor compound mapping in mint. Food Chem. 2021;365:130522. https://doi.org/10.1016/j.foodchem.2021.130522.
138. Olofsson JK, Tyler T, Dunning LT, Hjertson M, Rühling Å, Hansen AJ. Morphological and genetic evidence suggest gene flow among native and naturalized mint species. Am J Bot. 2024;111(2):e16280. https://doi.org/10.1002/ajb2.16280.
139. Kofidis G, Bosabalidis A, Kokkini S. Seasonal variation of essential oils in a linalool-rich chemotype of Mentha spicata grown wild in Greece. J Essent Oil Res. 2004;16:469–72. https://doi.org/10.1080/10412905.2004.9698773.
140. Gill LS, Lawrence BM, Morton JK. Variation in Mentha arvensis L. (Labiatae). I. The north American populations. Bot J Linn Soc. 1973;67:213–32. https://doi.org/10.1111/j.1095-8339.1973.tb01739.x.
141. Patonay K, Szalontai H, Radácsi P, Zámboriné-Németh É. Chemotypes and their stability in Mentha longifolia (L.) L.-A comprehensive study of five accessions. Plants. 2021;10(11):2478. https://doi.org/10.3390/plants10112478.
142. Khaldari I, Afshoon E, Nik SH. Phytohormonal elicitation triggers oxidative stress and enhances menthol biosynthesis through modulation of key pathway genes in Mentha piperita L. Sci Rep. 2025;15:30495. https://doi.org/10.1038/s41598-025-16667-1.

143. Fuchs LK, Holland AH, Ludlow RA, Coates RJ, Armstrong H, Pickett JA, Harwood JL, Scofield S. Genetic manipulation of biosynthetic pathways in mint. Front Plant Sci. 2022;13:928178. https://doi.org/10.3389/fpls.2022.928178.
144. Frey M, Gohr ST, Köllner TG, Bathe U, Lackus ND, Padilla-Gonzalez F, Ro DK, O'Connor SE, Degenhardt J, Tissier A. Biosynthesis of biologically active terpenoids in the mint family (Lamiaceae). Nat Prod Rep. 2025;42(11):1887–908. https://doi.org/10.1039/d5np00026b.
145. Croteau R, Gershenzon J. Genetic control of monoterpene biosynthesis in mints (Mentha: Lamiaceae). In: Ellis BE, Kuroki GW, Stafford HA, editors. Genetic engineering of plant secondary metabolism. Recent advances in Phytochemistry, vol. 28. Boston: Springer; 1994. https://doi.org/10.1007/978-1-4615-2544-8_8.
146. Abbott S, Fairbanks DJ. Experiments on plant hybrids by Gregor Mendel. Genetics. 2016;204(2):407–22. https://doi.org/10.1534/genetics.116.195198.
147. Hedley CL, Ambrose MJ. Designing "leafless" plants for improving yields of dried pea crops. Adv Agron. 1981;34:225–77.
148. Feng C, Chen B, Hofer J, et al. Genomic and genetic insights into Mendel's pea genes. Nature. 2025;642:980–9. https://doi.org/10.1038/s41586-025-08891-6.
149. https://www.johninnesfoundation.org.uk/timeline/.
150. Skinulienė L, Marcinkevičienė A, Dorelis M, Bogužas V. The effect of long-term crop rotations for the soil carbon sequestration rate potential and cereal yield. Agriculture. 2024;14:483. https://doi.org/10.3390/agriculture14030483.
151. Chaudhari N, Roper SD. Taste: from peripheral receptors to perception. Curr Opin Physio. 2021;21:44–7. https://doi.org/10.1016/j.cophys.2021.04.001.
152. Töle JC, Behrens M, Meyerhof W. Taste receptor function. Handb Clin Neurol. 2019;164:173–85. https://doi.org/10.1016/B978-0-444-63855-7.00011-3.
153. Li C, Li Y, Sun Q, Abdurehim A, Xu J, Xie J, Zhang Y. Taste and its receptors in human physiology: a comprehensive look. Food Front. 2024;5:1512–33. https://doi.org/10.1002/fft2.407.
154. Diepeveen J, Moerdijk-Poortvliet TCW, van der Leij FR. Molecular insights into human taste perception and umami tastants: a review. J Food Sci. 2022;87(4):1449–65. https://doi.org/10.1111/1750-3841.16101.
155. van der Werf A, Rovithi M, Langius JAE, van Schueren MAE, Verheul HMW. Insight in taste alterations during treatment with protein kinase inhibitors. Eur J Cancer. 2017;86:125–34. https://doi.org/10.1016/j.ejca.2017.09.006.
156. Shi Z, Xu W, Wu L, et al. Structural and functional characterization of human sweet taste receptor. Nature. 2025;645:801–8. https://doi.org/10.1038/s41586-025-09302-6.
157. D'Urso O, Drago F. Pharmacological significance of extra-oral taste receptors. Eur J Pharmacol. 2021;910:174480. https://doi.org/10.1016/j.ejphar.2021.174480.
158. Martens K, Steelant B, Bullens DMA. Taste receptors: the gatekeepers of the airway epithelium. Cells. 2021;10(11):2889. https://doi.org/10.3390/cells10112889.
159. Jaggupilli A, Singh N, Jesus VC, Duan K, Chelikani P. Characterization of the binding sites for bacterial acyl Homoserine lactones (AHLs) on human bitter taste receptors (T2Rs). ACS Infect Dis. 2018;4(7):1146–56. https://doi.org/10.1021/acsinfecdis.8b00094.
160. Kouakou YI, Lee RJ. Interkingdom detection of bacterial quorum-sensing molecules by mammalian taste receptors. Microorganisms. 2023;11(5):1295. https://doi.org/10.3390/microorganisms11051295.
161. Mori S, Nakamura N, Fuchigami A, Yoshimoto S, Sakakibara M, Ozawa T, Aoki J, Inoue A, Sumida H, Ando H, Nakamura M. Intracellular TAS2Rs act as a gatekeeper for the excretion of harmful substances via ABCB1 in keratinocytes. FASEB Bioadv. 2024;6(10):424–41. https://doi.org/10.1096/fba.2024-00074.
162. Jeon TI, Seo YK, Osborne TF. Gut bitter taste receptor signalling induces ABCB1 through a mechanism involving CCK. Biochem J. 2011;438(1):33–7. https://doi.org/10.1042/BJ20110009.

163. Robey RW, Pluchino KM, Hall MD, Fojo AT, Bates SE, Gottesman MM. Revisiting the role of ABC transporters in multidrug-resistant cancer. Nat Rev Cancer. 2018;18(7):452–64. https://doi.org/10.1038/s41568-018-0005-8.
164. Taha MA, Shortess CJ, Noonan MJ, Harrod CC, Barham MJR, et al. Validation study: evaluating phenotypic expression of bitter taste receptors. J Otolaryngol Rhinol. 2021;7:110.
165. Gaida MM, Mayer C, Dapunt U, Stegmaier S, Schirmacher P, Wabnitz GH, Hänsch GM. Expression of the bitter receptor T2R38 in pancreatic cancer: localization in lipid droplets and activation by a bacteria-derived quorum-sensing molecule. Oncotarget. 2016;7(11):12623–32. https://doi.org/10.18632/oncotarget.7206.
166. John DA, Babu GR. Lessons from the aftermaths of green revolution on food system and health. Front Sustain Food Syst. 2021;5:644559. https://doi.org/10.3389/fsufs.2021.644559.
167. Bailey-Serres J, Parker JE, Ainsworth EA, Oldroyd GED, Schroeder JIC. Genetic strategies for improving crop yields. Nature. 2019;575(7781):109–18. https://doi.org/10.1038/s41586-019-1679-0.
168. Fenibo EO, Ijoma GN, Nurmahomed W, Matambo T. The potential and green chemistry attributes of biopesticides for sustainable agriculture. Sustainability. 2022;14:14417. https://doi.org/10.3390/su142114417.
169. Svensson Grape E, Rooth V, Nero M, et al. Structure of the active pharmaceutical ingredient bismuth subsalicylate. Nat Commun. 2022;13:1984. https://doi.org/10.1038/s41467-022-29566-0.
170. Bierer DW. Bismuth subsalicylate: history, chemistry, and safety. Rev Infect Dis. 1990;12(1):3–8.
171. Madjar RM, Vasile Scăețeanu G. An overview of heavy metal contamination in water from agriculture: origins, monitoring, risks, and control measures. Sustainability. 2025;17:7368. https://doi.org/10.3390/su17167368.
172. May & Baker Ltd.- Bromoxynil: N-Octanoyl Ester of Bromoxynil (M & B 10,731) Selective Herbicide. Pest articles & news summaries. Section C Weed Control. 1965;11(3):309–12. https://doi.org/10.1080/05331856509432409.
173. Siddik MAB, Francis P, Rohani MF, Azam MS, Mock TS, Francis DS. Seaweed and seaweed-based functional metabolites as potential modulators of growth, immune and antioxidant responses, and gut microbiota in fish. Antioxidants. 2023;12:2066. https://doi.org/10.3390/antiox12122066.
174. Panda D, Mondal S, Mishra A. Liquid biofertilizers from seaweeds: a critical review. In: Ranga Rao A, Ravishankar GA, editors. Sustainable global resources of seaweeds, vol. 1. Cham: Springer; 2022. https://doi.org/10.1007/978-3-030-91955-9_26.
175. Ahuja I, Dauksas E, Remme JF, Richardsen R, Løes AK. Fish and fish waste-based fertilizers in organic farming - with status in Norway: a review. Waste Manag. 2020;115:95–112. https://doi.org/10.1016/j.wasman.2020.07.025.
176. Madende M, Hayes M. Fish by-product use as biostimulants: an overview of the current state of the art, including relevant legislation and regulations within the EU and USA. Molecules. 2020;25(5):1122. https://doi.org/10.3390/molecules25051122.
177. Dhar M, Jasrotia R, Langer S. Using fish waste and by-products for manufacturing organic fertilizers and manures. In: Maqsood S, Naseer MN, Benjakul S, Zaidi AA, editors. Fish waste to valuable products. Sustainable materials and technology. Singapore: Springer; 2024. https://doi.org/10.1007/978-981-99-8593-7_16.
178. Ghaly AE, Dave D, Budge S, Brooks MS. Fish spoilage mechanisms and preservation techniques: review. Am J Appl Sci. 2010;7:859–77.
179. Tomaś N, Myszka K. Current advances in the concept of quorum sensing-based prevention of spoilage of fish products by pseudomonads. Appl Sci. 2022;12:6719. https://doi.org/10.3390/app12136719.
180. Coolahan M, Whalen KE. A review of quorum-sensing and its role in mediating inter-kingdom interactions in the ocean. Commun Biol. 2025;8:179. https://doi.org/10.1038/s42003-025-07608-9.

181. Bobadilla LK, Tranel PJ. Predicting the unpredictable: the regulatory nature and promiscuity of herbicide cross resistance. Pest Manag Sci. 2024 Feb;80(2):235–44. https://doi.org/10.1002/ps.7728.
182. Sen MK, Bhattacharya S, Bharati R, Hamouzová K, Soukup J. Comprehensive insights into herbicide resistance mechanisms in weeds: a synergistic integration of transcriptomic and metabolomic analyses. Front Plant Sci. 2023;14:1280118. https://doi.org/10.3389/fpls.2023.1280118.
183. Duke SO. Overview of herbicide mechanisms of action. Environ Health Perspect. 1990;87:263–71. https://doi.org/10.1289/ehp.9087263.
184. Duke SO, Dayan FE. The search for new herbicide mechanisms of action: is there a 'holy grail'? Pest Manag Sci. 2022;78(4):1303–13. https://doi.org/10.1002/ps.6726.
185. Dayan FE, Duke SO. Natural compounds as next-generation herbicides. Plant Physiol. 2014;166(3):1090–105. https://doi.org/10.1104/pp.114.239061.
186. Win P-P, Park H-H, Kuk Y-I. Control efficacy of natural products on broadleaf and grass weeds using various application methods. Agronomy. 2023;13:2262. https://doi.org/10.3390/agronomy13092262.
187. Dyer WE. Stress-induced evolution of herbicide resistance and related pleiotropic effects. Pest Manag Sci. 2018 Aug;74(8):1759–68. https://doi.org/10.1002/ps.5043.
188. Bonny S. Genetically modified herbicide-tolerant crops, weeds, and herbicides: overview and impact. Environ Manag. 2016;57(1):31–48. https://doi.org/10.1007/s00267-015-0589-7.
189. Reynier E, Rubin E. Glyphosate exposure and GM seed rollout unequally reduced perinatal health. Proc Natl Acad Sci U S A. 2025;122(3):e2413013121. https://doi.org/10.1073/pnas.2413013121.
190. Benbrook C. Impacts of genetically engineered crops on pesticide use in the US—the first sixteen years. Environ Sci Eur. 2012;24:24.
191. Lacroix R, Kurrasch DM. Glyphosate toxicity: in vivo, in vitro, and epidemiological evidence. Toxicol Sci. 2023;192(2):131–40. https://doi.org/10.1093/toxsci/kfad018.
192. Steinberg P, van der Voet H, Goedhart PW, Kleter G, Kok EJ, Pla M, Nadal A, Zeljenková D, Aláčová R, Babincová J, Rollerová E, Jaďuďová S, Kebis A, Szabova E, Tulinská J, Líšková A, Takácsová M, Mikušová ML, Krivošíková Z, Spök A, Racovita M, de Vriend H, Alison R, Alison C, Baumgärtner W, Becker K, Lempp C, Schmicke M, Schrenk D, Pöting A, Schiemann J, Wilhelm R. Lack of adverse effects in subchronic and chronic toxicity/carcinogenicity studies on the glyphosate-resistant genetically modified maize NK603 in Wistar Han RCC rats. Arch Toxicol. 2019;93(4):1095–139. https://doi.org/10.1007/s00204-019-02400-1.
193. Sammons RD, Gaines TA. Glyphosate resistance: state of knowledge. Pest Manag Sci. 2014;70(9):1367–77. https://doi.org/10.1002/ps.3743.
194. Dong H, Huang Y, Wang K. The development of herbicide resistance crop plants using CRISPR/Cas9-mediated gene editing. Genes. 2021;12(6):912. https://doi.org/10.3390/genes12060912.
195. Berestetskiy A. Modern approaches for the development of new herbicides based on natural compounds. Plants. 2023;12(2):234. https://doi.org/10.3390/plants12020234.
196. Zeidler O. Verbindungen von chloral mit brom- und chlorbenzol. Ber Dtsch Chem Ges. 1874;7(2):1180–1.
197. Gelaye Y, Negash B. The role of baculoviruses in controlling insect pests: a review. Cogent Food Agric. 2023;9(1) https://doi.org/10.1080/23311932.2023.2254139.
198. Oberemok VV, Laikova KV, Gal'chinsky, N.V., et al. DNA insecticide developed from the Lymantria dispar 5.8S ribosomal RNA gene provides a novel biotechnology for plant protection. Sci Rep. 2019;9:6197. https://doi.org/10.1038/s41598-019-42688-8.
199. Palma L, Muñoz D, Berry C, Murillo J, Caballero P. Bacillus thuringiensis toxins: an overview of their biocidal activity. Toxins. 2014;6(12):3296–325. https://doi.org/10.3390/toxins6123296.

200. Abbas MST. Genetically engineered (modified) crops (bacillus thuringiensis crops) and the world controversy on their safety. Egypt J Biol Pest Control. 2018;28:52. https://doi.org/10.1186/s41938-018-0051-2.
201. Trewavas A. Toxins and genetically modified food. Lancet. 2000;355(9207):931.
202. Pavela R, Kovaříková K, Novák M. Botanical antifeedants: an alternative approach to pest control. Insects. 2025;16(2):136. https://doi.org/10.3390/insects16020136.
203. Mwamula AO, Kabir MF, Lee D. A review of the potency of plant extracts and compounds from key families as an alternative to synthetic nematicides: history, efficacy, and current developments. Plant Pathol J. 2022;38(2):53–77. https://doi.org/10.5423/PPJ.RW.12.2021.0179.
204. Pereira AA, Yaverino-Gutierrez MA, Monteiro MC, Souza BA, Bachheti RK, Chandel AK. Precision fermentation in the realm of microbial protein production: state-of-the-art and future insights. Food Res Int. 2025;200:115527. https://doi.org/10.1016/j.foodres.2024.115527.

Chapter 9
Chemistry and Products from Nature. Food Processing for Value and Bioactivity

Abstract The development of products of high intrinsic value. The links between disease and health and dietary components. The concept of food as medicine as suggested by Epicurus. Particular non-communicable disease states including cancer, cardiovascular disease and metabolic diseases (diabetes). Protective factors in the diet. The importance to Norfolk of barley, brewing and fermentation. Compounds that stimulate the growth of organisms in a specific fashion—stimbiotics. The chemistry of beer and its ingredients. The bitterness and biological properties of hops—monoterpenes. The distinction between correlative claims for diet and health and causality. The primary and other functions of taste receptors. The link between Alzheimer's disease and bitter receptors. The production of chocolate in Norwich. The principles of causality as discussed by Hume and Locke.

Keywords Processing · Value chain · Nutrition · Non-communicable disease (NCD) · Epicurus · Cancer · Diet · Brewing · Stimbiotics · Chocolate

This chapter will deal with various aspects of processing that are of relevance to Norfolk and also make use of a range of chemical techniques whose genesis and development is an extension of the scientific principles that have been developed over many years in Norfolk. The personal, institutional and societal assimilation of chemistry is part of the fabric of the county.

The development of food processing is a way of taking raw agricultural materials and progressing them along the value chain. It also requires an awareness of how materials respond and the potential for adding value in a number of ways. In addition to nutrition and the provision of macronutrients (carbohydrate, protein, lipids and fats) and micronutrients (essential vitamins and minerals), many of the added value attributes (taste, flavour, colour, health-promoting properties) are attributable to the presence of secondary metabolites and natural products. Although Norfolk, as has been mentioned, is an arable and marine county, processing and manufacturing of food has included high value products. This is in part due to the geographical and geological attributes of the county and partially because transport links favoured high value products rather than bulk commodities. This

B. Hanley, *The Chemistry of a County*,
https://doi.org/10.1007/978-3-032-21491-1_9

has been accompanied by a change in the way that food and nutrition are treated and considered [1]. Initially food intake recommendations were based upon nutritional adequacy and the prevention of malnutrition and deficiency diseases. As science progressed and the management of infectious diseases became less of an all-absorbing priority (the advent of the 'magic bullet' advocated by Ehrlich [2] and the growth in chemotherapy as an approach to curing infective diseases together with improvements in sanitation had an impact on prevention and cure), the importance of chronic and non-communicable diseases (NCDs) began to be realised. This included both chemotherapeutic approaches—for example the use of drugs such as statins to prevent hypercholesteremia and thereby reduce the incidence of cardiovascular disease [3]—and the use of diet beyond dealing with malnutrition and nutritional inadequacy (low sodium and saturated fat levels, for example) to reduce disease incidence [4] and to reduce the impact of risk factors became a realisable objective. This has created a paradigm whereby prevention (or the reduction in severity and consequences) of disease is considered to be at least as important as the treatment of the clinical symptoms of NCDs [5]. Underpinning, the concept of food and health is the idea of 'food sufficiency' as suggested by Epicurus [6]. Indeed Epicurus believed that medicine must treat diseases of the body as philosophy should treat maladies of the soul [7]. The improvement of health and protection from NCDs is, therefore, part of the tradition of Epicurean philosophy. The concept of 'food as medicine' has increased in popularity [8] however it is often supported by an evidence base that is mechanistically meagre and simply advocating a lower fat, more plant-based macronutrient diet [9] is overly simplistic. In addition, studies often lack statistical rigour [10]—although, in terms of general health it can show some positive results, however in terms of some specific markers of health endpoints, vegetarian diets were also associated with harmful outcomes on one-carbon metabolism markers and lower concentrations of vitamin B-12 than omnivorous diets [11]. In order to contextualise the information, the global incidences of various NCDs have been estimated [12]. The most significant NCD endpoints were found to those resulting from cardiovascular diseases, cancers, diabetes, kidney diseases, and mental disorders. In the case of cardiovascular disease, as noted previously, a number of dietary and clinical interventions can be envisioned. The suite of diseases defined as cancers are ubiquitous and are associated, on the one hand, with exposure to carcinogens and, on the other, with postulated dietary protective factors. This duality and the nature of the cancer process has attracted considerable attention. The incidence of tobacco-related cancers remains high in some parts of the globe and is associated with the inhalation of compounds that are known mutagens and carcinogens. In Europe, by contrast, prostate and breast cancers were the most common forms of the disease in men and women respectively [13] and the link to specific exposures is more difficult to confirm. The concept of dietary prevention of cancer is one that has attracted considerable attention. The simplistic view is that cancer prevention can reflect the physiological duality mentioned previously. Reducing exposure to carcinogenic substances is one factor while preventing the initiation or acceleration along the pathway leading to tumour formation is the other. The European Prospective

Investigation into Cancer and Nutrition (EPIC) investigation has generated considerable amounts of data [14] however many of these studies are correlative rather than demonstrating causality and, as shall be discussed later (Sect. 9.7) the link (and the effectiveness of any potential intervention) should be treated with caution. Data from the EPIC investigation has also suggested a potential link between food processing and some types of cancer [15] however the presence of confounders must be considered alongside apparent correlative evidence. The links between cancer incidence at certain sites and the intake of specific dietary components (for example, mycotoxins, benzo(a)pyrene) can be demonstrated using a range of *in vitro* and *in vivo* model systems that have a mechanistic underpinning and illustrate specific molecular events in the cancer pathway starting with a specific event (for example a mutation) together with evidence of exposure in specific organs. Similar mechanistically based studies on the cancer prevention effects of specific dietary components are more difficult to investigate using reductionist techniques because of the multiple stages in the cancer process and the effect of dietary components on inter-dependent pathways with significant confounders [16]. For example, a number of phytochemicals have been implicated in cancer treatment and prevention [17] however a substantial number of these are supported by observational and epidemiological data rather than by mechanisms that have been proven. A number of clinical and pre-clinical studies have been reported [18] however the putative mechanisms have often invoked generic effects where the link to cancer is at best tentative such as prevention of oxidative damage to cellular components or modification of a range of cell signalling pathways with a range of effects and outcomes—many of which are unproven in the context of the development of cancers. In addition, phytochemicals are often present in low concentrations and amounts and basic, absorption, distribution, metabolism, excretion (ADME) data (as required for pharmacological agents) is often not available hence the amount of active agent present at the biological target is unknown. Some advances have been made using nanotechnologies to optimise uptake [19] however this remains a challenging area and, despite the use of physicochemical models to determine uptake and/or absorption [20], data can still be difficult to obtain. The role of obesity in cancer is one area where evidence is accumulating and where long-term interventions have been advocated [21]. In contrast, most randomised controlled trials (RCTs) investigating the link between specific dietary components and health (including cancer-related) endpoints failed to demonstrate even an association [22]. The reductionist approach of attempting to link cancer prevention with specific isolated dietary components using, in essence, a pharmacological approach has been, in some cases, abandoned in favour of a more holistic, whole diet approach [23]. Further studies are required on the biological effects of dietary components, their biotransformation products, both *ex vivo* and after *in vivo* and *in situ* metabolism by constitutive and inducible phase I and phase II enzyme systems and the modulation (for example up and down regulation) of those systems themselves by dietary components [24].

9.1 Barley, Brewing and Fermentation

Fermentation is one of the oldest technologies known. It is defined as anaerobic fermentation. While for the fermentative organism, it is a way of manipulating redox balance and deriving energy, it is also an important technique for food preservation through the accumulation of side products such as acids and alcohols which prevent the growth of pathogens and spoilage organisms and provide positive organoleptic properties. It is these side products of the process that are responsible for many of the attributes of fermentation that considered to be of greatest value. In addition, the competition for nutrients by the fermenting species is an important mechanism for preventing the growth of some spoilage organisms [25]. In addition to preventing the growth of spoilage organisms, a number of products of the fermentative process have been suggested to contribute to, in particular, gastrointestinal health [26] and fermentative bacteria also produce a range of compounds with different types of biological activity [27] in addition to antimicrobial activity [28]. In addition, there are also a range of products which promote the growth of physiologically beneficial bacteria in the gut microbiome. These products are associated with *in situ* fermentation in the gut and were termed *prebiotics* by Gibson and Roberfroid [29]. The concept was expanded [30] to components that give rise to the development of a healthy, fermentative gut microbiome. If the definition of a prebiotic can be extended to include any molecule that may increase the population or stimulate the growth of beneficial bacteria, then the production of quorum sensing (QS) molecules might also be considered a prebiotic activity [31] however since prebiotics have to be substrates and used selectively then QS molecules cannot be defined as true prebiotics and another term should be sought. One possibility is to refer to molecules which stimulate the growth of specific species as *stimbiotics*.

The production of alcoholic beverages through fermentation is a simple procedure however many of the techniques that have been developed have relied upon advances in other areas of fermentation [32]. In essence, the technology has been developed to take a range of raw materials (for example, cereals, legumes, vegetables, roots and tubers, fish sauce and pastes, sausage, milk to produce a range of products which, in addition to being preserved and resistant to spoilage, also have particular desirable taste and flavour properties. Apart from its intoxicating effect, beer also contains a number of functional components (deriving in part from the specific starting materials used) that can be beneficial to health and it is, therefore in the same category as other health-promoting fermented foods [33]. In the case of barley, which was extensively grown in Norfolk, production of beer was an obvious step in the utilisation of the crop. In addition barley water (prepared by extracting barley with boiling water, filtering it and allowing it to cool) is a common beverage in many cultures. Many of the health-promoting properties of traditional, malted barley-derived beer are also found in barley itself and are a consequence of the presence of a range of phytochemicals, with the addition of hops (which will be considered in more detail in Sect. 9.3). The health factors that might be attributed to beer are complicated both by the compounds derived from the three main active ingredients (barley, hops and yeast) and by potential synergies between different components (for example, phenolic compounds and ethanol) [34]. Barley has been particularly associated with glucose and lipid metabolism [35] and the potential of barley as a source of functional ingredients for human health has been reviewed [36]. The development of beer (brewed with hops) and ales (brewed without hops but with other herbs instead) for medicinal purposes can be traced back thousands of years [37]. In addition, the combination of the brewing process (including boiling to extract sugars and flavour compounds) and the low pH of the product together with the presence of ethanol, reduced microbial (particularly pathogen) levels, contributed to enhanced sanitation among a rapidly growing population (such as happened in sixteenth century Norwich) and thereby increased the safety of the beverage consumed by the public [38] compared to drinking (often microbially contaminated) fresh water. In addition, the use of beer as a safe water source during long sea journeys gave an incentive for marine regions such as Norfolk (and others) to produce palatable beer for sea voyages. The importance of commercial brewing and beer was particularly evident from the activities of the Hanseatic League and this in turn emphasised the importance of beer (from barley) in Northern Europe compared to wine (from grapes) in Southern Europe. Indeed, Hamburg acquired the name "Brauhaus der Hanse" (the brewery of the Hanseatic League). The chemistry of beer (and cider) production will be detailed in Sect. 9.2. Another driving force for the production of both barley and beer in medieval Norfolk was the growth in the numbers of churches and other religious institutions. One of the most significant of these was the cathedral and (Benedictine) priory at Norwich [39] and it was noted that about two thirds of the malted barley supplied to the cathedral and priory was converted into beer. The arable nature of farming in Norfolk and the production of high-quality barley continued to have an impact on the production of beer in Norwich and Norfolk well into the nineteenth century and, through a process of consolidation, gave rise to the significant brewers of Victorian Norwich (Bullards, Steward & Patteson, Morgans and Young, Crawshaw and Youngs together with Lacons in Yarmouth). Many brewers accessed their most important ingredient (water) from their own bore holes. A large number of boreholes are shown to be present in Norwich in the survey data from the British Geological Survey and many of these are clustered along the Wensum valley. The ecosystems of chalk (such as might be found in Norwich) have been studied using boreholes [40] and this has yielded significant information about groundwater ecology and the presence of invertebrates and bacteria.

9.2 The Chemistry of Beer and Cider

The chemistry of beer is intrinsically linked to three key components:

- Malted barley (*Hordeum vulgare*)
- Hops (*Humulus lupulus*)
- Yeast (usually *Saccharomyces cerevisiae*)

The contribution of each of these will be considered from the standpoint of chemistry and the molecules that each contains.

The process of malting is essential to the production of beer and to many of its taste and flavour characteristics [41]. The malting process involves immersing the barley seed in water thereby allowing germination to take place. The germination leads to the release of enzymes that break down the grain cell wall and starch is released. Amylases are then released which convert the starch into fermentable sugars. Once this process is completed to the required degree (some proteolysis also takes place), the grain is kiln dried in stages until it has the correct colour and flavour characteristics. While physicochemical properties of the process (levels of moisture, temperature and time of drying/roasting) are important factors, the barley must also have certain defined characteristics (particularly starch and protein content) to reproducibly produce malted grains of the required characteristics. Numerous papers and reviews have been written about the malting and extraction of barley [42]. The possibility of breeding (including gene editing) to improve barley quality can include a number of targets including nitrogen and carbohydrate content and disease resistance and resilience to changing growing conditions [43]. As research into the most important factors involved in the malting process has continued, targets for breeding programmes such as modulation of water uptake by the grain have been identified [44] and metabolic/genomic prediction to improve the brewing characteristics of barley has been investigated [45]. As noted above, malting is influenced by the starting material in the grain (particularly the relative amounts of protein and starch) and by the mixture of enzymes. The correlation between intrinsic enzyme activity and malting characteristics has been investigated [46] and the use of exogenous enzyme preparations to facilitate the malting process has been demonstrated [47] however traditional methods continue to operate. The next stage in the brewing process is mashing in which crushed malt is mixed with water and the mixture maintained at the optimal temperature for the hydrolysis of starch to fermentable sugars by amylases (63C-68C). The sugar-rich liquor is then collected and taken to the next stage (addition of hops) before fermentation. The three enzymic processes in mashing involve proteolysis, cytolysis, and amylolysis and these degrade proteins, cell walls and starch respectively. The link between barley malting and mashing has been investigated since these are interdependent processes [48] and the relationship between, for example, particle size and the efficiency of the mashing process has been studied [49]. Balancing starch gelatinisation and hydrolysis through the manipulation of factors such as particle size are crucial to the efficiency of the mashing process [50].

The quality of the barley is an important factor in the brewing process and selection of varieties with optimal characteristics has already been noted. One aspect of the growth of barley (and, indeed other crop plants) has been the effect of the root exudates of different varieties on the soil microbiome and this has been investigated in Norwich. By selecting two different cultivars of barley, it was found that cultivar-driven *Pseudomonas* selection was driven by root exudate composition, with the greater abundance of hexose sugars secreted from the roots of one variety attracting microbes better adapted to growth on these metabolites and *vice versa* for the other variety [51]. This suggests that the root exudates may be demonstrating a true prebiotic effect with the selective use of specific substrates by particular microbial species which in turn will affect the growth of the plant. This is a further example of the importance of sugar composition to the whole brewing process from plant growth to processing and fermentation.

Once a sugar-rich extract has been obtained, the next stage in the brewing process is the addition of hops. A range of plant-based flavour compounds have, at various times, been used to provide flavour for beer. The mixture of herbs (which could include coriander, wild rosemary, heather, ginger, spruce, juniper, yarrow and bog myrtle) is referred to as gruit. The use of hops as the predominant flavoring agent was known in Flanders and Holland in the late fourteenth century and it was exported, among other places, to England and was also encouraged by the adoption of the Reinheitsgebot in Bavaria in 1516. The first import of Low Countries "beere" into England seems to have come in the fourteenth century, when James Dodynessone of Amsterdam paid a toll on beer at Great Yarmouth in 1361–62 and beer was imported from the Low Countries into other East coast ports in England at about this time [52]. Wine was imported *via* London in the fourteenth century and is detailed in the London Customs Accounts however there is no mention of beer. By the end of the fourteenth century Great Yarmouth was importing 40–80 barrels of beer a month. Indeed, much of the brewing of beer (containing hops in contrast to English ale which contained other herbs) in the fifteenth and into the sixteenth century was carried out by immigrants using imported hops. For indigenous brewers, the introduction of a tax on hops in England led to the abolition (by Act of Parliament) of the use of non-hop agents in beer in 1710. The apparent 'rediscovery' of Mendel's work on pea breeding resulted in the development of breeding programmes for hops at the beginning of the twentieth century. However, the breeding of hops with specific desirable characteristics significantly predates knowledge of how to carry out systematic breeding [53]. Accidental crosses between cultivated and wild hops were another route for the development of plants with novel properties. More modern strategies including those based on genotyping have been used to construct hop plants with defined, desirable characteristics [54] and marker assisted breeding is also used [55]. The correlation of genetic analysis with chemical constituents has been carried out in order to establish some knowledge of the diversity among wild hops [56] and this type of base knowledge can provide raw material for breeding programmes. Much of the interest in hops lies in the production of bitter oils that are produced by the female part of the plant. In addition to agronomic performance, it is the yield of secondary metabolites that create the aroma and taste characteristics of hops that underpins much of the interest in optimising hop varietal development. Hops contain a number of secondary metabolites—some of which have medicinal properties (see Sect. 9.3). The chemical transformations of natural products from hops—with specific reference to the brewing process—have been reviewed [57]. In addition to their flavour profile, some of the compounds present in hops have antimicrobial and anti-oxidative properties and therefore have a role in preventing

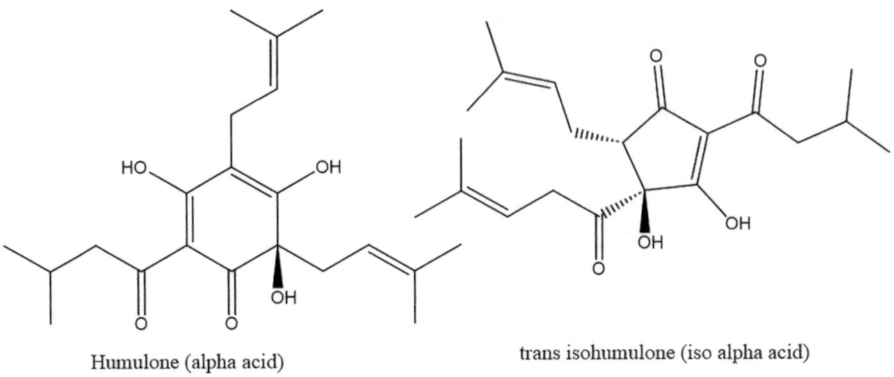

Fig. 9.1 Conversion of hop alpha acids into iso alpha acids

spoilage of beer. Hop alpha acids (also called humulone, adhumulone, and cohumulone) are converted by heating into iso alpha acids (Fig. 9.1) and these compounds, which are more water soluble than the parent compounds, impart bitterness to the beer.

While hop beta acids (lupulones) are less bitter than the alpha acids, they make a larger contribution to the aroma of beer. Different hop varieties have different proportions of bittering and aroma compounds and therefore are used in differing amounts in different types of beers. While the hop acids and other secondary natural products are present in the hop plant and do not require tissue damage for their generation (unlike the volatile secondary metabolites of plants of the genus *Brassica* or those of the genus *Allium*), glycosylated aroma precursors have been found in hops [58] and these may be released through the action of glycosidases or by acid-catalysed hydrolysis. One such compound is the glycoside of β-damascenone (Fig. 9.2) [59].

Other studies on the factors that affect the formation of terpene alcohols (and therefore impact the flavour properties of beer [60]) have been carried out and these indicate that glycosidase activity derived from yeast during fermentation may be an important factor in the accumulation of different compounds [61]. In addition to the acids mentioned above, the terpenoids, including humulene, myrcene, geraniol and linalool (and their glycosides) contribute to hop aroma (see above). Some of these compounds are also biotransformed by yeast during the brewing process (the monoterpene alcohols geraniol and linalool were biotransformed in a test system containing brewing yeast however in this study the major terpenoids were unaffected [62]). The planned manipulation of flavour and odour compounds by yeasts remain a viable approach [63]. The importance of the chemical constituents of hops (and aspects of the brewing process) has been reviewed from a historical perspective [64]. The effect of hop components on yeast metabolism has been studied comparatively less however the metabolomics of different hopping practices on yeast have been investigated using spectroscopic techniques and an effect on purine metabolism was noted [65]. The use of hops in brewing is carried out for a combination of

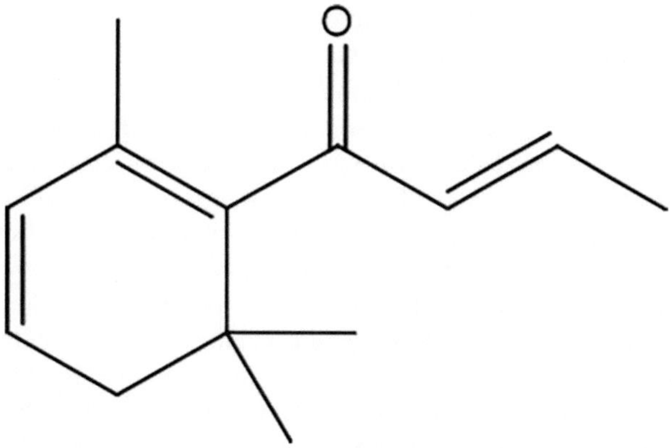

Fig. 9.2 The structure of β-damascenone

reasons including preservation and (more recently) health benefits for the consumer. The major driver for the use of hops, however, remains the effects they and their components have on the sensory properties of beer [66]. Other aspects of the properties of hops will be considered in more detail in Sect. 9.3.

In addition to malted barley and hops, the third important component of beer (apart from water) is yeast. The link between beer and intoxication is not as overt as might be assumed [38]. Indeed, as noted previously, beer was an important source of a microbiologically safe liquid (compared to water) and contained a number of vitamins and nutrients. The importance of beer for sailors has already been noted and, unsurprisingly, much of the work done on developing brewing (including selecting the optimal yeast strains) was carried out in Northern European, brewing, maritime countries. The Carlsberg Research facility for the investigation of the science of the brewing process was set up in 1876 and work there led to the isolation and purification of yeast strains suitable for brewing. The first pure strain lager yeast was *Saccharomyces carlsbergensis* which was isolated and characterised in 1883. This enabled consistent beer to be brewed and the use of a pure strain derived from a single yeast cell and allowed light, lager-type beer to be brewed at relatively low temperatures. More details of the characteristics of the main types of yeast will be given below. Later the genome sequence of *Saccharomyces carlsbergensis* was determined [67]. Before the advent of the isolation and characterisation of specific strains of yeast, fermentation was a consequence of spontaneous inoculation by wild type organisms. This is still the basis of the formation of lambic-style beers [68] which are still brewed in the traditional way in Belgium [69] although culture-dependent methods are also used. The formation of specific metabolites during the fermentation and maturation of lambic beers has been studied and the links between metabolite production and specific microorganism growth has been delineated [70]. Apart from the example of lambic beers, the control and characteristics of beer brewing requires the use of pure, specific yeast strains with distinct characteristics which leads to the production of a consistent beer [71]. This is what Emil Chr Hansen achieved at the Carlsberg Institute in 1883. Incidentally, the pH scale was also introduced at the Carlsberg laboratory by the Danish chemist, Søren Peter

Lauritz Sørensen [72]. Significant advances in the study of fermentation were made by Pasteur who was perhaps the first to implicate fermentation with a living species [73]. The yeast fermentation process involves the conversion of pyruvate into ethanol. This process is carried out under anaerobic (or low oxygen) levels (Fig. 9.3).

In simple terms two types of yeast are used in brewing: *S. cerevisiae* is a top-fermenting yeast which is used to make ales while *S. pastorianus* (a hybrid of *S. cerevisiae* and *S. eubayanus*) is a bottom-fermenting yeast used in lager brewing and is the modern name given to *Saccharomyces carlsbergensis*. One of the more interesting aspects of the use of nutritional interventions using beer was the study of strains of yeast that were significant producers of thiamine [74]. This may have provided an opportunity for a nutritional intervention [75] since thiamine deficiency is one of the causes of cognitive decline (termed Wernicke-Korsakoff syndrome (WKS)) caused by chronic alcohol abuse [76] and fortification of beer with thiamine was actively discussed [75]. A more detailed investigation of the factors that influence vitamin production and excretion into the medium (beer) by yeast will be necessary if any potential for claims that beer could be a 'functional food' will be required. While ethanol is toxic, the potential for the production of beer with functional properties [77] or the development non-alcoholic 'beer' with functional properties by selection of certain strains of yeast has been considered [78]. In addition, there is a substantial history of use of spent yeast and yeast extracts as functional foods [79].

The making of cider was an important aspect of the use of produce—particularly in parts of Norfolk. The earliest written account of apples in England comes from a thirteenth century Norfolk document, describing how a Norfolk farmer (Robert Evermere) paid his annual rent with '200 Pearmains and 4 hogsheads of Pearmain cider.' [80]. The importance of orchards in the landscape of Norfolk has been noted [81]. A number of local varieties of apple were developed in Norfolk including Happisburgh, Horsford Prolific, Hubbard's Pearmain, Norfolk Royal, Norwich Pippin, Dutch Mignonne, Norfolk Beefing and Winter Majetin. A number of the varieties which were later adapted, originated in Europe and, in particular, in Holland [82]. Cider was used as payment for farm labourers in Norfolk as early as the thirteenth century. In addition to the economic value of cider, there has also been a consideration of the beneficial aspects of cider and apple juice consumption—particularly in the context of cardiovascular health [83] Much of the interest in the potential health benefits of cider have concentrated on the effects of cider vinegar [84] rather than cider itself and there seems to be an association with glucose

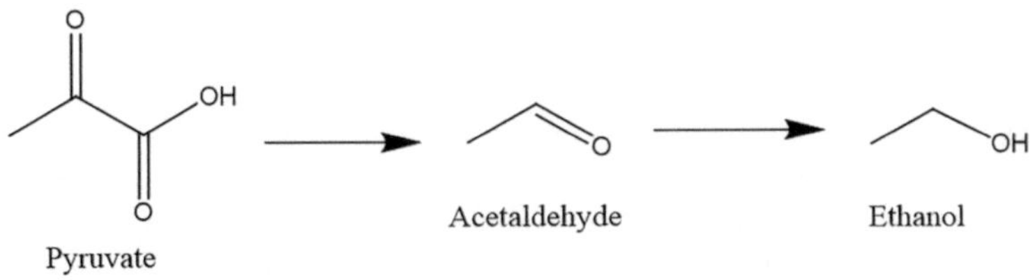

Fig. 9.3 The conversion of pyruvate to ethanol

metabolism. It is unclear what the potential mechanism of action of the vinegar might be although delayed gastric emptying and inhibition of amylase activity have both been suggested and these may impact upon glucose metabolism. It is, perhaps, more likely that any physiological effect can be ascribed to the acetic acid content rather than any apple-derived phytochemical [85] and an effect of acetic acid on post-prandial hyperglycemia has been suggested [84]. The positive health effects of different apple products have been reviewed and a number of associations suggested [86] and the physiological effects of a range of fermented apple products (generally de-alcoholised) have also been reviewed [87]. In addition, the use of probiotic bacteria (particularly *Lactobacillus* species) in the fermentation of apple juice rather than yeast has been suggested to promote healthy properties [88].

9.3 Hops and Health

As noted previously, hops have been associated with a number of health-promoting effects. Among other mechanisms, some of these may be associated directly with an interaction between bioactive bitter compounds in hops and the extra-oral bitter taste receptors but other associations and causalities have been suggested. The interaction specifically with bitter receptors will be noted here but considered in more detail in Sect. 9.4.

The development of hops and hop extracts for medicinal uses has a long history [89] and while some bioactive principles have been identified [90], many of the studies are correlative and observational rather than exploring the mechanistic basis of health promoting effects. The effects of hops have been grouped according to perceived endpoints. These include antioxidative activity [91], prevention of anxiety and stress [92], anti-inflammatory and antimicrobial effects [93]. These have been attributed to a range of phytochemicals present in hops including bitter compounds, prenyl flavonoids and a range of phenolic compounds [94]. The development of a more refined health message associated with hops can be considered in several different ways, depending on the nature of the claim, the availability of biological assays and the application of isolative chemistry techniques to crude extracts. Studies can either be chemistry or biological effect led (or a combination of the two)

- Claims associated with extracts containing a number of components and for which a dose: response relationship cannot be established.
- Claims based upon reductionist chemistry—the isolation of individual compounds (or groups of compounds) often with common structural motifs.
- Associative or correlative claims using either mixtures or individual compounds for which a clinical endpoint can be measured. Such an approach is bioassay led and treats chemistry and the identity of the active agents as a 'black box'. While correlative evidence can be fairly compelling, it often does not form the basis for

the development of optimal efficacy or for identifying particular structure:activity relationships.
- Claims based on a mechanism that can be proven and for which causality can be demonstrated. Such approaches are often the result of the application of reductionism to biological systems (for example, measuring agonist or antagonist activity against a particular isolated receptor) and, while they can form the basis for optimisation of activity, the precise nature of the biological target and its relationship to whole system clinical outcomes must be demonstrated.

Examples of how to probe these four modes of activity for hops will help to illustrate more generally the strengths and weaknesses of these approaches and their applicability under different circumstances. There is, unsurprisingly, some overlap between the approaches and one pathway (for example, reductionist chemistry) may lead to another (for example mechanism-based information). The use of hop extracts for medicinal/health uses has a considerable record. The first group to be considered will be multi-component extracts. One of the main drivers has been the extraction of used (spent) hops since these contain a number of bioactive constituents [95]. The biological targets for these hop extracts include hormonal, metabolic, inflammatory, and epigenetic pathways. A number of specific effects are associated with prenylated flavonoids and with chalcones and these will be discussed in more detail later when a more reductionist chemistry approach is considered. There is also significant interest in the examination of complete hop extracts for the control of menopausal symptoms [96] which implies an interaction with the estrogen receptor [96] although in the absence of specific receptor binding information, this remains a plausible hypothesis rather than a proven mechanism. In addition, tests on whole extracts are, as noted previously, of limited value. In addition to effects at a receptor level, hop extracts (as noted previously) also have general antioxidative properties [91]. The second category includes the isolation and purification of specific hop extract components. For example, some attempts have been made to implicate specific compounds in hops (for example, xanthohumol) with the clinical (anti-cancer) properties of hop extracts however the mechanism of action may involve the stimulation of formation of free radicals and the induction of apoptosis in cancer cells [97]. The third category are associative or correlative claims using either mixtures or individual compounds for which a clinical endpoint can be measured. An example of this is that hop extracts have been postulated to affect gut hormone secretion (although this may be a consequence of binding to a bitter receptor in the gut—see Sect. 9.4) [98]. Hop extract has also been found to reduce body fat levels [99] and also has specific anti-microbial and anti-biofilm activity which may have clinical consequences [100] and also anti-viral activity [101]. More generally, the antimicrobial activity of hop extracts has been reviewed [102]. Finally, hop extracts have been implicated in improved sleep patterns by hypothesised binding of components to the γ-aminobutyric acid (GABA) binding site on the GABA A receptor [103], however this falls short of the mechanistic and causal relationship mentioned in the fourth category above. Attempts to correlate sleep induction with specific groups of hop compounds by varietal variation in the observed effects [104] or attempts to establish a dose: response relationship have not been wholly convincing [105].

As noted previously, attempts to correlate the biological activity of hop extracts with specific components have met with limited success. A more significant pharmacological model would make use of specific isolated hop components being tested in isolation and this is the approach outlined in the fourth approach mentioned previously. This type of approach does not permit the study of synergistic effects. While effects on specific signalling pathways have been noted including ERK1/2 phosphorylation, regulation of AP-1 activity, PI3K-Akt, and nuclear factor NF-κB, many of these effects are associated with specific compounds. The primary pharmacologically active components of hops include, in particular, bitter acids and specific prenylflavonoids such as xanthohumol and 8-prenylnaringenin. The bitter acids from hops have a range of biological activities [106] including activity to counter inflammatory and metabolic diseases. In addition, iso-alpha acids derived from hops which contain a β-tricarbonyl moiety suppress microglial activation and improve memory impairment induced by inflammation or Alzheimer's disease pathology and may thereby prevent cognitive decline [107]. The suggestion that inflammation may be a core cause of Alzheimer's disease [108] provides a rationale for the possible treatment of the disease with specific anti-inflammatory compounds from hops. The anti-inflammatory mechanism of action of a range of hop acids has been studied and the possibility of synergistic effects with compounds with different modes of action noted [109]. This combines reductionist chemistry with a specific mode of biological activity. A key requirement for the study of phytochemicals is the development of a robust method of analysis and this has been achieved for hop acids [110]. The study of the biosynthesis of hop bitter acids has enabled biotechnological methods for their production to be developed [111] and this will facilitate the production and testing of specific components. As noted previously, the biological activity of crude hop extracts can, in part, be associated with the presence of prenylflavonoids such as xanthohumol and 8-prenylnaringenin. Xanthohumol is a prenylated chalcone (Fig. 9.4) and has been studied from a biosynthetic and bioactivity standpoint [112].

As with many phytochemicals, there have only been limited human studies and, apart from fairly generic claims associated with inflammation and anti-oxidative properties (for example) definitive human studies, including dose response investigations, are limited [113]. The other compound that has been studied as an isolated entity is 8-prenylnaringenin. The primary activity associated with this compound is estrogenicity but it is not structurally similar to estradiol (Fig. 9.5).

The nature and consequences of the binding of the molecule to one of the subtypes of estrogen receptor will be discussed later. The presence of 8-prenylnaringenin (also a hop-derived estrogen) has been demonstrated in a number of foods [114] however the levels commonly found in foodstuffs is below the level of pharmacological significance. Its isolation and characterisation is, however an example of a bioassay-guided fractionation [115]. A number of methods, particularly mass spectrometry-based, have been developed for reliable quantitative determination of 8-prenylnaringenin and related prenylated flavonoids however the interconversion

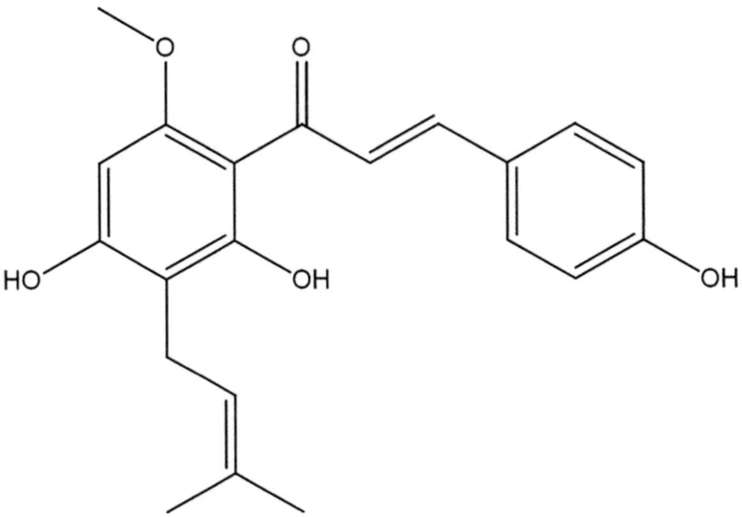

Fig. 9.4 Structure of xanthohumol

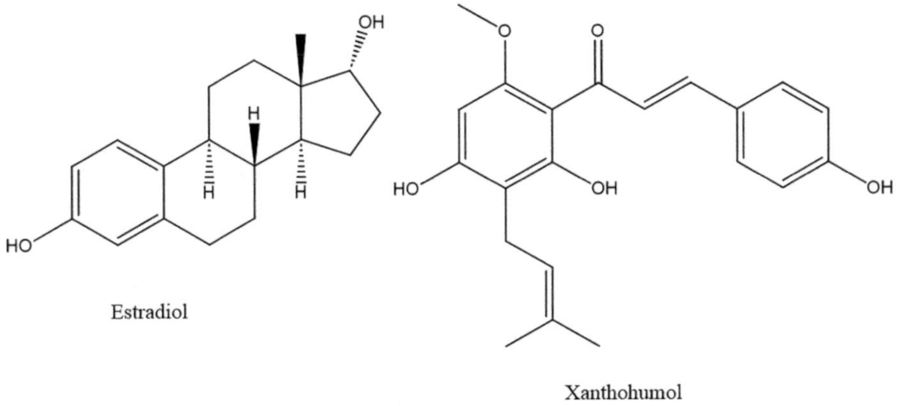

Fig. 9.5 Structures of estradiol and xanthohumol

and isomerisation of chalcones during the brewing process can lead to reports of different amounts of the compound being present in beer [116] and, in addition, it is metabolised in the human liver [117] and therefore, even when a defined phytochemical with a known mode of primary action is known, the clinical consequences of its ingestion can be challenging to deconvolute.

Some of the studies already mentioned can be categorised as associative or correlative claims using either mixtures or individual compounds for which a clinical endpoint can be measured. More examples that illustrate these types of investigation will be examined. The most common such studies are population-level interventions. Moderate consumption of both beer and wine have been found not to be deleterious to health and, indeed, may provide some protective effects [118] with a positive effect on endothelial function proposed [119] however it is unclear if this effect is potentiated by alcohol or by some of the other components present. Some of the effects may be associated with the presence of xanthohumol [120] however beer (and hop extracts) is a complex mixture of bioactive ingredients and it remains unclear if the application of reductionist techniques will lead to the identification of a primary causative agent or if the indications and studies will favour a correlative effect and possible synergies. One of the simplest approaches to try to simplify a major confounder is to remove the effect of alcohol since it is a highly bioactive compound that is metabolised (oxidised) in the liver and to concentrate on links between disease/health outcomes and hop constituents [121]. In addition to the chemically characterised prenylated chalcones, some of the major, biologically characterised groups of compounds from hops are antioxidants [91].

Since both hops and the products derived from them (beer or hop extracts) are mixtures containing complex chemical agents with a wide spectrum of biological activities, demonstrating causality can be challenging. The bioactivity-guided isolation of 8-prenylnaringenin has been mentioned however it is not possible to identify a single compound that is responsible for all of the activity of hops in a given physiological context [122]. One interesting, mechanistically derived, molecular effect of hops is the inhibition of the hydroxylation of estradiol at the 4-position which leads to the formation of the presumptive proximal carcinogen (an o-quinone). It was also found that 8-prenylnaringenin was a significant inhibitor of the hydroxylase enzyme while xanthohumol had no effect [123]. This inhibition of oxidative estrogen metabolism may, in a mechanistic sense along with inhibition of aromatase activity, account for the specific anti-proliferative and anti-cancer properties ascribed to hops [124]. The molecular activity of specific hop components (enzyme inhibition) can be interpreted in the context of a clinical/physiological endpoint [125]. Another example of a molecular event, but with a less defined clinical outcome is the binding of hop compounds—particularly 8-prenylnaringenin—preferentially to the estrogen receptor-alpha. This has been studied [126] and attempts made to correlate the receptor binding with clinical outcomes. Definitive studies linking mechanisms to clinical endpoints are relatively rare [127] and explaining the difference between the effects of mammalian and phytoestrogens in terms of physiological endpoints remains problematic. It is, however interesting that while many phytoestrogens bind preferentially to estrogen receptor—beta, 8-prenylnaringenin from hops binds preferentially to estrogen receptor—alpha [128].

While there has been significant interest in 8-prenylnaringenin, because of its differential binding properties, the consequences of the binding of 6-prenylnaringenin (Fig. 9.6) to the estrogen receptor has also been noted [129] and this has been associated with the prevention of the inhibition of detoxification of estrogen and to hence may give rise to a beneficial effect.

The health-promoting effects of hops, apples and barley has not traditionally been a major driver for the production of beer and cider. Indeed the dangers of 'the demon drink' have been at the forefront of health policy for much of the twentieth century [130]. The use of fermentation, in part for preservation and in part to

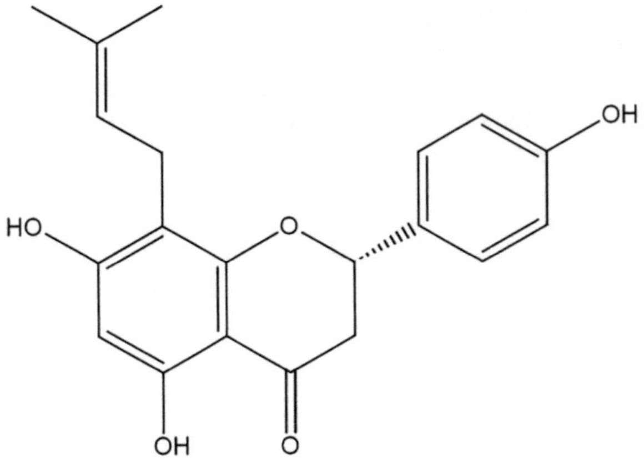

6-Prenylnaringenin

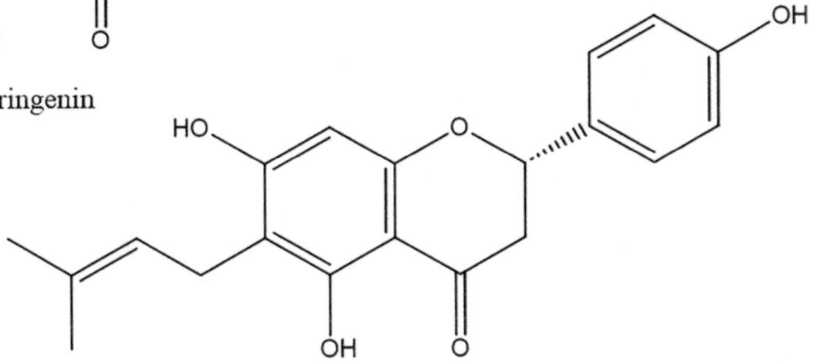

8-Prenylnaringenin

Fig. 9.6 The structures of 6-prenylnaringenin and 8-prenylnaringenin

optimise the production of the side product (ethanol) led to the commercial exploitation of alcoholic beverages. The processing and, to an extent, the inherent properties of the ingredients helped to endure that microbiologically safe beverages were produced, and any health benefit was somewhat incidental. Nonetheless the association of alcohol with wellbeing was an important aspect of the advertising and marketing of beer [131, 132]. More recently the potential effects of beer in the positive modulation of the gut microbiome have been suggested to be a potential causative mechanism [133] for beneficial outcomes.

The correlative and causative approaches to understanding the health-promoting advantages of dietary components follows from a better understanding of the molecular basis for health. The example of the bitter receptor will be used to demonstrate the progression from the (reductionist) study of the binding of components to receptors to clinical endpoints and a progression from sensory characteristics to health (and not simply avoidance due to an adverse taste). This will lead to a discussion of the production of chocolate in Norwich—despite cacao (unlike barley and hops) not being an indigenous crop in Norfolk—and finally to a discussion of correlation and causality in disease and health as mediated by diet.

9.4 The Taste Profile of Natural Products: Taste Receptors' Function and Health

Several natural products have a bitter taste and, it was previously supposed that this was associated with the prevention of predation of the producing organism. The production of compounds that induce taste aversion would seem to be a useful property. This concept has been somewhat challenging since, while an evolutionary advantage can be assumed for the producing organism, it relies, for its effectiveness, upon the bitter taste being detected by the consuming species and it is unclear how enhanced sensitivity to bitterness through the development of bitter receptors would be of value to a consuming species—unless there was a correlation between bitterness and toxicity. Bitter taste would therefore be a signal that a particular molecule is harmful. In addition, bitter receptors do not just occur at sites where food is ingested, and extra-oral bitter receptors seem to have various functions apart from activating the bitter taste transduction pathway. Before investigating further, the concept of bitterness and its potential physiological purposes, it is important to have an understanding of how bitter substances (and compounds with other taste properties) are detected, how taste receptors work in different environments and how the production of taste compounds is modulated by the physiological impact they have? This suggests some degrees of analogy with the chemistry of compounds produced by living species. Natural products may have different roles/benefits for producer organisms compared with consuming species. The development of the different types of chemistry that facilitated an understanding of crop development or the formation of pigments, gave rise to different products and adaptations in Norfolk compared to where they may have been first developed. For example, the use of fodder crops such as turnip led to a different crop rotation system in Norfolk compared to that developed in Northern Europe [134]. Similarly, the nature of the chalk springs (particularly high calcium levels) in Norfolk and therefore the availability of calcium in water [135] changed both local health indices, the characteristics of dyestuffs, in particular the use of madder as a staining agent [136], and the separate development of the brewing industries [137]. The analogy between the effect of the local conditions on the chemistry that was able to be carried out in different industries (agronomy, dystuffs and brewing) and the evolution and importance of (in particular) extra-oral bitter receptors (in addition to their role in taste sensation) in different environments is analogous. A taste receptor situated in one organ in a body will have a very different biological function in a different organ. The chemistry that develops and underpins particular industries will be different in Norfolk (with a particular tradition of expertise in certain types of chemistry on an individual, institutional and societal level) compared to other places with different traditions. A consideration of taste receptors will be reflective of these differences.

Taste receptors are of different types and a detailed description is beyond the scope of this chapter however salient aspects of G-protein coupled receptors (GPCRs) will be discussed in order to provide context. Taste receptors enable organisms to distinguish five different tastes (sweet, sour, salt, umami and bitter). In the case of salt and sour taste, the detection of sodium and hydrogen ions respectively are the crucial aspects. The mechanism of how salt is perceived, while appearing to be simple remains (detection of a particular ion) somewhat obscure in mechanistic detail [138] and, for example, both sodium-dependent and -independent pathways mediate attraction and aversion. As will become clear, even the apparently simplest of taste sensations is complex and provides, therefore, an analogy with the complexity of even apparently simple chemistry. In addition to sodium ions, some other dietary components (for example, some peptides) can enhance salt perception. Furthermore, the counter ion (chloride) is also an important factor in the detection of salt taste [139] and a receptor with a specificity for chloride ion has been suggested to be involved [140]. At a further level of organisation, taste cells are divided into four types (I, II, III and IV) with different properties and specificities. Type II cells are implicated in the detection of sweet, umami and bitter taste while sour taste is detected by type III cells. The cellular types responsible for detection of salt taste remains unknown although the sodium-dependent salt taste receptor is a sodium channel whereas the nonspecific salt taste receptor is a type vanilloid receptor-1 nonselective cation channel, TRPV1 [141]. The potential role of transmembrane channel-like 4 (TMC4) in the involvement of chloride perception of salt taste has been investigated [142] (also see above). Salt taste perception is suggested to be mediated in two ways—amiloride-sensitive and amiloride-insensitive reception, based on the difference in responsiveness to amiloride, which is an inhibitor of the epithelial Na^+ channel (ENaC). This suggests the presence of at least two mechanisms for the detection of salt taste. The link between salt intake and the risk of developing cardiovascular disease (CVD) has been well documented and, while the direct link between salt intake and blood pressure in salt-sensitive individuals has been given a predominant emphasis, other mechanisms have also been suggested to be involved [143]. The precise mechanistic details are not required for the link between excessive salt intake and CVD to be invoked to encourage the food industry to consider salt reduction strategies as an important formulation challenge with an impact on public health at a population level [144]. It has been suggested that approximately one third of individuals in a population are 'salt sensitive' and the effect of salt intake on blood pressure is abnormally exacerbated [145]. While selective intervention may be more effective the high proportion of salt sensitive individuals suggests that population level interventions are the simplest way of engineering a public health benefit. While single nucleotide polymorphisms (SNPs) in specific taste receptor genes have been found to have a correlation with psychophysical responses to sweet, fat, umami, and salt taste, it is unclear if specific polymorphisms are associated with other physiological responses (for example, the link between blood pressure and salt taste sensitivity). In the case of sour taste, the receptor involved (OTOP 1) is, like the sodium receptor, also an ion channel [146] but in this case it is selectively permeable to protons [147]. The link between salt intake and blood pressure has stimulated studies of the link between other dietary components and physiological consequences including the possibility that omega-3 fatty acids (particularly found in some types of fish) may help to reduce blood pressure. Several mechanisms have been suggested [148] although these are not linked to salt intake. Nonetheless this may suggest that diet (enhanced fish consumption) could help to reduce the CVD risk factors prevalent among seafarers [149], although the risk remains high—probably due to a range of lifestyle choices [150] and deconvoluting the link between diet and health is even more complicated than the mechanism of salt perception. There is no evidence that extra-oral salt receptors might be implicated. Genomic linkage map analysis of an isolated population suggests a link between chromosome 1p36 and systolic blood pressure [151] and while this may suggest some genetic involvement, it is insufficient to explain what the mechanism of any increase in risk or susceptibility might be. Gut microbiome-mediated associations have been noted between elevated levels of trimethylamine –N-oxide (TMAO) and the development of CVD [152] and the impact of a marine diet on circulating levels of TMAO has been noted [153] which may also contribute to elevated CVD incidence in marine areas such as Norfolk, however it should also be noted that fish from relatively shallow coastal waters have significantly less TMAO in muscle tissue compared to fish from deep sea fishing areas [154] so the complexity of any causative relationship between dietary components and health status remains difficult to prove.

The sweet receptor is conceptually more complex than the salt or sour receptors a heterodimer composed of two GPCRs (T1R2 and T1R3). Structural and functional characterisation of the human sweet taste receptor has been carried out [155]. Umami taste is also mediated by a heterodimer of two GPCRs (T1R1 and T1R3). The mechanism of taste transduction employed by GPCR taste receptors has been elucidated [156]. Both sweet and umami taste receptors have been found extra-orally (in particular in the gastrointestinal (GI) tract) and this may suggest an additional, non-gustatory role for this class of GPCRs. In the case of sweet receptors, they are found in the pancreas and brain as well as in the oral cavity and the GI tract [157]. The presence of other 'taste' receptors in these tissues may suggest a range of physiological functions for compounds that bind to them in addition to their sensory function. In the case of the sweet receptor and ligands which bind to it, such additional functions may include regulation of adipogenesis and bone biology [158] as well as glucose absorption and insulin release. The activities of various GPCRs, as mediated by exogenous ligand binding, suggests a complex interrelationship between living organisms and their environment. The extra-oral umami taste receptor also seems to be implicated in insulin regulation and the binding of umami-associated amino acids (glutamate and arginine) to the heterodimer composed of T1R1 and T1R3 has been demonstrated in pancreatic cells *in vitro* [159] and exposure to these amino acids also resulted in a significant increase in the secretion of insulin.

The most complex of the taste receptors are those associated with the detection of bitter compounds. These are also GPCRs but multiple forms of them with specificities for different ligands exist. From a pharmacological and physiological standpoint, the extra-oral consequences of the location of various bitter receptors have therefore attracted most attention and scientific investigation. The importance and function of bitter receptors may vary depending on their location. Those in the GI tract have been studied and found to be implicated in a number of physiological effects [160] including, in particular, the secretion of taste hormones and gastric acid [161]. In addition bitter compounds, acting through interactions with bitter receptors in the GI tract, can also modulate the innate immune response [162] and may, thereby, confer resistance to pathogens. In addition, polymorphisms in bitter receptors can affect both infection susceptibility and immune responses [163] and the impact of disease states modulated by infection such as cystic fibrosis, chronic rhinosinusitis, dental caries, and periodontitis. While the mechanism of action of different bitter compounds (*via* interaction with bitter receptors) may vary, indeed several pathways may operate, the detection of quorum sensing molecules produced by pathogens with subsequent induction of an innate immune response has been noted [164]. Bitter receptors are found in Paneth and goblet cells and are implicated in the release of antimicrobial peptides and glycoproteins, thereby providing a direct defence against pathogens [165]. Compounds produced by *Pseudomonas aeruginosa*, bind to bitter receptors in macrophages and airway epithelial cells leading to Ca^{2+}-dependent activation of endothelial nitric oxide (NO) synthase which then leads to formation of nitric oxide which has antimicrobial properties [166]. The effect of bitter compounds is also modulated by the stress-induced upregulation of production of a heat shock protein (HSP 90) which serves as an innate immune modulator by regulating NO production downstream of the bitter receptor T2R signalling [167]. Another physiologically important localisation of bitter receptors is in airway epithelial cells where they may be involved in a relaxation effect on airway smooth muscle [168] by a distinct signalling mechanism. It is known that bitter receptors can be involved in cholecystokinin (CCK) secretion from enteroendocrine cells and this has been shown to activate the ATP-binding cassette B1 (ABCB1) in intestinal cells [169]. The product of the ABCB1 gene (P-glycoprotein) is an efflux transporter and may be responsible for multidrug resistance [170]. The specific bitter taste receptor involved in this response (T2R38) binds compounds containing a thiourea (NC = S) moiety [171] which is found in the natural compound present in many *Brassica* vegetables, goitrin (Fig. 9.7).

Expression of T2R38 in pancreatic cancer cells [172] and its involvement in the expression of ABCB1 might suggest that, since *Brassica* vegetables contain an agonist of T2R38, the intake of these vegetables may contribute to the occurrence of multidrug resistance of tumour cells to chemotherapy. This conjecture will not affect advice concerning the prevention of pancreatic cancer in the first place [173] but may be relevant to post-operative treatment with drugs however further studies on the link between diet and cancer prognosis are required [174].

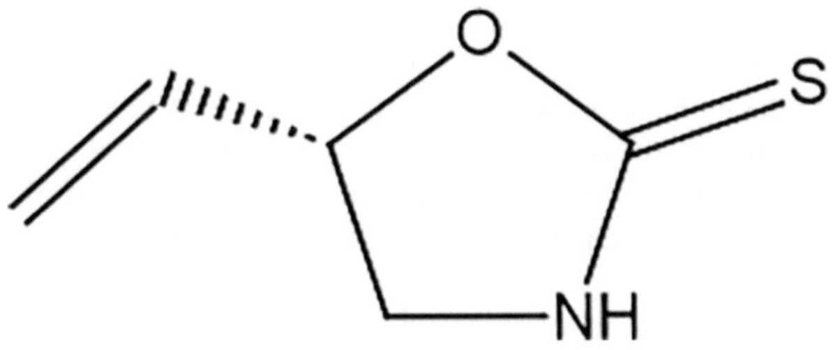

Fig. 9.7 The structure of goitrin

Another potential implication is the interaction between bitter receptors and microbial quorum sensing molecules which may also have some relevance to the treatment of Alzheimer's disease (AD). In order to deconvolute this possible causal link, it is necessary to understand some of the roles of different factors in the progression of AD. While the pathology of AD is considered to be strongly associated with the accumulation of beta amyloid plaques and neurofibrillary (tau) tangles in the brain [175] which result in neuronal cell death, the situation is complicated by both plaques and tangles being composed of endogenous elements which have a number of normal physiological functions. For example, the tau protein has an important role to play in the binding and stabilisation of microtubules [176] and amyloid beta has been implicated in neuroprotection in the ageing brain [177] and as an inducible antimicrobial factor and part of the innate immune defence system [178]. The role of inflammation in the development of pathogenesis of AD has been the subject of considerable speculation [179] and the presence of systemic [180] and local [181] infection have been implicated as triggers for neural inflammation. In the case of neural inflammation, the induction of expression of amyloid beta by infection and the formation of a biofilm around a bacterial species thereby preventing its removal by the formation of fibrils that can trap pathogens. An effect of amyloid beta on innate immunity has also been postulated (and mentioned previously) [182]. If the production of a biofilm prevents the effective removal of pathogens by amyloid beta then biofilm disruption may be postulated to permit pathogen removal and prevent the adverse accumulation of amyloid with an increased possibility of concomitant plaque formation. Both biofilms and pathogens have been associated with AD [183]. Indeed an effect of the interaction of bacterial biofilms derived from oral bacteria with amyloid beta has been proposed as a viable mechanism to accelerate neural plaque formation [184]. In summary, pathogen infection drives amyloid beta production in neural cells by a combination of increased inflammation and the production of biofilms to prevent expulsion of microbial species. Amyloid protein then co-aggregates with the bacterial biofilm to expedite plaque formation and increase downstream neuroinflammation. The next stage in the hypothesis requires the participation of bitter receptors and their extra-oral location. Bitter taste receptors have been found to be present in the brain [185]. In addition to binding to bitter compounds, some bitter receptors (T2Rs) contain binding sites for certain microbial quorum sensing molecules and this can provoke an antimicrobial innate immune response [186]. This activation of the bitter receptor at certain sites may prevent biofilm formation through induction of phagocytosis and removal of bacteria. In summary, binding of the microbial quorum sensing molecule, N-(3-oxododecanoyl)-L-homoserine lactone (AHL-12) [187] prevents biofilm formation and facilitates removal of pathogens.

The role of taste-active compounds in the physiological responses of consuming organisms clearly includes physiological responses that are not solely a function of taste aversion. It is known, for example, that some toxic compounds are not bitter and that many bitter compounds have negligible toxicity while having important health benefits [188]. These benefits seem to include activation of the innate immune system and other benefits that are mediated to a large extent through the presence of extra-oral taste receptors at a range of locations and with a range of functions [189]. Further investigations are likely to yield more information regarding the modes of action of natural bioactive compounds and, in probably many cases, these will be different to current perceptions. The classification of bitter taste receptors, in particular, is perhaps more biologically relevant if they are considered to be a specialised GPCR rather than part of the sensory detection system.

9.5 The Chemistry of Chocolate and its Manufacture in Norwich

Chocolate (derived from cacao) contains a number of bioactive compounds that have been implicated in various physiological processes. The development of chocolate (and other industries) in the nineteenth century is linked to social awareness (alternatives to alcohol and even to tea and coffee) and to particular religious and social movements—a number of chocolate manufacturers were involved in social programmes such as prison reform (Fry), social reform (Cadbury at Bourneville) social conditions (Rowntree). Chocolate manufacture is linked to the production of mineral waters in Norwich as water bottlers such as A J Caley expanded into drinking chocolate manufacture which gave them an additional market opportunity in the winter months. The popularity of mineral waters increased as spa towns grew in popularity and this led to the bottling, transport and sale of natural mineral waters in Europe in the mid-sixteenth century, with mineral water marketed as from Spa in Belgium, from Vichy in France, from Ferrarelle in Italy and as Apollinaris in Germany. They were sold as health drinks through pharmacies. In the UK, due to the geology present, thermal mineral water sources are relatively rare. Most mineral waters contain dissolved inorganic solutes [190] however the mineral content varies depending on the source of the water. Unlike spa towns such as Bath, Buxton and Harrogate, Norwich has no history of bathing in underground spring water for health purposes. The use of marine water for health-promoting bathing (thalassotherapy) was widely adopted in the eighteenth century [191] and this increased the popularity of coastal resorts [192] such as Great Yarmouth as an alternative to the spa towns. The more general use of spas and mineral waters to promote health has been the subject of review [193]. The popularity of bathing in water from a spa has been recounted [194] and there seems to have been a distinct focus on rheumatic diseases [195] and sanatoria for the treatment of tuberculosis were opened [196] however the lack of controlled clinical trials necessarily limits the effective use of spa therapy. In addition, compared to pharmaceutical interventions, it is expensive and labour intensive. In Europe, there has been considerable interest in the potential of natural and artificial mineral water [197] alongside advances in analytical methodologies which enabled the correlation of specific mineral mixtures with defined physiological endpoints [198]. The presence of chalk springs in Norfolk encouraged the development of spring and carbonated water bottlers since they could make claims based on the high mineral content of the water. Any assessment of mineral waters required that they are analysed accurately and this was considered to be a significant challenge to the chemist [199] and it was not until the eighteenth century that the concept of mineral waters containing some 'essence' derived from an earthly 'vital spirit' was discarded and mineral water was confirmed to be water that contained a number of dissolved components. The development of reproducible analytical methods and the debunking of the 'vital spirit' hypothesis enabled the preparation of artificial spring waters which were claimed to mimic waters from some of the most popular sources. One of the most successful enterprises was that founded by Jacob Schweppe, a jeweller in Geneva towards the end of the eighteenth century and building on the work done by Priestley on 'fixed air'. In the case of Albert Jarman Caley, he opened a small chemist's business in London Street, Norwich in 1857. Caley continued to bottle and sell mineral water before sinking two wells at Chapelfield in Norwich which provided him with a source of pharmaceutical mineral water (1880) and enabled him to open his 'Fleur de Lys' factory in 1890. By 1898, A J Caley and Son was a manufacturer of mineral water, ginger beer and chocolate and was established as a limited company. By 1915, Caley was marketing a mineral water, Sirus, which was described as being similar in composition and effect as the German mineral water Appolinaris, a natural spring water from Bad Neuenahr in Germany and was the European market leader. It is generally accepted that Caley's diversified into chocolate (drink) manufacture to create a source of revenue during the winter when mineral water was less popular (even for medicinal purposes). The transition to making solid chocolate was aided by an invention from the Netherlands. Dutch chocolate maker Coenraad Johannes van Houten and his father Casparus developed a method which used a hydraulic press to remove the fat, or cocoa butter, from ground and roasted cocoa beans, creating cocoa powder. In 1828, Coenraad Johannes patented the process, which became known as Dutching. After that, in 1847 the English company J.S. Fry & Sons developed the first chocolate for eating rather than drinking. This process was adopted by Caley in Norwich and formed the basis of his business. Unlike agriculture, beer making or weaving, chocolate making does not involve exploiting a locally available resource [200]. The connection of chocolate with health has been less direct than that of mineral water where a health benefit was a major driver from the outset. For chocolate, its primary role (due to the presence of xanthines such as caffeine and theophylline) was as a stimulant before other constituents were investigated [201]. These components will be described in more detail in Sect. 9.6. The use of cacao spread from South America, through Spain to the rest of Europe. While the popularity of it was firstly a consequence of its health-promoting properties and its use as a stimulant, by the twentieth century the nutritional properties of chocolate was being increasingly recognised and it became, over the years, a luxury item with an increasing appreciation of its link to health and wellness [202]. It should be noted, however, that, in a non-clinical setting, chocolate consumption is linked to a better standard of living and therefore a causative link is difficult to prove. Overall the link between chocolate consumption and health is weak—if it exists at all [203]. Perhaps the most speculative aspect of health/behaviour and the link to chocolate consumption is the apparent link, on a population level, between the chocolate consumption of a country and the awarding of Nobel prizes [204]. Such associations should, perhaps encourage critical consideration of the difference between causality and correlation.

9.6 Natural Products from Cacao and their Biological Properties

The growth, cultivation and use of cacao (*Theobroma cacao*) in South America has been studied using archaeogenomic approaches and the pre-Columbian origin of the crop traced [205]. The genomic variation in populations of the plant is a signifier of the development of varieties that are favoured for use by human populations. In Mesoamerica and Central America, the ancient domestication of *T. cacao* has been particularly well documented, in part owing to the tree's cultural significance to the ancient and current cultures of these regions however the details of the spread of the most popular (and aromatic) varieties remain somewhat obscure. The initial spread of cacao was a consequence of the presence of the stimulants caffeine (typically about 0.071 mg/g in cocoa products) and theobromine (at a higher level of 0.695 mg/g) (Fig. 9.8).

Caffeine and theobromine are botanically comparatively rare with the latter only found in only nineteen other plants [206, 207]. In addition, cacao contains polyphenols (classified as either flavonoids or non-flavonoids) and these are responsible for many of the biological properties reported for products from cacao. While the main bioactive compounds present in cacao have been extensively studied [208], the extraction and use of the cocoa bean only uses about 10% of the plant material and uses for the unused parts (husk, shell and pulp) are the subject of investigation

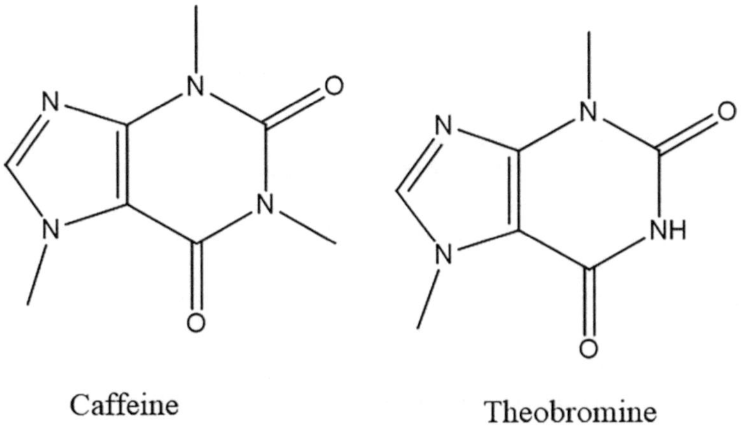

Fig. 9.8 The structures of caffeine and theobromine

[209]. The proportions of the different bioactive polyphenols and the xanthine alkaloids varies both with the variety of cacao, the growing conditions and the processing carried out [210] and these will impact upon the observed biological properties after consumption. Some of the most significant effects have been ascribed to antioxidative and anti-inflammatory pathways with some mechanistic details suggested [211]. These are relatively recent studies and the value of the crop predates modern investigations. The transatlantic cocoa trade in the eighteenth century was largely controlled by Spain [212] and this, in turn, was driven by consumer demand and the perceived stimulatory properties of the drink. A drink made from raw cacao beans is bitter tasting (caffeine and theobromine are known bitter agents) The components responsible affect mood and level of alertness [213] and the development of the cocoa trade is linked to the increasing use of sugar as a sweetening agent and the link between sugar and chocolate continues [214]. The development of sweetened products from cacao changed the drivers for consumption from stimulatory and other health properties to indulgence in the 17th and early 18th centuries [215]. In terms of trade, the Netherlands became one of the largest importers of cocoa pods in the nineteenth century. This combined with the development of new methods for processing cacao beans by the Dutch chemist Coenraad Johannes van Houten (15 March 1801–27 May 1887) mentioned previously, led to the production of cocoa butter and cocoa powder that could be more effectively transported. The proximity of Norfolk to the Netherlands may have been beneficial to trade in this respect. The change in focus from health to indulgence was accelerated by the production of solid chocolate by Fry in Bristol—a technology that spread elsewhere and enabled chocolate to be marketed and sold as a commodity to individuals rather than requiring the preparation of a beverage, by manufacturers such as Cadbury (Bourneville), Rowntree (York), Fry (Bristol), Mackintosh (Halifax) and Caley (Norwich) and this removed the need for a drinking establishment to be the sole source. In addition, it enabled the commodification of chocolate for the benefit of consumers. Chocolate bars became an important component of the rations for soldiers since they were a portable source of energy and, because of the xanthine content, produced improved concentration. Chocolate rations were provided by manufacturers such as Cadbury and Caley (who also produced a cocoa paste emergency ration that could either be eaten directly or used to make a drink).

9.7 Causality and Correlation in Diet-Related Disease and Health

The difference between causality and correlation lies at the core of scientific thought and the philosophy of science. David Hume (and others) have emphasised the importance of experience and observation in building a picture of the world. An extreme version of this is the 'tabula rasa' concept that was described by Locke. The thinking that the human mind is a blank slate that is gradually filled by experiences and this is what creates personal identity (consciousness) [216]. Hume was interested in the concept of causality. In that respect he was, like Locke, an empiricist. He considered that experience resulted in expectation and in the concept of a 'necessary connection' between things. The naturally sceptical and reductionist consequence of Hume's empiricism was that there was no such thing as 'innate knowledge' and that all knowledge came from experience. In addition, Hume and his idea of necessary connection implies that one thing can be causally linked to another. One consequence of this is that things can be described in terms of their constituent parts even if those parts, in the particular conjunction of interest at a moment in time, have not been fully described in terms of their interaction and their constituents. This enabled Hume (and others) to identify where gaps in knowledge might exist even if the nature of the entities that will fill those gaps are not known. This is analogous to Mendeleev's construction of the Periodic Table with gaps where undiscovered elements will fit when they are isolated. This approach suggests that things are the sum of their parts (this is an important criticism of vitalism in the context of organic molecules) however the molecules that make up a compound are not simply the *blend* of their constituents with properties that are the average of those of its constituents. If they were then dimethyl ether (C_2H_6O) would be identical to ethanol (C_2H_6O) since they are constitutionally identical. This confusion is further increased if more complex isomers are considered, or enantiomers are included. The question of causality in the context of molecules is a challenge on a molecular level for chemists but it can at least be partially addressed by amending a pure reductionist approach, where molecules are thought of as being composed of their smallest viable entities (atoms) and are, instead, thought of in terms of discrete functional groups with similar properties. In this consideration, similarities are found between, for example, organic acids that all contain a -COOH functionality. Molecules can then be causally related according to their constituent parts, structure and function. Coherent reductionism is not synonymous with the simplest possible way of looking at molecules. Correlation (if A happens then B also happens) can be considered to contain, within it, the expectation of limited causality. For causality to exceed simple correlation, there must be, not only the observation of something changes as something else changes but also a sense of the interrelationship of the ideas that underpin the events that are subject to change and this brings a predictive capability. There is, even with an understanding of how things are related and hence why change in one translates into change in another, no absolute and invariable causation within the concept of correlation. For proof of causality, it is necessary that the relationship between things can be used to predict consequences of change (in either direction). A further complication, alluded to above, exists when trying to provide definitions of causality and correlation in the context of chemistry and molecular interactions. This is particularly the case for causality that is implied from correlation [217]. At one extreme, the interaction of molecules with living systems can be considered to be an inter-molecular exchange (for example, receptor/agonist interactions) which leads to a change in the structure and function of the biological entity and this then has distinct biological consequences for the whole organism (signal transduction). In order for such a chain of events to function, each link in the chain of causality must respond in a wholly predictable way to the previous event and thus a complete chain of causality can be constructed. Biological systems are both complex and are not subject to a complete division between states. This means that the core causal chain can become diluted and dissipated. Complete transfer between links in the chain do not invariably occur. In addition, some of the links are acted upon by multiple other causal pathways of variable importance and hence may respond in unexpected ways. Even with a single, bimolecular (agonist/receptor) interaction as a start point, the consequences further down the causality chain may not be completely predictable. In the case of diet-related disease and health, the situation is further confounded by the lack (in most cases) of a distinct starting point and the large number of possible molecular confounders that occur in the diet. Some researchers have tried to create a predictable biological system or a distinct, measurable biological event and have then determined the effects of complete diets on this event. Others have adopted a more classically reductionist approach by trying to isolate the most biologically reactive chemical species from a complex mixture and postulated that the effect of this particular component will be the most significant contributor to the effect on the biological system. One example of the former approach is the measurement of arterial stiffness as a predictor of cardiovascular morbidity and mortality [218] and the impact of different diets on arterial stiffness has been studied [219] however multiple pathways are associated with the risk of developing cardiovascular disease [220] and having a focus on a single parameter may not be as useful as initially hypothesised. Fractionation of chemical components in a mixture based upon the measurement of a biological endpoint has formed the basis of the isolation of active constituents (activity-guided fractionation) and has been effectively used as a tool for pharmacological investigations [221] however while this approach is useful for the investigation and isolation of compounds for medicinal uses, it does not often define the biological consequences of ingestion of a whole diet—or many of its components in that setting. The correlation of diet with disease outcomes or health is only practical on a population rather than an individual (physiological effect or individual compound) basis. Philosophically (and practically) then, causality is a subset of correlation with predictive power and the distinction between causation and correlation in a molecular sense when dealing with whole organisms and the impact of mixtures or even individual components in a diet, is not possible.

Food processing for value and enhanced functionality has been a feature of the development of technology in Norfolk. This has involved interventions at the agronomy/raw materials stage and, to a lesser extent optimisation of products to provide additional attributes. Agronomy and food processing has formed the basis for other types of (small scale, generally) industry and provided wealth from such diverse sources as weaving, brewing, condiment manufacture, chocolate development and light engineering. The underlying technology has included fermentation, breeding and manipulation of plants and microbes and molecular sciences including chemistry. The driver has often been the creation of high value rather than bulk commodities and the accompanying intellectual investment has enabled Norfolk to remain at the centre of certain niche areas. The change in emphasis from trading and collaborating with Europe to a London-centric and US-facing philosophy has been to the detriment of East Anglia as a whole and Norfolk (with its links to the Hanseatic League and to Flanders) in particular. The loss of strong links with Europe combined with the failure of UK infrastructure to develop an East-West (rather than South-North) axis has led to Norfolk, in particular, being increasingly isolated. The natural association between Norwich and Norfolk and Europe has been disrupted by the isolationist tendencies of small groups whose obsession with the perception of sovereignty has diminished both current opportunities and future possibilities. This will reduce both wealth and knowledge creation and, for perhaps the first time, disrupt the flow of intellectual power and opportunity that underpinned the pre- and post-industrial wealth of Norwich and Norfolk through chemistry and its application. These themes will be examined more closely in the next and concluding chapter.

References

1. Cannon G, Leitzmann C. Food and nutrition science: the new paradigm. Asia Pac J Clin Nutr. 2022;31(1):1–15. https://doi.org/10.6133/apjcn.202203_31(1).0001.
2. Strebhardt K, Ullrich A. Paul Ehrlich's magic bullet concept: 100 years of progress. Nat Rev Cancer. 2008;8(6):473–80. https://doi.org/10.1038/nrc2394.
3. Endo A. A historical perspective on the discovery of statins. Proc Jpn Acad Ser B Phys Biol Sci. 2010;86(5):484–93. https://doi.org/10.2183/pjab.86.484.
4. Badimon L, Chagas P, Chiva-Blanch G. Diet and cardiovascular disease: effects of foods and nutrients in classical and emerging cardiovascular risk factors. Curr Med Chem. 2019;26(19):3639–51. https://doi.org/10.2174/0929867324666170428103206.
5. Parkinson MD, Stout R, Dysinger W. Lifestyle medicine: prevention, treatment, and reversal of disease. Med Clin North Am. 2023 Nov;107(6):1109–20. https://doi.org/10.1016/j.mcna.2023.06.007.
6. Hémar-Nicolas V, Hedegaard L. Food sufficiency, an approach rooted in the ethics of Epicurus: analysis framework and research agenda. Recherche Appl Mark. 2023;38(2):2–24. https://doi.org/10.1177/20515707231168595.
7. Luke S. DeHart, improving public health from an epicurean perspective. J Public Health. 2015;37(4):741–3. https://doi.org/10.1093/pubmed/fdu106.
8. Downer S, Berkowitz SA, Harlan TS, Olstad DL, Mozaffarian D. Food is medicine: actions to integrate food and nutrition into healthcare. BMJ. 2020;369:m2482. https://doi.org/10.1136/bmj.m2482.
9. Reddy KR, et al. Food is medicine: the time is now. Am J Med. 2024;137(12):1180–3.
10. Seligman HK, Angell SY, Berkowitz SA, Elkind MSV, Hager K, Moise N, Posner H, Muse J, Odoms-Young A, Ridberg R, Troxel AB, Yaroch AL, Volpp KG. A systematic review of "food is medicine" randomized controlled trials for noncommunicable disease in the United States: a scientific statement from the American Heart Association. Circulation. 2025;152(4):e32–46. https://doi.org/10.1161/CIR.0000000000001343.
11. Oussalah A, Levy J, Berthezène C, Alpers DH, Guéant JL. Health outcomes associated with vegetarian diets: an umbrella review of systematic reviews and meta-analyses. Clin Nutr. 2020;39(11):3283–307. https://doi.org/10.1016/j.clnu.2020.02.037.
12. Li J, Pandian V, Davidson PM, Song Y, Chen N, Fong DYT. Burden and attributable risk factors of non-communicable diseases and subtypes in 204 countries and territories,

1990-2021: a systematic analysis for the global burden of disease study 2021. Int J Surg. 2025;111(3):2385–97. https://doi.org/10.1097/JS9.0000000000002260.
13. Elmadani M, Mokaya PO, Omer AAA, Kiptulon EK, Klara S, Orsolya M. Cancer burden in Europe: a systematic analysis of the GLOBOCAN database. BMC Cancer. 2025;25(1):447. https://doi.org/10.1186/s12885-025-13862-1.
14. Ubago-Guisado E, Rodríguez-Barranco M, Ching-López A, Petrova D, Molina-Montes E, Amiano P, Barricarte-Gurrea A, Chirlaque MD, Agudo A, Sánchez MJ. Evidence update on the relationship between diet and the most common cancers from the European prospective investigation into cancer and nutrition (EPIC) study: a systematic review. Nutrients. 2021;13(10):3582. https://doi.org/10.3390/nu13103582.
15. Kliemann N, et al. Food processing and cancer risk in Europe: results from the prospective EPIC cohort study. Lancet Planetary Health. 2023;7(3):e219–32.
16. Chapela S, Locatelli J, Saettone F, Forte CA, Memoli P, Cucalon G, Frias-Toral E. The role of nutrition in cancer prevention: the effect of dietary patterns, bioactive compounds, and metabolic pathways on cancer development. Food Agric Immunol. 2025;36(1):2490003. https://doi.org/10.1080/09540105.2025.2490003.
17. Rudzińska A, Juchaniuk P, Oberda J, Wiśniewska J, Wojdan W, Szklener K, Mańdziuk S. Phytochemicals in cancer treatment and cancer prevention-review on epidemiological data and clinical trials. Nutrients. 2023;15(8):1896. https://doi.org/10.3390/nu15081896.
18. Lekhak N, Bhattarai HK. Phytochemicals in cancer chemoprevention: preclinical and clinical studies. Cancer Control. 2024;31:10732748241302902. https://doi.org/10.1177/10732748241302902.
19. Puranik N, Yadav SK, Song M. Pharmacological mechanisms of action of phytochemicals in cancer prevention and treatment. In: Puranik N, editor. Nano-formulation of dietary phytochemicals for cancer management. Singapore: Springer; 2024. https://doi.org/10.1007/978-981-97-8005-1_6.
20. Selby-Pham SNB, Miller RB, Howell K, et al. Physicochemical properties of dietary phytochemicals can predict their passive absorption in the human small intestine. Sci Rep. 2017;7:1931. https://doi.org/10.1038/s41598-017-01888-w.
21. Sauter ER, Butera G, Agurs-Collins T. Long-term randomized controlled trials of diet intervention reports and their impact on cancer: a systematic review. Cancers. 2024;16(19):3296. https://doi.org/10.3390/cancers16193296.
22. Ilerhunmwuwa NP, Khader AHSA, Smith C, Scheffer Cliff ER, Booth CM, Hottel E, Aziz M, Lee-Smith W, Goodman A, Chakraborty R, Mohyuddin GR. Dietary interventions in cancer: a systematic review of all randomized controlled trials. JNCI J Natl Cancer Inst. 2024;116(7):1026–34. https://doi.org/10.1093/jnci/djae051.
23. Lampe JW. Diet and cancer prevention research: from mechanism to implementation. J Cancer Prev. 2020;25(2):65–9. https://doi.org/10.15430/JCP.2020.25.2.65.
24. Hodges RE, Minich DM. Modulation of metabolic detoxification pathways using foods and food-derived components: a scientific review with clinical application. J Nutr Metab. 2015;2015:760689. https://doi.org/10.1155/2015/760689.
25. Siedler S, Rau MH, Bidstrup S, Vento JM, Aunsbjerg SD, Bosma EF, McNair LM, Beisel CL, Neves AR. Competitive exclusion is a major bioprotective mechanism of lactobacilli against fungal spoilage in fermented Milk products. Appl Environ Microbiol. 2020;86(7):e02312. https://doi.org/10.1128/AEM.02312-19.
26. Mukherjee A, Breselge S, Dimidi E, Marco ML, Cotter PD. Fermented foods and gastrointestinal health: underlying mechanisms. Nat Rev Gastroenterol Hepatol. 2024;21(4):248–66. https://doi.org/10.1038/s41575-023-00869-x.
27. Pyo Y, Kwon KH, Jung YJ. Probiotic functions in fermented foods: anti-viral, immunomodulatory, and anti-cancer benefits. Foods. 2024;13(15):2386. https://doi.org/10.3390/foods13152386.
28. Harms H, König GM, Schäberle TF. Production of antimicrobial compounds by fermentation. Methods Mol Biol. 2017;1520:49–61. https://doi.org/10.1007/978-1-4939-6634-9_3.

29. Gibson GR, Roberfroid MB. Dietary modulation of the human colonic microbiota. Introducing the concept of prebiotics. J Nutr. 1995;125:1401–12.
30. Gibson GR, Probert HM, Loo JV, Rastall RA, Roberfroid MB. Dietary modulation of the human colonic microbiota: updating the concept of prebiotics. Nutr Res Rev. 2004;17(2):259–75. https://doi.org/10.1079/NRR200479.
31. Li H, He B, Ma N, Liu C, Cai K, Zhang X, Ma X. Quorum sensing of bifidobacteria: research and progress. Microbiol Res. 2025;294:128102. https://doi.org/10.1016/j.micres.2025.128102.
32. Ray RC, Paramithiotis S, Thekkangil A, Nethravathy V, Rai AK, Martin JGP. Food fermentation and its relevance in the human history. In: Martin JGP, De Dea Lindner J, Pereira M, Ray RC, editors. Trending topics on fermented foods. Cham: Springer; 2024. https://doi.org/10.1007/978-3-031-72000-0_1.
33. Zeng Y, Ahmed HGM, Li X, Yang L, Pu X, Yang X, Yang T, Yang J. Physiological mechanisms by which the functional ingredients in beer impact human health. Molecules. 2024;29(13):3110. https://doi.org/10.3390/molecules29133110.
34. Ambra R, Pastore G, Lucchetti S. The role of bioactive phenolic compounds on the impact of beer on health. Molecules. 2021;26(2):486. https://doi.org/10.3390/molecules26020486.
35. Zhang J, Deng H, Bai J, Zhou X, Zhao Y, Zhu Y, McClements DJ, Xiao X, Sun Q. Health-promoting properties of barley: a review of nutrient and nutraceutical composition, functionality, bioprocessing, and health benefits. Crit Rev Food Sci Nutr. 2023;63(9):1155–69. https://doi.org/10.1080/10408398.2021.1972926.
36. Zeng Y, Pu X, Yang J, Du J, Yang X, Li X, Li L, Zhou Y, Yang T. Preventive and therapeutic role of functional ingredients of barley grass for chronic diseases in human beings. Oxidative Med Cell Longev. 2018;2018:3232080. https://doi.org/10.1155/2018/3232080.
37. Kuropatnicki AK. Medicinal properties of beer: from ancient to early modern times. Acta Med Hist Adriat. 2024;22(1):83–110. https://doi.org/10.31952/amha.22.1.4.
38. Raihofer L, Zarnow M, Gastl M, Hutzler M. A short history of beer brewing: alcoholic fermentation and yeast technology over time: alcoholic fermentation and yeast technology over time. EMBO Rep. 2022;23(12):e56355. https://doi.org/10.15252/embr.202256355.
39. Slavin P. Bread and ale for the brethren: the provisioning of Norwich cathedral priory 1260–1536, studies in regional and local history. Hatfield: University of Hertfordshire Press; 2012.
40. Sorensen JP, Maurice L, Edwards FK, Lapworth DJ, Read DS, Allen D, Butcher AS, Newbold LK, Townsend BR, Williams PJ. Using boreholes as windows into groundwater ecosystems. PLoS One. 2013;8(7):e70264. https://doi.org/10.1371/journal.pone.0070264.
41. Rani H, Bhardwaj RD. Quality attributes for barley malt: "the backbone of beer". J Food Sci. 2021;86:3322–40. https://doi.org/10.1111/1750-3841.15858.
42. Kumar V, Chaturvedi SK, Singh GP. Brief review of malting quality and frontier areas in barley. Cereal Res Commun. 2023;51:45–59. https://doi.org/10.1007/s42976-022-00292-z.
43. Meng G, Rasmussen SK, Christensen CSL, Fan W, Torp AM. Molecular breeding of barley for quality traits and resilience to climate change. Front Genet. 2023;13:1039996. https://doi.org/10.3389/fgene.2022.1039996.
44. O'Lone CE, Juhász A, Nye-Wood M, Moody D, Dunn H, Ral JP, Colgrave ML. Advancing sustainable malting practices: Aquaporins as potential breeding targets for improved water uptake during controlled germination of barley (Hordeum vulgare L.). J Agric Food Chem. 2024;72(17):10149–61. https://doi.org/10.1021/acs.jafc.4c00884.
45. Guo X, Sarup P, Jahoor A, et al. Metabolomic-genomic prediction can improve prediction accuracy of breeding values for malting quality traits in barley. Genet Sel Evol. 2023;55:61. https://doi.org/10.1186/s12711-023-00835-w.
46. Cai K, Wu X, Yue W, Liu L, Song X, Ge F, Wang Q, Wang J. Enzyme activities shape malting quality standards. Food Sci Nutr. 2024;13(1):e4702. https://doi.org/10.1002/fsn3.4702.
47. Grujić O. Application of a commercial enzyme preparation in the barley malting process. J Inst Brew. 1998;104:249–53. https://doi.org/10.1002/j.2050-0416.1998.tb00997.x.

48. Wenwen Y, Tao K, Gidley MJ, Fox GP, Gilbert RG. Molecular brewing: molecular structural effects involved in barley malting and mashing. Carbohydr Polym. 2019;206:583–92. https://doi.org/10.1016/j.carbpol.2018.11.018.
49. Yin Tan W, Li M, Devkota L, Attenborough E, Dhital S. Mashing performance as a function of malt particle size in beer production. Crit Rev Food Sci Nutr. 2023;63(21):5372–87. https://doi.org/10.1080/10408398.2021.2018673.
50. De Schepper CF, Courtin CM. Reassessing the importance of barley starch and amylolytic enzyme properties in malting and brewing. Compr Rev Food Sci Food Saf. 2024;23(6):e70069. https://doi.org/10.1111/1541-4337.70069.
51. Pacheco-Moreno A, Bollmann-Giolai A, Chandra G, Brett P, Davies J, Thornton O, Poole P, Ramachandran V, Brown JKM, Nicholson P, Ridout C, DeVos S, Malone JG. The genotype of barley cultivars influences multiple aspects of their associated microbiota via differential root exudate secretion. PLoS Biol. 2024;22(4):e3002232. https://doi.org/10.1371/journal.pbio.3002232.
52. Pajic M. 'Ale for an Englishman is a natural drink': the Dutch and the origins of beer brewing in late medieval England. J Mediev Hist. 2019;45(3):285–300. https://doi.org/10.1080/03044181.2019.1612182.
53. Scot R. Perfite platforme of a hoppe garden. London: Henrie Denham; 1576.
54. Salem KFM, Ibrahim AA. Advances in common hop (Humulus lupulus L.) breeding strategies. In: Al-Khayri JM, Jain SM, Penna S, editors. Biodiversity and genetic improvement of medicinal and aromatic plants I. Advances in plant breeding strategies, vol. 9. Cham: Springer; 2025. https://doi.org/10.1007/978-3-031-81846-2_7.
55. McAdam EL, Freeman JS, Whittock SP, Buck EJ, Jakse J, Cerenak A, Javornik B, Kilian A, Wang CH, Andersen D, Vaillancourt RE, Carling J, Beatson R, Graham L, Graham D, Darby P, Koutoulis A. Quantitative trait loci in hop (Humulus lupulus L.) reveal complex genetic architecture underlying variation in sex, yield and cone chemistry. BMC Genomics. 2013;14:360. https://doi.org/10.1186/1471-2164-14-360.
56. Patzak J, Nesvadba V, Krofta K, Henychova A, Marzoev AI, Richards K. Evaluation of genetic variability of wild hops (Humulus lupulus L.) in Canada and the Caucasus region by chemical and molecular methods. Genome. 2010;53(7):545–57. https://doi.org/10.1139/g10-024.
57. Steenackers B, De Cooman L, De Vos D. Chemical transformations of characteristic hop secondary metabolites in relation to beer properties and the brewing process: a review. Food Chem. 2015;172:742–56. https://doi.org/10.1016/j.foodchem.2014.09.139.
58. Caffrey A, Ebeler SE. The occurrence of glycosylated aroma precursors in Vitis vinifera fruit and Humulus lupulus hop cones and their roles in wine and beer volatile aroma production. Foods. 2021;10(5):935. https://doi.org/10.3390/foods10050935.
59. Chevance F, Guyot-Declerck C, Dupont J, Collin S. Investigation of the β-Damascenone level in fresh and aged commercial beers. J Agric Food Chem. 2002;50:3818–21. https://doi.org/10.1021/jf020085i.
60. Dietz C, Cook D, Wilson C, Oliveira P, Ford R. Exploring the multisensory perception of terpene alcohol and sesquiterpene rich hop extracts in lager style beer. Food Res Int. 2021;148:110598. https://doi.org/10.1016/j.foodres.2021.110598.
61. Sharp DC, Steensels J, Shellhammer TH. The effect of hopping regime, cultivar and β-glucosidase activity on monoterpene alcohol concentrations in wort and beer. J Inst Brew. 2017;123:185–91. https://doi.org/10.1002/jib.418.
62. King AJ, Dickinson JR. Biotransformation of hop aroma terpenoids by ale and lager yeasts. FEMS Yeast Res. 2003;3:53–62.
63. Buiatti S, Tat L, Natolino A, Passaghe P. Biotransformations performed by yeasts on aromatic compounds provided by hop—a review. Fermentation. 2023;9:327. https://doi.org/10.3390/fermentation9040327.
64. Ting PL, Ryder DS. The bitter, twisted truth of the hop: 50 years of hop chemistry. J Am Soc Brew Chem. 2017;75(3):161–80. https://doi.org/10.1094/ASBCJ-2017-3638-01.

65. Spevacek AR, Benson KH, Bamforth CW, Slupsky CM. Beer metabolomics: molecular details of the brewing process and the differential effects of late and dry hopping on yeast purine metabolism. J Inst Brew. 2016;122:21–8. https://doi.org/10.1002/jib.291.
66. Dietz C, Cook D, Huismann M, Wilson C, Ford R. The multisensory perception of hop essential oil: a review. J Inst Brew. 2020;126:320–42. https://doi.org/10.1002/jib.622.
67. Walther A, Hesselbart A, Wendland J. Genome sequence of Saccharomyces carlsbergensis, the world's first pure culture lager yeast. G3 (Bethesda). 2014;4(5):783–93. https://doi.org/10.1534/g3.113.010090.
68. De Keersmaecker J. The mystery of Lambic beer. Sci Am. 1996;275(2):74–80.
69. Spitaels F, Wieme AD, Janssens M, Aerts M, Daniel HM, Van Landschoot A, De Vuyst L, Vandamme P. The microbial diversity of traditional spontaneously fermented lambic beer. PLoS One. 2014;9(4):e95384. https://doi.org/10.1371/journal.pone.0095384.
70. De Roos J, Vandamme P, De Vuyst L. Wort substrate consumption and metabolite production during Lambic beer fermentation and maturation explain the successive growth of specific bacterial and yeast species. Front Microbiol. 2018;9:2763. https://doi.org/10.3389/fmicb.2018.02763.
71. Maicas S. The role of yeasts in fermentation processes. Microorganisms. 2020;8(8):1142. https://doi.org/10.3390/microorganisms8081142.
72. Sörensen SPL. Enzyme studies. 2nd report. On the measurement and the importance of hydrogen ion concentration during enzymatic processes. Biochem Z. 1909;21:131–304.
73. Barnett JA. A history of research on yeasts 2: Louis Pasteur and his contemporaries, 1850-1880. Yeast. 2000;16(8):755–71. https://doi.org/10.1002/1097-0061(20000615)16:8<755::AID-YEA587>3.0.CO;2-4.
74. Rocchi R, Wolkers-Rooijackers JCM, Liao Z, Tempelaars MH, Smid EJ. Strain diversity in Saccharomyces cerevisiae thiamine production capacity. Yeast. 2023;40:628–39. https://doi.org/10.1002/yea.3906.
75. Jedamzik T, Juckel G. Fortifying alcoholic beverages with thiamine: lessons from history and future opportunities. Drug Alcohol Depend Rep. 2025;16:100358. https://doi.org/10.1016/j.dadr.2025.100358.
76. Langlais PJ. Alcohol-related thiamine deficiency: impact on cognitive and memory functioning. Alcohol Health Res World. 1995;19(2):113–21.
77. Habschied K, Živković A, Krstanović V, Mastanjević K. Functional beer—a review on possibilities. Beverages. 2020;6:51. https://doi.org/10.3390/beverages6030051.
78. Vaštík P, Brunner J, Jung R, Klempová T, Furdíková K, Šmogrovičová D, Dostálek P. Functional non-alcoholic beer fermented with potential probiotic yeasts. Beverages. 2025;11:140. https://doi.org/10.3390/beverages11050140.
79. Rakowska R, Sadowska A, Dybkowska E, Świderski F. Spent yeast as natural source of functional food additives. Rocz Panstw Zakl Hig. 2017;68(2):115–21.
80. Archaeol J 2019;5:djvu/402.
81. Dallas P, Barnes G, Williamson T. Orchards in the landscape: a Norfolk case study. Landscapes. 2015;16(1):26–43. https://doi.org/10.1179/1466203515Z.00000000039.
82. Cornille A, Gladieux P, Smulders MJ, Roldán-Ruiz I, Laurens F, Le Cam B, Nersesyan A, Clavel J, Olonova M, Feugey L, Gabrielyan I, Zhang XG, Tenaillon MI, Giraud T. New insight into the history of domesticated apple: secondary contribution of the European wild apple to the genome of cultivated varieties. PLoS Genet. 2012;8(5):e1002703. https://doi.org/10.1371/journal.pgen.1002703.
83. Vallée Marcotte B, Verheyde M, Pomerleau S, Doyen A, Couillard C. Health benefits of apple juice consumption: a review of interventional trials on humans. Nutrients. 2022;14(4):821. https://doi.org/10.3390/nu14040821.
84. Jafarirad S, Elahi MR, Mansoori A, Khanzadeh A, Haghighizadeh MH. The improvement effect of apple cider vinegar as a functional food on anthropometric indices, blood glucose and lipid profile in diabetic patients: a randomized controlled clinical trial. Front Clin Diabetes Healthc. 2023;4:1288786. https://doi.org/10.3389/fcdhc.2023.1288786.

85. Santos HO, de Moraes WMAM, da Silva GAR, Prestes J, Schoenfeld BJ. Vinegar (acetic acid) intake on glucose metabolism: a narrative review. Clin Nutr ESPEN. 2019;32:1–7. https://doi.org/10.1016/j.clnesp.2019.05.008.
86. Zhang Y, Zeng M, Zhang X, Yu Q, Zeng W, Yu B, Gan J, Zhang S, Jiang X. Does an apple a day keep away diseases? Evidence and mechanism of action. Food Sci Nutr. 2023;11(9):4926–47. https://doi.org/10.1002/fsn3.3487.
87. Guiné RPF, Barroca MJ, Coldea TE, Bartkiene E, Anjos O. Apple fermented products: an overview of technology. Prop Health Effects Processes. 2021;9:223. https://doi.org/10.3390/pr9020223.
88. Li H, Liang J, Han M, Wang X, Ren Y, Wang Y, Huang J, Li S, Liu C, Wang Z, Yue T, Gao Z. Sequentially fermented dealcoholized apple juice intervenes fatty liver induced by high-fat diets via modulation of intestinal flora and gene pathways. Food Res Int. 2022;156:111180. https://doi.org/10.1016/j.foodres.2022.111180.
89. Koetter U, Biendl M. Hops (Humulus lupulus): a review of its historic and medicinal uses. HerbalGram. 2010;87:44–57.
90. Astray G, Gullón P, Gullón B, Munekata PE, Lorenzo JM. Humulus lupulus L. as a natural source of functional biomolecules. Appl Sci. 2020;10:5074. https://doi.org/10.3390/app10155074.
91. Zugravu CA, Bohiltea RE, Salmen T, Pogurschi E, Otelea MR. Antioxidants in hops: bioavailability, health effects and perspectives for new products. Antioxidants. 2022;11(2):241. https://doi.org/10.3390/antiox11020241.
92. Kyrou I, Christou A, Panagiotakos D, Stefanaki C, Skenderi K, Katsana K, Tsigos C. Effects of a hops (Humulus lupulus L.) dry extract supplement on self-reported depression, anxiety and stress levels in apparently healthy young adults: a randomized, placebo-controlled, double-blind, crossover pilot study. Hormones. 2017;16(2):171–80. https://doi.org/10.14310/horm.2002.1738.
93. Carbone K, Gervasi F. An updated review of the genus Humulus: a valuable source of bioactive compounds for health and disease prevention. Plants. 2022;11(24):3434. https://doi.org/10.3390/plants11243434.
94. Piekara J, Piasecka-Kwiatkowska D, Hołaj H, Jędryczka M, Daniel Daramola O, Dwiecki K. Hop (Humulus lupulus L.) phenolic compounds profile depends on cultivar and plant organ maturity. Molecules. 2025;30:2365. https://doi.org/10.3390/molecules30112365.
95. Bolton JL, Dunlap TL, Hajirahimkhan A, Mbachu O, Chen SN, Chadwick L, Nikolic D, van Breemen RB, Pauli GF, Dietz BM. The multiple biological targets of hops and bioactive compounds. Chem Res Toxicol. 2019;32(2):222–33. https://doi.org/10.1021/acs.chemrestox.8b00345.
96. Aghamiri V, Mirghafourvand M, Mohammad-Alizadeh-Charandabi S, Nazemiyeh H. The effect of hop (Humulus lupulus L.) on early menopausal symptoms and hot flashes: a randomized placebo-controlled trial. Complement Ther Clin Pract. 2016;23:130–5.
97. Zhang B, Chu W, Wei P, Liu Y, Wei T. Xanthohumol induces generation of reactive oxygen species and triggers apoptosis through inhibition of mitochondrial electron transfer chain complex I. Free Radic Biol Med. 2015;89:486–97.
98. Walker EG, Lo KR, Pahl MC, Shin HS, Lang C, Wohlers MW, Poppitt SD, Sutton KH, Ingram JR. An extract of hops (Humulus lupulus L.) modulates gut peptide hormone secretion and reduces energy intake in healthy-weight men: a randomized, crossover clinical trial. Am J Clin Nutr. 2022;115(3):925–40. https://doi.org/10.1093/ajcn/nqab418.
99. Morimoto-Kobayashi Y, Ohara K, Ashigai H, Kanaya T, Koizumi K, Manabe F, Kaneko Y, Taniguchi Y, Katayama M, Kowatari Y, Kondo S. Matured hop extract reduces body fat in healthy overweight humans: a randomized, double-blind, placebo-controlled parallel group study. Nutr J. 2016;15:25. https://doi.org/10.1186/s12937-016-0144-2.
100. Di Lodovico S, Menghini L, Ferrante C, Recchia E, Castro-Amorim J, Gameiro P, Cellini L, Bessa LJ. Hop extract: an efficacious antimicrobial and anti-biofilm agent against multidrug-

resistant staphylococci strains and Cutibacterium acnes. Front Microbiol. 2020;11:1852. https://doi.org/10.3389/fmicb.2020.01852.

101. Di Sotto A, Checconi P, Celestino I, Locatelli M, Carissimi S, De Angelis M, Rossi V, Limongi D, Toniolo C, Martinoli L, et al. Antiviral and antioxidant activity of a hydroalcoholic extract from Humulus lupulus L. Oxidative Med Cell Longev. 2018;2018:5919237. https://doi.org/10.1155/2018/5919237.
102. Bocquet L, Sahpaz S, Rivière C. An overview of the antimicrobial properties of hop. In: Mérillon J-M, Riviere C, editors. Natural antimicrobial agents. Cham: Springer; 2018. p. 31–54.
103. Lee H, Chung SH, Kwon DJ, Nam MJ, Choi JH, Suh HJ, Choi HS, Han SH. Sleep-enhancing effect of Hongcheon-hop (Humulus lupulus L.) extract containing xanthohumol and humulone through GABA A receptor. J Ethnopharmacol. 2025;338(2):119019. https://doi.org/10.1016/j.jep.2024.119019.
104. Min B, Ahn Y, Cho HJ, Kwak WK, Jo K, Suh HJ. Chemical compositions and sleep-promoting activities of hop (Humulus lupulus L.) varieties. J Food Sci. 2023;88(5):2217–28. https://doi.org/10.1111/1750-3841.16544.
105. Franco L, Sánchez C, Bravo R, Rodriguez A, Barriga C, Juánez JC. The sedative effects of hops (Humulus lupulus), a component of beer, on the activity/rest rhythm. Acta Physiol Hung. 2012;99(2):133–9. https://doi.org/10.1556/APhysiol.99.2012.2.6.
106. Van Cleemput M, Cattoor K, De Bosscher K, Haegeman G, De Keukeleire D, Heyerick A. Hop (Humulus lupulus)-derived bitter acids as multipotent bioactive compounds. J Nat Prod. 2009;72(6):1220–30. https://doi.org/10.1021/np800740m.
107. Ano Y, Ohya R, Yamazaki T, Takahashi C, Taniguchi Y, Kondo K, Takashima A, Uchida K, Nakayama H. Hop bitter acids containing a β-carbonyl moiety prevent inflammation-induced cognitive decline via the vagus nerve and noradrenergic system. Sci Rep. 2020;10(1):20028. https://doi.org/10.1038/s41598-020-77034-w.
108. Kinney JW, Bemiller SM, Murtishaw AS, Leisgang AM, Salazar AM, Lamb BT. Inflammation as a central mechanism in Alzheimer's disease. Alzheimer's Dement. 2018;4:575–90. https://doi.org/10.1016/j.trci.2018.06.014.
109. Van Cleemput M, Heyerick A, Libert C, Swerts K, Philippé J, De Keukeleire D, Haegeman G, De Bosscher K. Hop bitter acids efficiently block inflammation independent of GRalpha, PPARalpha, or PPARgamma. Mol Nutr Food Res. 2009;53(9):1143–55. https://doi.org/10.1002/mnfr.200800493.
110. Taniguchi Y, Matsukura Y, Taniguchi H, Koizumi H, Katayama M. Development of preparative and analytical methods of the hop bitter acid oxide fraction and chemical properties of its components. Biosci Biotechnol Biochem. 2015;79(10):1684–94. https://doi.org/10.1080/09168451.2015.1042832.
111. Zhang G, Zhang N, Yang A, et al. Hop bitter acids: resources, biosynthesis, and applications. Appl Microbiol Biotechnol. 2021;105:4343–56. https://doi.org/10.1007/s00253-021-11329-4.
112. Awouafack MD, Lee YE, Morita H. Xanthohumol: recent advances on resources, biosynthesis, bioavailability and pharmacology. In: Xiao J, editor. Handbook of dietary flavonoids. Cham: Springer; 2023. https://doi.org/10.1007/978-3-030-94753-8_52-1.
113. Langley BO, Ryan JJ, Hanes D, Phipps J, Stack E, Metz TO, Stevens JF, Bradley R. Xanthohumol microbiome and signature in healthy adults (the XMaS Trial): safety and tolerability results of a phase I triple-masked, placebo-controlled clinical trial. Mol Nutr Food Res. 2021;65(8):e2001170. https://doi.org/10.1002/mnfr.202001170.
114. Pohjanvirta R, Nasri A. The potent phytoestrogen 8-prenylnaringenin: a friend or a foe? Int J Mol Sci. 2022;23(6):3168. https://doi.org/10.3390/ijms23063168.
115. Milligan S, Kalita J, Heyerick A, Rong H, De Cooman L, De Keukeleire D. Identification of a potent phytoestrogen in hops (Humulus lupulus L.) and beer. J Clin Endocrinol Metab. 1999;84:2249. https://doi.org/10.1210/jcem.84.6.5887.

116. Stevens JF, Taylor AW, Clawson JE, Deinzer ML. Fate of xanthohumol and related prenylflavonoids from hops to beer. J Agric Food Chem. 1999;47(6):2421–8. https://doi.org/10.1021/jf990101k.
117. Nicolic D, et al. Metabolism of 8-prenylnaringenin, a potent phytoestrogen from hops (Humulus lupulus), by human liver microsomes. Drug Metabol Disp. 2004;32(2):272–9.
118. de Gaetano G, Costanzo S, Di Castelnuovo A, Badimon L, Bejko D, Alkerwi A, Chiva-Blanch G, Estruch R, La Vecchia C, Panico S, Pounis G, Sofi F, Stranges S, Trevisan M, Ursini F, Cerletti C, Donati MB, Iacoviello L. Effects of moderate beer consumption on health and disease: a consensus document. Nutr Metab Cardiovasc Dis. 2016;26(6):443–67. https://doi.org/10.1016/j.numecd.2016.03.007.
119. Spaggiari G, Cignarelli A, Sansone A, Baldi M, Santi D. To beer or not to beer: a meta-analysis of the effects of beer consumption on cardiovascular health. PLoS One. 2020;15(6):e0233619. https://doi.org/10.1371/journal.pone.0233619.
120. Neumann HF, Frank J, Venturelli S, Egert S. Bioavailability and cardiometabolic effects of xanthohumol: evidence from animal and human studies. Mol Nutr Food Res. 2022;66:2100831. https://doi.org/10.1002/mnfr.202100831.
121. Karabín M, Hudcová T, Jelínek L, Dostálek P. Biologically active compounds from hops and prospects for their use. Compr Rev Food Sci Food Saf. 2016;15(3):542–67. https://doi.org/10.1111/1541-4337.12201.
122. Iniguez AB, Zhu MJ. Hop bioactive compounds in prevention of nutrition-related noncommunicable diseases. Crit Rev Food Sci Nutr. 2021;61(11):1900–13. https://doi.org/10.1080/10408398.2020.1767537.
123. Hemachandra LP, Madhubhani P, Chandrasena R, Esala P, Chen SN, Main M, Lankin DC, Scism RA, Dietz BM, Pauli GF, Thatcher GR, Bolton JL. Hops (Humulus lupulus) inhibits oxidative estrogen metabolism and estrogen-induced malignant transformation in human mammary epithelial cells (MCF-10A). Cancer Prev Res. 2012;5(1):73–81. https://doi.org/10.1158/1940-6207.CAPR-11-0348.
124. Monteiro R, Faria A, Azevedo I, Calhau C. Modulation of breast cancer cell survival by aromatase inhibiting hop (Humulus lupulus L.) flavonoids. J Steroid Biochem Mol Biol. 2007;105(1-5):124–30. https://doi.org/10.1016/j.jsbmb.2006.11.026.
125. Generali D, Berardi R, Caruso M, Cazzaniga M, Garrone O, Minchella I, Paris I, Pinto C, De Placido S. Aromatase inhibitors: the journey from the state of the art to clinical open questions. Front Oncol. 2023;13:1249160. https://doi.org/10.3389/fonc.2023.1249160.
126. Keiler AM, Zierau O, Kretzschmar G. Hop extracts and hop substances in treatment of menopausal complaints. Planta Med. 2013;79(7):576–9. https://doi.org/10.1055/s-0032-1328330.
127. Chadwick LR, Pauli GF, Farnsworth NR. The pharmacognosy of Humulus lupulus L. (hops) with an emphasis on estrogenic properties. Phytomedicine. 2006;13(1-2):119–31. https://doi.org/10.1016/j.phymed.2004.07.006.
128. Štulíková K, Karabín M, Nešpor J, Dostálek P. Therapeutic perspectives of 8-prenylnaringenin, a potent phytoestrogen from hops. Molecules. 2018;23(3):660. https://doi.org/10.3390/molecules23030660.
129. Hitzman RT, Dunlap TL, Howell CE, Chen SN, Vollmer G, Pauli GF, Bolton JL, Dietz BM. 6-Prenylnaringenin from hops disrupts ERα-mediated downregulation of CYP1A1 to facilitate estrogen detoxification. Chem Res Toxicol. 2020;33(11):2793–803. https://doi.org/10.1021/acs.chemrestox.0c00194.
130. Petticrew M, Shemilt I, Lorenc T, Marteau TM, Melendez-Torres GJ, O'Mara-Eves A, Stautz K, Thomas J. Alcohol advertising and public health: systems perspectives versus narrow perspectives. J Epidemiol Community Health. 2017;71(3):308–12. https://doi.org/10.1136/jech-2016-207644.
131. Duncan M. Cheers! Selling health and happiness: advertising alcohol in New Zealand, c.1900–1945. Open Access Te Herenga Waka-Victoria University of Wellington. Thesis. 2020. https://doi.org/10.26686/wgtn.17148500.

132. Crawford R. "Drink beer regularly – it's good for you (and us)": selling tooth's beer in a depressed market. Soc Hist Alcohol Drugs. 2007;21(2):160–82.
133. Quesada-Molina M, Muñoz-Garach A, Tinahones FJ, Moreno-Indias I. A new perspective on the health benefits of moderate beer consumption: involvement of the gut microbiota. Meta. 2019;9(11):272. https://doi.org/10.3390/metabo9110272.
134. Timmer CP. The turnip, the new husbandry, and the English agricultural revolution. Q J Econ. 1969;83(3):375–95.
135. Pop MS, Cheregi DC, Onose G, Munteanu C, Popescu C, Rotariu M, Turnea MA, Dogaru G, Ionescu EV, Oprea D, Iliescu MG, Minea M, Stanciu LE, Silişteanu SC, Oprea C. Exploring the potential benefits of natural calcium-rich mineral waters for health and wellness: a systematic review. Nutrients. 2023;15(14):3126. https://doi.org/10.3390/nu15143126.
136. Puchtler H, Meloan SN, Terry MS. On the history and mechanism of alizarin and alizarin red S stains for calcium. J Histochem Cytochem. 1969;17(2):110–24. https://doi.org/10.1177/17.2.110.
137. Liu T, Wang Q, Shi G, Zhan Y, Shen Y, Zhao Y, Tian J, Liu Y, Ma C, Yao Z. Optical fingerprint database of beer brewing water based on supramolecular probes and its application in the evaluation of the water source. Langmuir. 2025;41(43):29367–76. https://doi.org/10.1021/acs.langmuir.5c04373.
138. Taruno A, Gordon MD. Molecular and cellular mechanisms of salt taste. Annu Rev Physiol. 2023;85:25–45. https://doi.org/10.1146/annurev-physiol-031522-075853.
139. Murphy C, Cardello AV, Brand J. Tastes of fifteen halide salts following water and NaCl: anion and cation effects. Physiol Behav. 1981;26(6):1083–95.
140. Kasahara Y, Narukawa M, Ishimaru Y, et al. TMC4 is a novel chloride channel involved in high-concentration salt taste sensation. J Physiol Sci. 2021;71:23. https://doi.org/10.1186/s12576-021-00807-z.
141. DeSimone JA, Lyall V. Taste receptors in the gastrointestinal tract III. Salty and sour taste: sensing of sodium and protons by the tongue. Am J Physiol Gastrointest Liver Physiol. 2006;291(6):1005–10. https://doi.org/10.1152/ajpgi.00235.2006.
142. Kasahara Y, Narukawa M, Takeuchi A, et al. Molecular logic of salt taste reception in special reference to transmembrane channel-like 4 (TMC4). J Physiol Sci. 2022;72:31. https://doi.org/10.1186/s12576-022-00856-y.
143. Han F, Li W, Duan N, Hu X, Yao N, Yu G, Qu J. Relationship between salt intake and cardiovascular disease. J Clin Hypertens. 2025;27(6):e70078. https://doi.org/10.1111/jch.70078.
144. Sood S, Methven L, Cheng Q. Role of taste receptors in salty taste perception of minerals and amino acids and developments in salt reduction strategies: a review. Crit Rev Food Sci Nutr. 2025;65(18):3444–58. https://doi.org/10.1080/10408398.2024.2365962.
145. Bailey MA, Dhaun N. Salt sensitivity: causes, consequences, and recent advances. Hypertension. 2024;81(3):476–89. https://doi.org/10.1161/Hypertesionaha.123.17959.
146. Chandrashekar J, Hoon MA, Ryba NJ, Zuker CS. The receptors and cells for mammalian taste. Nature. 2006;444(7117):288–94.
147. Liman ER, Kinnamon SC. Sour taste: receptors, cells and circuits. Curr Opin Physiol. 2021;20:8–15. https://doi.org/10.1016/j.cophys.2020.12.006.
148. Brosolo G, Da Porto A, Marcante S, Picci A, Capilupi F, Capilupi P, Bertin N, Vivarelli C, Bulfone L, Vacca A, Catena C, Sechi LA. Omega-3 fatty acids in arterial hypertension: is there any good news? Int J Mol Sci. 2023;24(11):9520. https://doi.org/10.3390/ijms24119520.
149. Sagaro GG, Angeloni U, Marotta C, Nittari G, Rezza G, Silenzi A, Battineni G, Amenta F. The magnitude of cardiovascular disease risk factors in seafarers from 1994 to 2021: a systematic review and meta-analysis. J Pers Med. 2023;13(5):861. https://doi.org/10.3390/jpm13050861.
150. Roberts SE, Jaremin B. Cardiovascular disease mortality in British merchant shipping and among British seafarers ashore in Britain. Int Marit Health. 2010;62:107–16.

151. Bellis C, Cox HC, Dyer TD, Charlesworth JC, Begley KN, Quinlan S, Lea RA, Heath SC, Blangero J, Griffiths LR. Linkage mapping of CVD risk traits in the isolated Norfolk Island population. Hum Genet. 2008;124(5):543–52. https://doi.org/10.1007/s00439-008-0580-y.
152. Tang WHW, Li XS, Wu Y, Wang Z, Khaw KT, Wareham NJ, Nieuwdorp M, Boekholdt SM, Hazen SL. Plasma trimethylamine N-oxide (TMAO) levels predict future risk of coronary artery disease in apparently healthy individuals in the EPIC-Norfolk prospective population study. Am Heart J. 2021;236:80–6. https://doi.org/10.1016/j.ahj.2021.01.020.
153. Wang Z, Tang WHW, O'Connell T, Garcia E, Jeyarajah EJ, Li XS, Jia X, Weeks TL, Hazen SL. Circulating trimethylamine N-oxide levels following fish or seafood consumption. Eur J Nutr. 2022;61(5):2357–64. https://doi.org/10.1007/s00394-022-02803-4.
154. Treberg JR, Driedzic WR. Elevated levels of trimethylamine oxide in deep-sea fish: evidence for synthesis and intertissue physiological importance. J Exp Zool. 2002;293:39–45. https://doi.org/10.1002/jez.10109.
155. Shi Z, Xu W, Wu L, Yue X, Liu S, Ding W, Zhang J, Meng B, Zhao L, Liu X, Liu J, Liu ZJ, Hua T. Structural and functional characterization of human sweet taste receptor. Nature. 2025;645(8081):801–8. https://doi.org/10.1038/s41586-025-09302-6.
156. Wu B, Eldeghaidy S, Ayed C, Fisk ID, Hewson L, Liu Y. Mechanisms of umami taste perception: from molecular level to brain imaging. Crit Rev Food Sci Nutr. 2021;62(25):7015–24. https://doi.org/10.1080/10408398.2021.1909532.
157. Yoshida R, Ninomiya Y. Mechanisms and functions of sweet reception in oral and extraoral organs. Int J Mol Sci. 2024;25(13):7398. https://doi.org/10.3390/ijms25137398.
158. Laffitte A, Neiers F, Briand L. Functional roles of the sweet taste receptor in oral and extraoral tissues. Curr Opin Clin Nutr Metab Care. 2014;17(4):379–85. https://doi.org/10.1097/MCO.0000000000000058.
159. Oya M, Suzuki H, Watanabe Y, Sato M, Tsuboi T. Amino acid taste receptor regulates insulin secretion in pancreatic β-cell line MIN6 cells. Genes Cells. 2011;16:608–16. https://doi.org/10.1111/j.1365-2443.2011.01509.x.
160. Behrens M, Lang T. Extra-Oral taste receptors-function, disease, and perspectives. Front Nutr. 2022;9:881177. https://doi.org/10.3389/fnut.2022.881177.
161. Liszt KI, Ley JP, Lieder B, Behrens M, Stoger V, Reiner A, et al. Caffeine induces gastric acid secretion via bitter taste signaling in gastric parietal cells. Proc Natl Acad Sci USA. 2017;114:E6260. https://doi.org/10.1073/pnas.1703728114.
162. Liszt KI, Wang Q, Farhadipour M, Segers A, Thijs T, Nys L, et al. Human intestinal bitter taste receptors regulate innate immune responses and metabolic regulators in obesity. J Clin Invest. 2022;132:e144828. https://doi.org/10.1172/JCI144828.
163. Rudolph E, Dychtenberg H, Pozniak A, Pundir P. Bitter taste receptors in bacterial infections and innate immunity. Immun Inflamm Dis. 2025;13(7):e70232. https://doi.org/10.1002/iid3.70232.
164. Medapati MR, Singh N, Bhagirath AY, Duan K, Triggs-Raine B, Batista EL Jr, Chelikani P. Bitter taste receptor T2R14 detects quorum sensing molecules from cariogenic Streptococcus mutans and mediates innate immune responses in gingival epithelial cells. FASEB J. 2021;35(3):e21375. https://doi.org/10.1096/fj.202000208R.
165. Sternini C, Rozengurt E. Bitter taste receptors as sensors of gut luminal contents. Nat Rev Gastroenterol Hepatol. 2025;22:39–53. https://doi.org/10.1038/s41575-024-01005-z.
166. Okda M, Spina S, Safaee Fakhr B, Carroll RW. The antimicrobial effects of nitric oxide: a narrative review. Nitric Oxide. 2025;155:20–40. https://doi.org/10.1016/j.niox.2025.01.001.
167. Carey RM, Hariri BM, Adappa ND, Palmer JN, Lee RJ. HSP90 modulates T2R bitter taste receptor nitric oxide production and innate immune responses in human airway epithelial cells and macrophages. Cells. 2022;11(9):1478. https://doi.org/10.3390/cells11091478.
168. Conaway S Jr, Huang W, Hernandez-Lara MA, Kane MA, Penn RB, Deshpande DA. Molecular mechanism of bitter taste receptor agonist-mediated relaxation of airway smooth muscle. FASEB J. 2024;38(14):e23842. https://doi.org/10.1096/fj.202400452R.

169. Jeon TI, Seo YK, Osborne TF. Gut bitter taste receptor signalling induces ABCB1 through a mechanism involving CCK. Biochem J. 2011;438(1):33–7. https://doi.org/10.1042/BJ20110009.
170. Leschziner GD, Andrew T, Pirmohamed M, Johnson MR. ABCB1 genotype and PGP expression, function and therapeutic drug response: a critical review and recommendations for future research. Pharmacogenomics J. 2007;7(3):154–79. https://doi.org/10.1038/sj.tpj.6500413.
171. Risso DS, Mezzavilla M, Pagani L, Robino A, Morini G, Tofanelli S, Carrai M, Campa D, Barale R, Caradonna F, Gasparini P, Luiselli D, Wooding S, Drayna D. Global diversity in the TAS2R38 bitter taste receptor: revisiting a classic evolutionary PROPosal. Sci Rep. 2016;6:25506. https://doi.org/10.1038/srep25506.
172. Gaida MM, Mayer C, Dapunt U, Stegmaier S, Schirmacher P, Wabnitz GH, Hänsch GM. Expression of the bitter receptor T2R38 in pancreatic cancer: localization in lipid droplets and activation by a bacteria-derived quorum-sensing molecule. Oncotarget. 2016;7(11):12623–32. https://doi.org/10.18632/oncotarget.7206.
173. Zhang G, Li Y, Sun Y. Cruciferous vegetables intake reduces pancreatic cancer risk: an updated systematic review with meta-analysis. Eur J Nutr. 2024;63(7):2421–35. https://doi.org/10.1007/s00394-024-03472-1.
174. Hardt L, Mahamat-Saleh Y, Aune D, Schlesinger S. Plant-based diets and cancer prognosis: a review of recent research. Curr Nutr Rep. 2022;11(4):695–716. https://doi.org/10.1007/s13668-022-00440-1.
175. Zheng Q, Wang X. Alzheimer's disease: insights into pathology, molecular mechanisms, and therapy. Protein Cell. 2025;16(2):83–120. https://doi.org/10.1093/procel/pwae026.
176. Parra Bravo C, Naguib SA, Gan L. Cellular and pathological functions of tau. Nat Rev Mol Cell Biol. 2024;25(11):845–64. https://doi.org/10.1038/s41580-024-00753-9.
177. Morley JE, Farr SA, Nguyen AD, Xu F. Editorial: what is the physiological function of amyloid-Beta protein? J Nutr Health Aging. 2019;23(3):225–6. https://doi.org/10.1007/s12603-019-1162-5.
178. Gosztyla ML, Brothers HM, Robinson SR. Alzheimer's amyloid-β is an antimicrobial peptide: a review of the evidence. J Alzheimer's Dis. 2018;62(4):1495–506. https://doi.org/10.3233/JAD-171133.
179. Akiyama H, Barger S, Barnum S, Bradt B, Bauer J, Cole GM, Cooper NR, Eikelenboom P, Emmerling M, Fiebich BL, Finch CE, Frautschy S, Griffin WS, Hampel H, Hull M, Landreth G, Lue L, Mrak R, Mackenzie IR, McGeer PL, O'Banion MK, Pachter J, Pasinetti G, Plata-Salaman C, Rogers J, Rydel R, Shen Y, Streit W, Strohmeyer R, Tooyoma I, Van Muiswinkel FL, Veerhuis R, Walker D, Webster S, Wegrzyniak B, Wenk G, Wyss-Coray T. Inflammation and Alzheimer's disease. Neurobiol Aging. 2000;21(3):383–421. https://doi.org/10.1016/s0197-4580(00)00124-x.
180. Lim SL, Rodriguez-Ortiz CJ, Kitazawa M. Infection, systemic inflammation, and Alzheimer's disease. Microbes Infect. 2015;17(8):549–56. https://doi.org/10.1016/j.micinf.2015.04.004.
181. Uwishema O, Mahmoud A, Sun J, Correia IFS, Bejjani N, Alwan M, Nicholas A, Oluyemisi A, Dost B. Is Alzheimer's disease an infectious neurological disease? A review of the literature. Brain Behav. 2022;12(8):e2728. https://doi.org/10.1002/brb3.2728.
182. Prosswimmer T, Heng A, Daggett V. Mechanistic insights into the role of amyloid-β in innate immunity. Sci Rep. 2024;14:5376. https://doi.org/10.1038/s41598-024-55423-9.
183. Chakravarthi ST, Joshi SG. An association of pathogens and biofilms with Alzheimer's disease. Microorganisms. 2021;10(1):56. https://doi.org/10.3390/microorganisms10010056.
184. Dumoulin D, Fulop T, Beauregard PB. The role of the biofilm of Porphyromonas gingivalis in driving amyloid beta secretion and aggregation. Alzheimers Dement. 2023;19:e073279. https://doi.org/10.1002/alz.073279.
185. Tuzim K, Korolczuk A. An update on extra-oral bitter taste receptors. J Transl Med. 2021;19:440. https://doi.org/10.1186/s12967-021-03067-y.
186. McMahon DB, Kuek LE, Johnson ME, Johnson PO, Horn RLJ, Carey RM, Adappa ND, Palmer JN, Lee RJ. The bitter end: T2R bitter receptor agonists elevate nuclear calcium and

induce apoptosis in non-ciliated airway epithelial cells. Cell Calcium. 2022;101:102499. https://doi.org/10.1016/j.ceca.2021.102499.
187. Gaida MM, Dapunt U, Hänsch GM. Sensing developing biofilms: the bitter receptor T2R38 on myeloid cells. Pathogens Dis. 2016;74(3):4. https://doi.org/10.1093/femspd/ftw004.
188. Nissim I, Dagan-Wiener A, Niv MY. The taste of toxicity: a quantitative analysis of bitter and toxic molecules. IUBMB Life. 2017;69:938–46. https://doi.org/10.1002/iub.1694.
189. Tong A, Yang H, Yu X, Wang D, Guan J, Zhao M, Li J. Mechanisms and novel therapeutic roles of bitter taste receptors in diseases. Theranostics. 2025;15(9):3961–78. https://doi.org/10.7150/thno.107406.
190. Smedley P. A survey of the inorganic chemistry of bottled mineral waters from the British Isles. Appl Geochem. 2010;25:1872–88.
191. Moss GA. Water and health: a forgotten connection? Perspect Public Health. 2010;130(5):227–32. https://doi.org/10.1177/1757913910379192.
192. Borsay P. Health and leisure resorts 1700–1840. In: Clark P, editor. The Cambridge urban history of Britain. Cambridge: Cambridge University Press; 2000. p. 775–804.
193. Porter R. The medical history of waters and spas. Introduction. Med Hist Suppl. 1990;10:vii–xii.
194. van Tubergen A, van der Linden S. A brief history of spa therapy. Ann Rheum Dis. 2002;61(3):273–5. https://doi.org/10.1136/ard.61.3.273.
195. Sukenik S, Flusser D, Abu-Shakra M. The role of spa therapy in various rheumatic diseases. Rheum Dis Clin N Am. 1999;25(4):883–97. https://doi.org/10.1016/s0889-857x(05)70108-3.
196. McCarthy OR. The key to the sanatoria. J R Soc Med. 2001;94(8):413–7. https://doi.org/10.1177/014107680109400813.
197. Nocco PB. Mineral water as a cure. Veroff Schweiz Ges Gesch Pharm. 2008;29:402.
198. Lennartson A. Bergman and the chemistry of mineral waters. In: Carl Wilhelm Scheele and Torbern Bergman. Perspectives on the history of chemistry. Cham: Springer; 2020. https://doi.org/10.1007/978-3-030-49194-9_17.
199. Coley NG. Physicians, chemists, and the analysis of mineral waters: "the most difficult part of chemistry". Med Hist Suppl. 1990;10:56–66. https://doi.org/10.1017/s002572730007099x.
200. Barišić V, Kopjar M, Jozinović A, Flanjak I, Ačkar Đ, Miličević B, Šubarić D, Jokić S, Babić J. The chemistry behind chocolate production. Molecules. 2019;24(17):3163. https://doi.org/10.3390/molecules24173163.
201. Wilson PK, Hurst WJ, editors. Wilson, in chocolate and health: chemistry, nutrition and therapy, vol. 1. London: The Royal Society of Chemistry; 2015. p. 1–27.
202. Zhao B, Gan L, Yu K, Männistö S, Huang J, Albanes D. Relationship between chocolate consumption and overall and cause-specific mortality, systematic review and updated meta-analysis. Eur J Epidemiol. 2022;37(4):321–33. https://doi.org/10.1007/s10654-022-00858-5.
203. Veronese N, Demurtas J, Celotto S, Caruso MG, Maggi S, Bolzetta F, Firth J, Smith L, Schofield P, Koyanagi A, Yang L, Solmi M, Stubbs B. Is chocolate consumption associated with health outcomes? An umbrella review of systematic reviews and meta-analyses. Clin Nutr. 2019;38(3):1101–8. https://doi.org/10.1016/j.clnu.2018.05.019.
204. Messerli FH. Chocolate consumption, cognitive function, and nobel laureates. N Engl J Med. 2012;367:1562–4. https://doi.org/10.1056/NEJMon1211064.
205. Lanaud C, Vignes H, Utge J, Valette G, Rhoné B, Garcia Caputi M, Angarita Nieto NS, Fouet O, Gaikwad N, Zarrillo S, Powis TG, Cyphers A, Valdez F, Olivera Nunez SQ, Speller C, Blake M, Valdez FJ, Raymond S, Rowe SM, Duke GS, Romano FE, Loor Solórzano RG, Argout X. A revisited history of cacao domestication in pre-Columbian times revealed by archaeogenomic approaches. Sci Rep. 2024;14(1):2972. https://doi.org/10.1038/s41598-024-53010-6.
206. Craig WJ, Nguyen TT. Caffeine and theobromine levels in cocoa and carob products. J Food Sci. 1984;49(1):302–3.

207. Coe SD, Coe MD. The true history of chocolate. 3rd ed. London: Thames & Hudson; 2013.
208. Razola-Díaz MDC, Aznar-Ramos MJ, Verardo V, Melgar-Locatelli S, Castilla-Ortega E, Rodríguez-Pérez C. Exploring the nutritional composition and bioactive compounds in different cocoa powders. Antioxidants. 2023;12(3):716. https://doi.org/10.3390/antiox12030716.
209. Soares TF, Oliveira MBPP. Cocoa by-products: characterization of bioactive compounds and beneficial health effects. Molecules. 2022;27(5):1625. https://doi.org/10.3390/molecules27051625.
210. Atanassova M, Aayush K. Chemical diversity of Theobroma cacao L. In: Mérillon JM, Rivière C, Lefèvre G, editors. Natural products in beverages. Reference series in phytochemistry. Cham: Springer; 2025. https://doi.org/10.1007/978-3-031-38663-3_3.
211. Magrone T, Russo MA, Jirillo E. Cocoa and dark chocolate polyphenols: from biology to clinical applications. Front Immunol. 2017;8:677. https://doi.org/10.3389/fimmu.2017.00677.
212. Valle Pavón G. The cocoa trade in New Spain in the 18th century. In: Oxford research encyclopedia of Latin American history. Oxford: Oxford University Press; 2021.
213. Martínez-Pinilla E, Oñatibia-Astibia A, Franco R. The relevance of theobromine for the beneficial effects of cocoa consumption. Front Pharmacol. 2015;6:30. https://doi.org/10.3389/fphar.2015.00030.
214. Mellor DD, Amund D, Georgousopoulou E, Naumovski N. Sugar and cocoa: sweet synergy or bitter antagonisms. Formulating cocoa and chocolate products for health: a narrative review. Int J Food Sci Technol. 2018;53(1):33–42. https://doi.org/10.1111/ijfs.13651.
215. Loveman K. The introduction of chocolate into England: retailers, researchers, and consumers, 1640-1730. J Soc Hist. 2013;47(1):27–46.
216. Nimbalkar N. John Locke on personal identity. Mens Sana Monogr. 2011;9(1):268–75. https://doi.org/10.4103/0973-1229.77443.
217. Gershman SJ, Ullman TD. Causal implicatures from correlational statements. PLoS One. 2023;18(5):e0286067. https://doi.org/10.1371/journal.pone.0286067.
218. Chowienczyk P. Role of arterial stiffness in cardiovascular disease JRSM. Cardiovasc Dis. 2012;1(4):012016. https://doi.org/10.1258/cvd.2012.012016.
219. Zupo R, Castellana F, Lisco G, Corbo F, Crupi P, Sardone R, Catino F, Perna S, Gesualdo L, Lozupone M, Panza F, Clodoveo ML. The effect of a Mediterranean diet on arterial stiffness: a systematic review. Nutrients. 2025;17(7):1192. https://doi.org/10.3390/nu17071192.
220. Roth GA, Mensah GA, Johnson CO, Addolorato G, Ammirati E, Baddour LM, Barengo NC, Beaton AZ, Benjamin EJ, Benziger CP, Bonny A, Brauer M, Brodmann M, Cahill TJ, Carapetis J, Catapano AL, Chugh SS, Cooper LT, Coresh J, Criqui M, DeCleene N, Eagle KA, Emmons-Bell S, Feigin VL, Fernández-Solà J, Fowkes G, Gakidou E, Grundy SM, He FJ, Howard G, Hu F, Inker L, Karthikeyan G, Kassebaum N, Koroshetz W, Lavie C, Lloyd-Jones D, Lu HS, Mirijello A, Temesgen AM, Mokdad A, Moran AE, Muntner P, Narula J, Neal B, Ntsekhe M, Moraes de Oliveira G, Otto C, Owolabi M, Pratt M, Rajagopalan S, Reitsma M, Ribeiro ALP, Rigotti N, Rodgers A, Sable C, Shakil S, Sliwa-Hahnle K, Stark B, Sundström J, Timpel P, Tleyjeh IM, Valgimigli M, Vos T, Whelton PK, Yacoub M, Zuhlke L, Murray C, Fuster V, GBD-NHLBI-JACC Global Burden of Cardiovascular Diseases Writing Group. Global burden of cardiovascular diseases and risk factors, 1990-2019: update from the GBD 2019 study. J Am Coll Cardiol. 2020;76(25):2982–3021. https://doi.org/10.1016/j.jacc.2020.11.010.
221. Kellogg JJ, Todd DA, Egan JM, Raja HA, Oberlies NH, Kvalheim OM, Cech NB. Biochemometrics for natural products research: comparison of data analysis approaches and application to identification of bioactive compounds. J Nat Prod. 2016;79(2):376–86. https://doi.org/10.1021/acs.jnatprod.5b01014.

Chapter 10
Conclusions and Future Opportunities

Abstract Explaining the interactions of humans with their environment through the lens of chemistry. Knowledge is limited by experience. The knowledge base can be expanded by travelling to seek new (or applicable) knowledge. There are examples of the transmission of knowledge to Norfolk (and to other places in England) from Europe *via* the Hanseatic League and from further afield. The knowledge of atoms, structures of molecules and bonding increased gradually. This led to proposals regarding concepts such as tautomerism and to the creation of increasingly sophisticated models. Knowledge is created and assimilated on a personal, institutional and societal level and can be transmitted between entities at each of these levels. There can be unintended and unanticipated consequences of the generation of knowledge and some of these will create fear and misgivings about progress. It is important that chemists (and scientists generally) carry out work that meets a perceived or potential need and has value. Embarking on a programme of work simply to demonstrate that it is possible or to just showcase scientific expertise is often meaningless and counterproductive. It also creates a barrier between science and society and thereby undermines the partnership referred to earlier.

Keywords Temporal linkages · Trade · Travel · Acquired knowledge · Quantum and Newtonian theories · Analysis/synthesis dyad · Tautomerism · Chemistry and logic

10.1 Chemistry as a Continuum

The common thread that illustrates the temporal linkages developed throughout this volume is proposed to be chemistry. In order to understand chemistry and its role, it is important to define again what chemistry is and how it affects the interactions of humans with their environment and the exploitation of that environment. In the case of Norfolk, the agricultural (land) and marine (sea) environments are the starting point. The basis of the interaction is the materials that are found naturally—fish and other marine species—and the components of the terrestrial system—plants and

B. Hanley, *The Chemistry of a County*,
https://doi.org/10.1007/978-3-032-21491-1_10

animals. Some of the very earliest ways in which these resources were exploited involved the gathering of them for food or, in the case of woad (*Isatis tinctoria*) and rose madder (*Rubia tinctorum*) as the source of an ornament (colour) after processing. Indeed making a colour from a plant is an early example of a transformative creative process which involves certain types of chemistry (extraction, reduction, oxidation) and optimisation to give the desired hue. Another significant natural resource that has influenced the development of Norfolk both in the short and in the longer term (and which has not so far been discussed in any detail) is the collection and burning of peat. Peat was found in deposits in wetlands adjacent to the Southern North Sea in England and in European coastal areas such as the Netherlands [1]. As with many human interventions that make use of a natural resource, the persistent impact of the cutting and gathering of peat on the landscape and on the subsequent growth of plants has been studied [2]. The consequence of human interference in the ecosystem (some of the technology for which was imported from the Netherlands) [3] together with rising sea levels, was the creation of the Norfolk Broads. The inland port of Norwich was accessible *via* various rivers and waterways and it was further elaborated by early medieval land drainage to create arable opportunities and this led to an improved (more reliable) river transport between Norwich and the coast. While improved trade was one consequence of greater numbers of navigable options, the establishment of links between Norfolk and Europe also resulted in an exchange of ideas. The spread of knowledge is considered to be circumscribed by experience (this was in part, the doctrine of Kant but was perhaps more fundamental to the 'tabula rasa' philosophy of Locke and the approaches of Freud) [4]. The corollary of the paradigm of knowledge by experience is the concept that things which cannot be experienced are not simply unknown but are actually unknowable. If something is not (or cannot be) experienced, then its existence is a matter for debate. These ideas, in addition to being part of a political philosophy [5] may also have an impact on some specific aspects of science and learning [6]. In the context of the spread of agronomic knowledge and the development of useable knowledge through practice—experience is crucial. The development of the Norfolk crop rotation system was only possible because of the travels of Coke, Townshend and others to Europe where they observed crop rotation in practice [7] and were able to import the necessary expertise and apply it to an English setting [8]. As discussed earlier, the development of the science of breeding plants for favourable traits has been a theme in agronomic practice—even if trait identification was not underpinned by a mechanistic knowledge of how genes operated. The development of 'accepted practice' was central to the philosophical writings of Hume and others and to the concept of acquired rather than innate knowledge [9]. This leads to the appeal of learning by experience (empiricism) rather than innate knowledge. The importance of the link between impressions and, in particular, the interrelationships between them, is considered by some to be a key aspect of the philosophy of Hume [10]. If knowledge is restricted to that which is factually established by experience, then the concept of probable truths is introduced—this acceptance of the validity of probable truths suggests that the description of a concept in words must correspond to ideas and the ideas themselves must be rooted in experience. There is, therefore, a chain of connected concepts—experience creates ideas which then give rise to the development of concepts. This then enables chemistry to be conceptualised as a mixture of empirical (and testable) experiences giving rise to ideas and therefore to concepts. Physics, on the other hand, can be a scientific discipline in which the key principles can be established by mathematical relationships rather than by observation, testing and inspection. The challenge to this purely *a priori* conceptual interpretation is that the models which are used to construct the mathematical relationships are not invariably completely coherent with one another. This means that the approach developed by Newton differs from classical mechanics in how it describes systems. Nonetheless, both models are, in essence similar and, if not precisely derivable from one another, are not completely divorced. Quantum theory, on the other hand has a completely different starting point and scale and produces very different sets of paradigms. This means that theories and knowledge developed using classical mechanics cannot be used to advise theories that are based upon quantum mechanics. By contrast, the development of theories in chemistry can be thought of as being additive. While the development of ideas about atoms lies at the core of chemistry, these are complemented by theories about how atoms combine through the formation of bonds and, eventually, structures. These theories are not mutually exclusive however, unlike those developed in physics. Their assumptions (atoms, bonds and structures) are not directly observable and hence the theories fit those behaviours which can be observed by implication even if extrapolation to what direct observation may reveal is not possible. There is no discontinuity in thought in chemistry and the addition of new knowledge adds to the accepted framework rather than requiring a complete reinterpretation of knowledge acquired previously. Another feature of chemistry is that atoms, molecules and structures are invariant with regards to time and place. Molecules do not spontaneously change their composition and structural characteristics. For example, quinine isolated from *Cinchona* species today [11] is molecularly and structurally identical to that isolated by Caventou and Pelletier [12]. This provides a consistency in the basic certainties about molecules and in the empirical results obtained from testing substances. It is not necessary to construct a new molecular tabula rasa every time a particular molecule is isolated. This applies particularly to organic molecules (natural products from living systems and also the identical molecules obtained by synthesis) but is also relevant to inorganic molecules. For example, the precise nature of a multi-atom inorganic molecule may vary [13] if it interacts with itself, and this affects the properties of the material. The question of absolute structure as it relates to biological properties and function can be more complex and is an example of how the conditions that are prevalent may influence the structure (or, more accurately, preferred conformation) of a molecule. The favoured conformation of a molecule within a biological system (for example at an interface) is often different to that observed in, for example, a crystal structure [14] and this has a significant impact on the observed biological properties. This does not apply solely to organic molecules and, as with the example of the multi-atom inorganic molecule given previously, there is an appreciation of the importance of inorganic complexes and their interactions with living systems [15]. Indeed, one of the first examples of a 'magic bullet' pharmacological agent developed by

Ehrlich was an organoarsenical compound marketed under the name of Salvarsan [16] and given the chemical designation of arsphenamine (Fig. 10.1).

Even the apparent chemical consistency of Salvarsan was illusionary (before its interaction with biological systems) as the original structure proposed by Ehrlich containing a chemically implausible As = As double bond (Fig. 10.1). This was later supplanted by other cyclic structures involving tetramers and pentamers of Arsenic (see Lloyd et al. [16] above). The plausibility of specific chemical structures may change as more knowledge is acquired. Another example of the development of greater understanding based upon knowledge acquisition occurs in the unsuccessful attempts by Perkin to synthesise quinine. He based his reaction scheme on the empirical formulae of his starting materials and product. These were structurally unrelated to one another and, rather than synthesising the complex natural product, quinine, the reaction led to the formation of the aniline dye called mauveine which, by historical recreation of the original synthesis, has been found to be a mixture of compounds rather than a single molecular type [17]. A third example of the development of structural knowledge being developed some time after the original discovery or synthesis of a molecule is the antacid marketed as Pepto Bismol (bismuth subsalicylate). The structural details of the molecule were only discerned over 100 years after its initial synthesis [18]. All three of these examples derived from the chemical synthesis of molecules whose medicinal or colouring properties were not dependent upon having knowledge of the precise molecular structure—although such information is normally the basis for quantitative structure activity relationship (QSAR) studies to optimise the biological activity of a series of compounds [19]. With stand-alone individual compounds there is less of an imperative to try to relate specific structural characteristics to functional outcomes and to thereby optimise the molecular and functional characteristics. This is illustrated by the three examples given.

There have, however, also been examples of the misassignment of structures in the area of natural product isolation and characterisation. One example of this is the structure of glucosinolates—the precursors of the compounds (isothiocyanates) responsible for the pungent flavour of mustard (Chapter Eight). The reaction to form the isothiocyanate involves the enzymic hydrolysis of a glucose molecule and

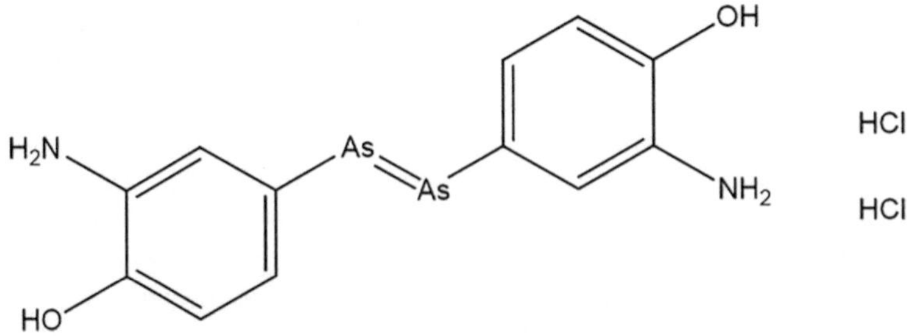

Fig. 10.1 Structure of arsephenamine

elimination of a sulphate group to give an isothiocyanate. This led to the structure shown (Fig. 10.2).

being proposed in the nineteenth century. This was later amended and the current structure (Fig. 10.3) confirmed by X-ray crystallography [20] and by synthesis [21].

Despite the development of thought around the roles of atoms, molecules and structures in chemistry in the seventeenth, eighteenth and nineteenth centuries, in particular, and the way that the progression of ideas shifted the focus of chemistry from isolation of pure elements (gases and metals in particular) to the isolation and limited structural characterisation of molecules from living systems (the beginnings

Fig. 10.2 The original proposed structure of glucosinolates

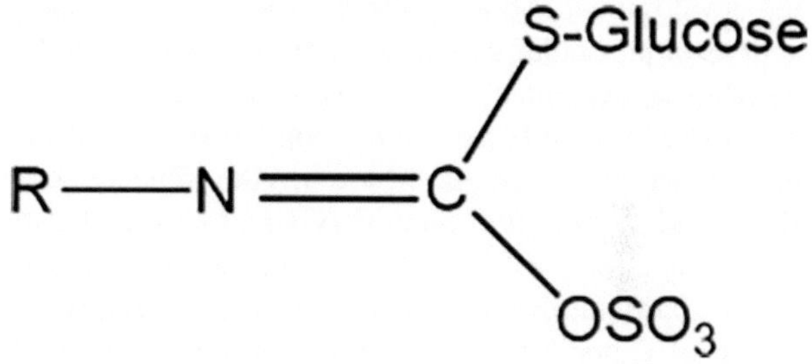

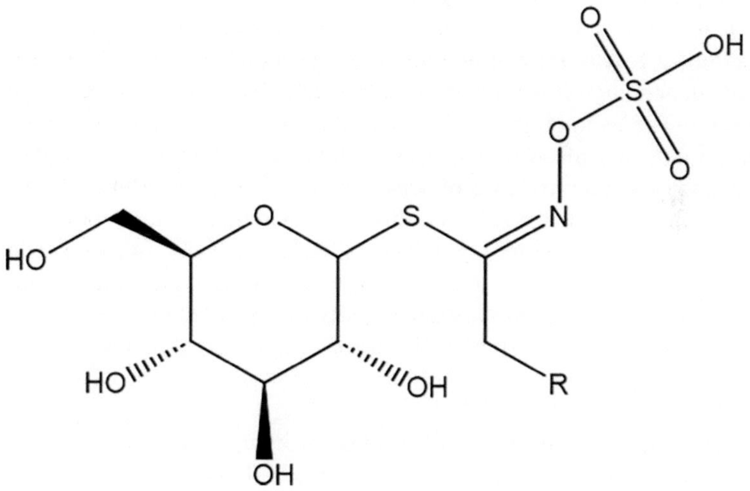

Fig. 10.3 Confirmed structure of glucosinolates

of natural product and organic chemistry—see above). The interrelated analysis/synthesis dyad, which was the gold standard to prove chemical structures, the study of how molecules are created in living systems and the use of 'living factories' to design and create molecules (synthetic biology) are all features that provide evidence of an intellectual progression within the common theme of chemistry. The development of methods to determine chemical structures which did not rely on decomposition and subsequent fragment analysis [22] to determine structural features represents a paradigm shift for chemistry however, the essential theme of chemistry represented by the investigation of increasingly complex molecules and interactions remained conceptually consistent even as the tools and techniques that could be used progressed. At one end of the theme lies the reductionism that led to the development and adoption of empirical formulae while at the other is the interactions of particular structural features that exist in different molecules. There are a number of examples of how these different approaches can affect the way that chemistry is considered within a logical and consistent framework. One of the simplest molecules—and one which was the subject of considerable interest, was water. Both from a structural chemistry standpoint and due to its nature as an essential component for public health [23], water was significant. Considerable scientific efforts, particularly in England [24] and France, were directed towards the investigation of water [25]. The proportions of hydrogen and oxygen in water became known and this would appear to give a simple formula—two hydrogen atoms for each oxygen atom. It would be wrong, however, to assume (from its bulk properties) that each and every molecule of water contains hydrogen and oxygen atoms in the fixed ratio of 2:1. Any sample of water contains a proportion of dissociated molecules and, depending on a number of factors, H^+, OH^-, H_3O^+ and other species are also present. The average proportions of hydrogen and oxygen in pure water are, indeed, 2:1 but this macroscopic average hides discontinuities at a microscopic level. In addition, the structure of water in living systems contains, on average, about 3% of non-bulk water [26] and this is affected by the other components present including, in particular, biomolecules including both macromolecules and smaller molecular species. In addition to biomolecules affecting water structure, the converse is also true. The stability and activity of biomolecules such as enzymes is affected by the water activity of their surroundings [27]. Another example of the difficulty in assigning a specific structure to an organic molecule is provided by keto-enol tautomerism (Fig. 10.4).

Keto-enol tautomerism in one of a range of interconvertible pairs of forms of molecules and is of particular importance in the biological activity of small molecules. It can occur in drug molecules but also in natural products such as curcumin (Fig. 10.5).

where tautomerism has been implicated in its biological activity [28] and has been extensively studied in drug development [29]. The definition of keto-enol tautomerism is a molecule in which a C=O double bond (keto form) is interconverted to an alkene (enol form) with the shift of a proton from the alpha carbon of the keto form to create a hydroxyl moiety (Fig. 10.4). In the case of β-diketones the molecules exist dominantly in the enol form. The formation of isomeric forms of organic

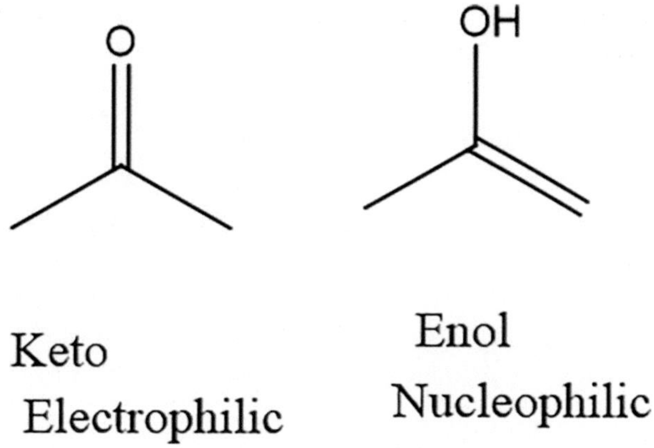

Fig. 10.4 Keto-enol tautomerism

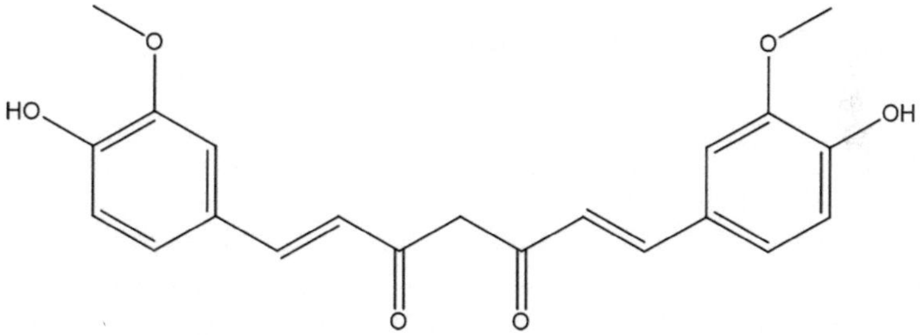

Fig. 10.5 The structure of curcumin

molecules (of which tautomerism is a specific example) is an important characteristic and clearly demonstrates why simple molecular formulae (the ratio of the different types of atoms) does not provide enough information to be able to define the properties of a compound. The spatial arrangement of groups with particular properties and how they are joined together (type of bonds formed) is also required. Even with such information, the interconversion of molecules between different forms (as is the case with keto-enol tautomerism) which is dependent upon the molecule itself and the conditions it is surrounded by, can further complicate a simplistic view of how molecules are best represented. The other factor to be considered is the reactivity of different tautomers. The enol form of many molecules can act as a nucleophile in many reactions, for example in acid-catalysed aldol condensations (Fig. 10.6).

In addition, tautomerism can provide a useful template for molecular simulation calculations [30] which, in addition to providing information about the interconversion between tautomeric forms, also provides evidence for the applicability of the nature of the chemical bond model used to describe the way molecules are assembled.

The iterative process by which knowledge in chemistry is gradually built is, as noted previously, in contrast to the largely mutually exclusive classical, Newtonian

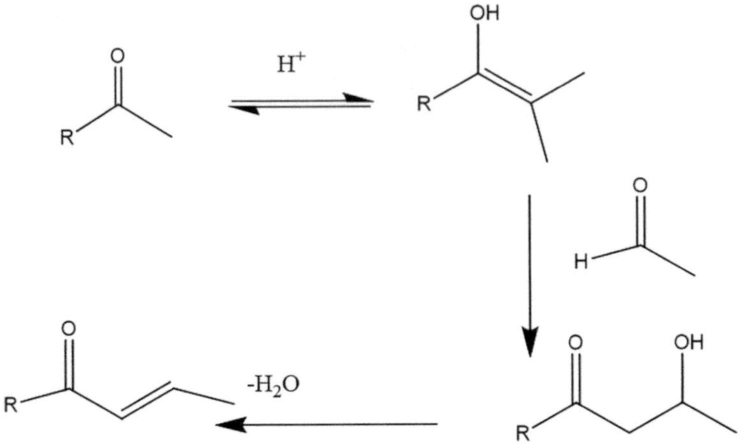

Fig. 10.6 The acid-catalysed aldol condensation reaction

and quantum models developed by physicists. The evidence base for chemistry develops through the accumulation of knowledge, the development of empirical testing and the application of reductionism. As a result, intermediate models are consistent with the level of data available and, as more sophisticated techniques are developed then models become more precise rather than being discarded in favour of completely different systems and paradigms. This process is seen in the development of models in a range of applied areas as well as in pure chemistry. For example, in the ultimate aim of improving agronomy through crop rotation, progress is measured (in increasing levels of detail) in terms of crop yield [31], constituents of soil [32] and through a molecular understanding of the soil microbiome [33]. These approaches are not mutually exclusive since each helps to build a better understanding different consequences of the crop rotation process. There is, therefore, a lineage from the Norfolk crop rotation system introduced initially in Belgium in the sixteenth century and later refined in Norfolk in the eighteenth century with the aim of improving crop yield and the quality of the agronomic environment to the most modern techniques for measuring improvements in the soil microbiome and microbial quorum sensing and signalling [34]. Similarly an understanding of natural products and how they interact, react and their biological properties has built up from crude extraction of mixtures of active principles, the fractionation and isolation of individual compounds and, more recently, the development of methods to measure the impact of mixtures of compounds in terms of synergies [35]. In each case, the useable knowledge is obtained through the application of the available techniques—reductionist isolation or measurement of multiple effects through eventually to a deep mechanistic understanding.

The progress in the development of chemistry has been driven, at least in part, by economic advantage as well as the accumulation of knowledge. When Perkin failed to synthesise quinine and made a new pigment instead, he did not discard his discovery as a 'failed experiment' but found an application that created a whole new area of chemistry (based on aniline dyes) and a significant commercial opportunity leading to the widescale production of synthetic dyes [36]. The importance of chemistry in the creation of wealth, particularly in the context of Norwich and Norfolk, will be examined in Sect. 10.2.

10.2 The Role of Chemistry in Wealth Creation

When considering the value of chemistry, it is important to include the non-tangible benefits as well as actual measurable products such as agricultural produce and manufactured products such as cloth. Properties that are associated with the chemistry of substances are part of chemical knowledge and determine how materials and situations acquire value. In addition to knowledge, there must also exist the determination to apply the knowledge to create something of value. The lack of this determination is actually a part of a lack of knowledge and any failure to reap the benefits of the application of specific knowledge is involuntary because it is the consequence of a lack of a particular attribute (application) (and hence a deficiency of knowledge). The application of knowledge leads to a change which in turn has the consequence of creating a new situation that can be responded to and which creates the opportunity for further change. The eventual actions at a specified point in time are a consequence of circumstance and application of knowledge however this is an ongoing iterative process and therefore actions are not endpoints in themselves but part of the process. In this way learning can be considered to be a process of the recollection of information or knowledge, and this can exist within an individual, a group (collective knowledge within an organisation) or society as a whole. It is clear, then, that there are different ways of benefitting (or failing to benefit) from knowledge. The knowledge of specific chemical processes can sometimes be applied to other situations. For example, the binding of ligands to receptors was a key aspect of the discovery and use of successive different histamine receptor antagonists [37] and this was also was dependent upon the discovery of histamine and its receptors and the incorporation into an assay [38]. The subsequent development of medicinal agents enabled a range of profitable physiological treatments to be developed—all based on the use of receptor-based screening technology. This is an example of the application of conceptual technology to a specific set of opportunities. This can lead to products with added value and hence the knowledge (technology) is both valuable in itself and its application is of value in economic terms. Secondly, the collective knowledge of an organisation can be applied—if an opportunity can be recognised and appreciated. For example, Pfizer developed deep tank fermentation for the production of citric acid. This was a commercially valuable process [39] and became an essential method for the production of penicillin. The institutional knowledge of the technique for the production of one microbial product was applied to a second valuable product [40]. Thirdly, societal-level knowledge will underpin advances and ensure that the benefits are appreciated and understood through knowledge, experience and familiarity. The awareness of textile weaving and a society with members that were skilled in those techniques contributed to the success of wire netting manufacture on a loom-based system in Norwich in the nineteenth century. The lost opportunity can arise if any of these criteria are missing or as noted above, if an involuntary failure occurs as a consequence of a lack of determination to carry an initiative forward. The logic of treating knowledge in this way was part of Platonic thought which was popular in the years before the Enlightenment. The examples of weaving and brewing have already been used to illustrate the development and application of chemistry knowledge under different circumstances and at different times and these will be revisited. Both are examples of the application of knowledge to raw materials to create additional value and are, to that extent, both situationally and intellectually specific when considered in combination.

Weaving to make cloth was established in Norfolk in medieval times and, in medieval Norwich, more people were employed in the textile industry than in any other. In conjunction with weaving, a range of ancillary trades were also commonly followed. In addition to producing wool for weaving, sheep are a source of lanolin which can be used as a general lubricant and for making cloth waterproof [41]. The beginnings of the Norfolk cloth industry began with the arrival of weavers from Flanders in the sixteenth century (certified by Royal proclamation in 1565). One of the technologies they brought was the combing of wool which led to the production of better quality and more resilient cloth produced from longer threads. This cloth was called Worsted Wool and was in great demand across Europe. Much of the weaving was carried out by individuals and techniques were passed on within families however the development of an apprentice system helped to formalise the techniques. In addition, by the 1650s, technology for the installation of sluices and other ways of managing water had been imported from Flanders and France and adapted to local conditions however it was the arrival of the Huguenots that led to the most significant advances in land management. Between 1627 and 1652 they reclaimed 40,000 acres of fenland. In addition to creating more arable land, this also improved navigability and hence helped the export trade. The arrival of refugee French Protestants—the Huguenots—in the seventeenth century also led to the development of new technology which combined the Flemish worsted wool (which had fallen out of favour) with silk to produce 'Norwich crape (or crepe)'. One of the leading crepe manufacturers in Europe was Grout and company in Norwich. The founder of the company, Joseph Grout (likely a descendant of a Huguenot immigrant) was born in Essex but chose Norwich as the centre of his business and by the nineteenth century over 3000 people were employed in crepe manufacture which was a successor to the weaving industry that produced worsted wool. Another feature of the weaving industry was the development of dyeing using rose madder and woad (imported from Hanseatic ports in Germany *via* King's Lynn)—both being processed to produce colours used in the textile industry. The individual and collective development of technology associated with weaving had a significant effect on the economy of Norfolk and Norwich from the time of the immigrant populations of the Flemish and the Walloons in the sixteenth century, followed by the influx of

Huguenots in the seventeenth century through to the nineteenth and even the twentieth centuries.

The development of brewing (particularly of beer) in Norwich and Norfolk has some parallels with that of weaving. The industry was founded on the production of a particular arable crop (barley rather than wool) combined with a local resource (water from chalk streams) to create profit and opportunities. Increases in crop (particularly cereal crop) yields have been found in England in the eighteenth and nineteenth centuries [42]. In addition, probate data specifically from East Anglia has been used to assess the impact of various changes to agronomic practices [43]. Crop yield data from the eighteenth century reflects more closely the impact of crop rotation and the elimination of a fallow period in agronomic practice [44]. The rapid increase in enclosure land compared to earlier, piecemeal strip farming also had a significant impact and resulted in increased yields. Once the best variety of barley has been selected (yield, disease resistance and the quality of the malted grain must all be considered) and the crop grown and harvested, the barley is malted (about I tonne of malt is prepared from 1.3 tons of barley). The malting process involves a number of steps

1. steeping in water with dry rests over 1–2 days
2. germination at 12–16 °C in a flow of humid air to maintain malt moisture at 42–46%;
3. kilning at between 50–85 °C with hot dry air and
4. storage and blending of malt to specification.

In addition to measures of key quality parameters such as microbial populations [45] and the effects of altering processing variables, the malting process has also been studied on a gene expression level [46] and this adds to the degree of control available to the maltster and beer maker. In addition to a supply of barley as a raw material, the main factors that influence the nature of the final product are the water used and the different aspects of the microbiology of the process [47]. In the case of water, the optimal compositional properties depends on the type of beer being produced [48] with very soft water (low levels of calcium and magnesium) being favoured for European lagers and the opposite (very hard water) being better for English bitter type beer. In the case of Norfolk, the very hard, chalk spring derived water will favour English-style beer production. In addition, the technologies developed for beer production are also beneficial to other fermentative and aseptic processes [49] and, in this way, the collective scientific (chemistry) knowledge of the society increases and can be applied in processes other than brewing beer. Another aspect of beer production was that it provided a safer supply of liquid than the water supply in the eighteenth century [50]. In addition, more recent research has identified a number of potential benefits associated with moderate beer consumption due to some of the components present [51], including a positive effect on cardiovascular and metabolic diseases [52]. The idea of combining health with beer is not intuitively obvious although the link has been used as a marketing tool and a possible association between beer consumption and the increased diversity of the gut microbiome has been suggested [53]. The creation of an added value product (beer) through the application of particular chemistry processes to raw materials (water and barley) is a clear example of the role of chemistry in creating enhanced value and in wealth creation. Such a model is not limited to the eighteenth and nineteenth centuries. The example of creation of economic value through the application of innovation in beer brewing has also been cited more recently [54]. In addition to the value associated with the primary product, waste material produced by the brewing process including post-fermentation lees [55], barley rootlets [56], spent grain [57] and hops [58] have all been investigated as a potential source of valuable material. This creation of added economic value and the reduction of waste that needs to be subject to disposal, is often a consequence of the development of a more complete understanding of the chemical constituents either released or remaining after a process to make the primary product has been carried out.

10.3 The Logic of Chemistry

The philosophical basis and the logic of chemistry has been debated and discussed by may people over many years. One aspect that is evident when considering chemistry at different points in time is the continuity of thought and approach and this lies at the core of this volume. Chemistry lies on an axis between trial and error on the one hand and learning through experience on the other. At the heart of chemistry lies the realisation that we cannot have absolute and certain knowledge of things that we cannot measure directly. This is a challenge for an empirically driven exercise such as chemistry. If we cannot directly inspect and test atoms and bonds, then how can empirically based conclusions be achieved. The answer to this may lie, in part, in what Hume described in the context of experience enabling expectation. While expectation can be anticipated based on experience, this does not mean that a cause/effect relationship invariably follows. Nonetheless expectation built on experience enables chemists to adapt solutions to different situations in the measurement of the properties and responses of substances and entities. There are some principles that guide the development of viable expectations. The concept of molecules and blends, for example, is key to the understanding of how to measure the properties of substances and to understand how to interpret these measurements. This is a defining principle of the concept of the properties of fundamental and macromolecular objects. First of all, however, it is appropriate to revisit the concept of mixtures and blends which was an important aspect of Aristolean thought (and later the development of the theories of mass and composition developed by Black and Lavoisier [59]). While both blends and unique combinations of the constituent elements of a compound are viable concepts, there exists a series of differences between them. Among the most significant is that a blend has the identifiable properties of its constituents, and the constituents are both discernible and separable, so the formation of a blend is, at least in principle, a reversible process. A combination in which a new entity is formed with distinct and novel properties is actually and philosophically different. The properties of the new entity are often not deducible directly (or by inference) from the properties of the constituent parts. In addition, the components are frequently not easily dissociable into their isolated, pre-combined, constituents. This can be further elaborated in the context of bond formation and energy diagrams however such explanations are both dependent upon a model of atoms and bonds which remains 'unknowable' (in the view of Hume) and are only measurable as macroscopic representations of microscopic events. This then introduces the concept of the possible link between the fundamental (microscopic) and macromolecular properties of substances. It has been suggested that the world is hierarchical and is structured such that fundamental (quantum) concepts or entities are real and can be demonstrated to operate at a basic level. Higher orders of things are necessarily more complex and are composed of entities that derived from the fundamental principles by inference or directly. This is the basis of the 'reduction' of chemistry to physics which, although superficially plausible has, nonetheless been rejected and criticised [60]. The reductionist argument about the link between chemistry and physics (and the proof of the existence of basic chemical entities) requires that two principles are invoked:

1. Atoms and molecules have to be reducible to fundamental entities that can be defined by physics.
2. The macromolecular (and measurable) properties of substances must be able to be described in the context of their constituent molecular species which are able to be defined as noted above. This then raises the further question of whether changes in the macromolecular properties in the chemical world can influence changes in the microscopic, quantum molecular world. In other words, is the link between macro and microscopic effects symmetric or asymmetric. If it is symmetric then this implies that both downward and upward causation are possible.

If these two sets of criteria cannot be met then the physics/chemistry reductionism hypothesis falls. In that case there has to be an alternative coherent narrative that creates the link between the macromolecular and microscopic aspects of chemistry. If it is not the case that all of chemistry is rendered intelligible by derivation from physics, then some other factor (learning through experience?) must be involved. The microstructure (as has been observed with water) is dependent upon context. While the macroscopic properties of water are those of a substance with the average constituents of H-O-H, on a molecular level, the existence of H^+ and OH^- (and other species) is postulated. The amounts of these that are present can be a function of the state of matter (they exist in liquid water, for example but not in water vapour) or other properties of the system such as the addition of ionic species. As observed previously, the same argument applies to organic molecules that are structurally able to exist in enol or keto forms. The proportion of each can be a function of context and can be influenced, for example, by redox control and electric field modulation [61]. The logic of chemistry is, therefore, a function of the empirically derived hypotheses regarding the formation of molecules and bonds and the understanding of the microscopic (fundamental) properties of chemical species. Within this, however, the higher level of chemistry is not wholly (asymmetrically) dependent upon the lower (fundamental) level of physics. The mutually exclusive nature of different interpretations of physics (classical, Newtonian and quantum) does not impact upon the validity of the totality of all chemical knowledge and, even if quantum mechanics is eventually replaced, the molecular and chemical bonding nature of chemistry will still be a valid representation that explains empirical observations and experiments and the historical theme of chemistry will still unite discoveries and knowledge acquired on an individual, collective and societal level.

10.4 The Fear of Progress and the Side Effects of Growth

The acquisition and development of new technology (particularly chemistry) can create both wealth and uncertainty. Examples of this include:

1. The production of synthetic colours and the replacement of traditional, natural colours
2. The introduction of novel crops and farming methods
3. The use of new technologies (particularly gene manipulation) to create crops with desirable properties.

The uncertainty can take many forms including, in particular, the replacement of existing technology (and the supplanting of traditional skills) and a fear of the new technology itself and the understanding of the potential consequences of its introduction.

As noted previously, the introduction of technology to extract and fix pigments so that dyeing textiles became viable was important to the weaving industry in Norfolk. The use of woad (*Isatis tinctoria*) as a source of blue dye was important in sixteenth century Europe. Indeed, even prior to that, woad was used as the source of a dyeing agent. Trade in woad was sufficiently well established (and profitable) in Thuringia (part of modern-day Germany) that a university was established in the most important town in the region, Erfurt, in 1392 and woad was extensively grown, harvested and processed. By the sixteenth century, the major threat to the cultivation and growth of the woad trade was the importation of another natural blue colour, indigo, from indican derived from plants of the genus Indigofera (primarily *Indigofera tinctoria* L. and *I. suffruticosa* Mill). This represented the replacement of one material (and technology) with another. In the sixteenth century, England was largely an exporter of raw wool. The development of textile production and the indigenous production of the woad-derived blue pigment greatly increased the value of English manufactured goods. One of the problems came from the early introduction of tariffs on imported (but not exported) goods. This led to a reduction in Crown revenues as exports with enhanced value were exported without financial benefit to the Crown. The solution adopted was to limit the cultivation of woad and to thereby control the production of finished textile goods ("No person was to break up ground for the present nor to sow woad within four miles of a market or clothing town."). Eventually this restriction was rescinded however the growing and processing of woad was still forbidden near to Royal castles and palaces, but this was because of the smell generated. Rose Madder was the principal alternative colouring material used for textiles. It comes from the root of the madder plant (*Rubia tinctorum),* which contains a natural dye called alizarin. This has been used for thousands of years to colour textiles. The development of a process to ensure the consistent and equivalent adhesion of rose madder to silk and to wool was carried out in Norwich by a Scottish chemist and has been referred to previously (Chap. 6). The main anthraquinones responsible for the colour are alizarin (1,2-dihydroxyanthraquinone), purpurin (1,2,4-trihydroxyanthraquinone) and pseudopurpurin (1,3,4-trihydroxyant hraquinone-2-carboxylic acid) and glycosides such as ruberythric acid. The proportions of these (and a range of other) compounds give rise to the different shades of rose madder. One of the consequences of the large-scale production of cloth (silk and wool) dyed with natural rose madder in Norwich was the discolouration of the river Wensum with effluent from the dyeworks. Following the development of different methods to extract pigments from plant material and the optimisation of methods for dyeing textiles, the next significant step came with the accidental production of mauve by W H Perkin while seeking a route for the synthesis of quinine [17]. This led to the rapid development of the synthetic dyestuffs industries in England, Germany and France. In the case of blue (indigo) and red (alizarin), synthetic methods were developed [62, 63] and these quickly replaced the traditional, plant-derived dyeing methods. In addition to having an impact upon industrial processes, the development of synthetic routes also changes agronomic and agricultural practice, trade in raw materials and the need to dispose of extracted (waste) plant material as well as highlighting issues of agronomic practice and competitiveness [64].

The development of novel varieties of crops with particular agronomic characteristics has previously been discussed (Chap. 3). The impact of different crops (including those included in crop rotation) has been referred to in passing in the discussion on quorum sensing in the soil microbiome (Chap. 4). While Norfolk was one of the most advanced agricultural counties in England by the beginning of the nineteenth century due to advances in agronomic practices and the quality of the available arable land, the agricultural practices had remained relatively unchanged since the introduction of crop rotation [65]. Nonetheless, the introduction of new varieties (particularly of grain crops) and the optimisation of crop protection practices resulted in an increase in yield during the 18th and the 19th centuries. For example, wheat yields increased by about a quarter between 1700 and 1800 due largely to the increase in useable nitrogen from legume crops and animal manure [66]. The introduction of synthetic fertilisers in the twentieth century provided a further boost to production [67]. While the efficient use of animals and the development of new machinery is considered responsible for a proportion of the improvement, better farm management and greater efficiency in using the workforce was likely the major factor. It has been noted that a smaller population was required to work the land and this made labour available for work in factories thereby sparking the industrial revolution [68]. While the development of more efficient agricultural practices has been largely welcomed due to the increased availability of cheap food, some concerns about 'factory farming' have been raised [69] including the adverse environmental impacts and the effect of crop monoculture on the soil microbiome [70]. Consequently less intensive approaches have been proposed [71] even if there may be negative impacts on crop yields in the short term. In addition, the impact of changes in one part of the economy of a country may be felt in others. For example, the development of mechanisation stimulated by the industrial revolution and the collapse of European trade had a severe effect on weaving in Norfolk [72]. While the improvements in agronomic practice meant that fewer people were required to produce sufficient food to provide for the whole population in agricultural counties such as Norfolk, therefore labour was available for mechanised industries to make use of to generate wealth through the efficiencies brought about by the introduction of factory working in the short term. This however contributed eventually to the rapid decline of the weaving industry in Norfolk since factory-produced textiles could be created more quickly and were cheaper than those produced by small scale or even individual working practices. Improvements in agricultural practices in initially created wealth for Norfolk landowners. Such improvements may take time to be impactful in terms of increases in crop yield and efficiency of production and the initial benefit for the wealthy few may not be sustained in the economy as a whole. There may also be unanticipated consequences. Improvements in agriculture eventually resulted in the loss of the weaving industry in Norfolk and, as a consequence, Norfolk and Norwich became a relatively poor region of England with low wages and little indigenous industry and a greater reliance on non-manufacturing (service) industries such as insurance and banking [73].

Another example of the unintended consequences of advances lies in the area of genetically modified (GM) crops. The initial target of genetic modification of crops was the improvement of crop quality. The first commercially grown GM food licenced for human consumption was the 'Flavr Savr' tomato. The problem that the technology sought to address was the need to harvest tomatoes while they were still unripe since the ripe fruit was unable to be transported without bruising and loss of fruit due to quality issues. Unfortunately unripe fruit was, after harvest, unable to develop a desirable flavour profile. The solution was to target an enzyme (polygalacturonase) that was involved in the breaking down of pectin (an important component of plant cell walls which is responsible for maintaining cell wall integrity) [74]. The breaking down of pectin during ripening was an important aspect of the softening of the fruit. By reducing the synthesis of polygalacturonase, fruit could be left to ripen on the vine while maintaining the structural integrity of the fruit enabling it to be transported. Later advances mediated by GM technology were targeted towards agronomic advances such as herbicide and pest resistance however in the case of the first application, the target was a benefit to the consumer and food manufacturer in terms of food quality (and sustainability with a reduction in the amount of product being wasted). A more recent example of a consumer-oriented modification of tomatoes will be introduced later. While the commercial production of GM tomato was relatively brief, it established, among other things, the concept of substantial equivalence—judging GM foods based on their compositional similarity to conventional counterparts rather than the process used to create them and this became a cornerstone of regulatory frameworks globally. There was little economic advantage in the production of GM tomatoes and its success was, therefore, dependent upon a high level of consumer acceptance of GM technology. A number of studies (of varying degrees of scientific validity) were produced over the years to suggest that GM was an unsafe technology [75] and this, together with a growing consumer awareness affected by environmental pressure groups, has limited the spread of GM technology. In the light of such concerns [76] it is perhaps surprising that scientists have persisted in developing foods with characteristics that are dependent upon GM. In some cases, the benefits ascribed to the modified products have been directly related to health. For example, golden rice contains transgenes that enable the plant to produce 20–30 μg of β-carotene per gram of edible rice [77]. Vitamin A deficiency in known to be an issue with some populations [78] however the uptake of GM rice has been low due to social rather than scientific issues [79]. Given the difficulty in popularising the use of a GM crop to solve a known nutritional deficiency, it is perhaps surprising that one of the other attempts to produce a biofortified crop used an unproven benefit—enhanced anthocyanin level—as a target benefit [80]. Polyphenols in general and anthocyanins in particular have been associated with a range of health benefits [81] however many studies have been associative rather than demonstrating causality and there are a number of dietary sources of these compounds. Nonetheless, the insertion of genes from snapdragon (*Antirrhinum majus*) into tomato resulted in the production of purple fruit with levels of anthocyanins comparable to that found in blackberries and blueberries [82]. The production of purple tomatoes through conventional plant breeding and manipulation of growing conditions has also been achieved [83] and while the GM purple tomatoes are available in the US for sale to home gardeners, the commercial future of a GM product with a relatively weak health claim remains unclear. It is possible that the GM, purple tomato will be another example of scientific hubris by molecular biologists who have produced something 'because they can' rather than for any particular translatable benefit for either the consumer or for industry.

References

1. Waller M, Kirby J. Coastal peat-beds and peatlands of the southern North Sea: their past, present and future. Biol Rev Camb Philos Soc. 2021;96(2):408–32. https://doi.org/10.1111/brv.12662.
2. Giller KE, Wheeler BD. Past peat cutting and present vegetation patterns in an undrained fen in the Norfolk Broadland. J Ecol. 1986;74(1):219–47.
3. Smith CT. Dutch peat digging and the origin of the Norfolk broads. Geogr J. 1966;132(1):69–72.
4. Petryszak NG. Tabula rasa–its origins and implications. J Hist Behav Sci. 1981;17(1):15–27. https://doi.org/10.1002/1520-6696(198101)17:1<15::aid-jhbs2300170104>3.0.co;2-3.
5. Krogerus M. The decision book: fifty models for strategic thinking. Roman Tschäppeler, Jenny Piening. 1st ed. New York: W.W. Norton & Co.; 2012. p. 86–7.
6. Thompson JP, et al. Known knowns, known unknowns, and unknown unknowns: can systems medicine provide a new approach to sepsis? Br J Anaesth. 2015;114(6):874–7.
7. Plumb JH. Sir Robert Walpole and Norfolk husbandry. Econ Hist Rev. 1952;5(1):86–9.

8. Campbell BMS, Overton M. A new perspective on medieval and early modern agriculture: six centuries of Norfolk farming c.1250-c.1850. Past Present. 1993;141:38–105.
9. Wang JJ, Feigenson L. Is empiricism innate? Preference for nurture over nature in people's beliefs about the origins of human knowledge. Open Mind. 2019;3:89–100. https://doi.org/10.1162/opmi_a_00028.
10. Clay G. Knowledge and sensory knowledge in Hume's treatise. In: Rutherford D, editor. Oxford studies in early modern philosophy, volume X. Oxford studies in early modern philosophy. Oxford: Oxford Academic; 2021. https://doi.org/10.1093/oso/9780192897442.003.0007.
11. Ochoa G, Armijos L, Figueroa JG, Jaramillo-Fierro X, Solano-Cueva N. Optimization of two methods for the rapid and effective extraction of quinine from Cinchona officinalis. Plants. 2025;14(3):364. https://doi.org/10.3390/plants14030364.
12. Pai-Dhungat JV. Caventou, Pelletier & history of quinine. J Assoc Physicians India. 2015;63(3):58.
13. Baruah JB. Inorganic molecular complexes: potential for growth of a new subject area in self-assembly. Top Curr Chem. 2020;378(2):30. https://doi.org/10.1007/s41061-020-0294-8.
14. Le Questel JY. Conformations and physicochemical properties of biological ligands in various environments. Int J Mol Sci. 2023;24(11):9630. https://doi.org/10.3390/ijms24119630.
15. Karges J, Stokes RW, Cohen SM. Metal complexes for therapeutic applications. Trends Chem. 2021;3(7):523–34. https://doi.org/10.1016/j.trechm.2021.03.006.
16. Lloyd NC, Morgan HW, Nicholson BK, Ronimus RS. The composition of Ehrlich's salvarsan: resolution of a century-old debate. Angew Chem Int Ed Engl. 2005;44(6):941–4. https://doi.org/10.1002/anie.200461471.
17. Cova TFGG, Pais AACC, Seixas de Melo JS. Reconstructing the historical synthesis of mauveine from Perkin and Caro: procedure and details. Sci Rep. 2017;7(1):6806. https://doi.org/10.1038/s41598-017-07239-z.
18. Grape ES, et al. Nat Commun. 2022;13:1984. https://doi.org/10.1038/s41467-022-29566-0.
19. Kwon S, Bae H, Jo J, et al. Comprehensive ensemble in QSAR prediction for drug discovery. BMC Bioinf. 2019;20:521. https://doi.org/10.1186/s12859-019-3135-4.
20. Marsh RE, Waser J. Refinement of the crystal structure of sinigrin. Acta Crystallogr Sect B Struct Crystallogr Crystal Chem. 1970;26(7):1030–7.
21. Benn MH, Ettlinger MG. The synthesis of sinigrin. Chem Commun. 1965;19:445.
22. Ivanov NN, Shulga DA, Palyulin VA. Decomposition of small molecules for fragment-based drug design. Biophysica. 2023;3:362–72. https://doi.org/10.3390/biophysica3020024.
23. Hardy A. Water and the search for public health in London in the eighteenth and nineteenth centuries. Med Hist. 1984;28(3):250–82. https://doi.org/10.1017/s0025727300035936.
24. West JB. Henry Cavendish (1731-1810): hydrogen, carbon dioxide, water, and weighing the world. Am J Physiol Lung Cell Mol Physiol. 2014;307(1):L1–6. https://doi.org/10.1152/ajplung.00067.2014.
25. Kambas L. Antoine-Laurent Lavoisier's 'Sur la nature de l'eau': an annotated English translation. Ann Sci. 2025;82(1):102–32. https://doi.org/10.1080/00033790.2023.2289531.
26. Lang X, Shi L, Zhao Z, Min W. Probing the structure of water in individual living cells. Nat Commun. 2024;15(1):5271. https://doi.org/10.1038/s41467-024-49404-9.
27. Anthonsen T, Sjursnes BJ. Importance of water activity for enzyme catalysis in non-aqueous organic systems. In: Gupta MN, editor. Methods in non-aqueous enzymology. Methods and tools in biosciences and medicine. Basel: Birkhäuser; 2000. https://doi.org/10.1007/978-3-0348-8472-3_2.
28. Leung MHM, Addicoat MA, Lincoln SF, Metha GF, Kee TW. Time-resolved keto-enol tautomerization of the medicinal pigment curcumin. Phys Chem Chem Phys. 2024;26(20):14970–9. https://doi.org/10.1039/d4cp01006j.
29. de Oliveira A, Paula A, Wegermann CA, Ferreira AMDC. Keto-enol tautomerism in the development of new drugs. Front Chem Biol. 2024; https://doi.org/10.3389/fchbi.2024.1400642.

30. Cucinotta CS, Ruini A, Catellani A, Stirling A. Ab initio molecular dynamics study of the keto-enol tautomerism of acetone in solution. ChemPhysChem. 2006;7(6):1229–34. https://doi.org/10.1002/cphc.200600007.
31. Mudare S, Jing J, Makowski D, He X, Liang Z, Sims Z, Wanger TC, Tilman D, Zhang F, Cong WF. Crop rotations synergize yield, nutrition, and revenue: a meta-analysis. Nat Commun. 2025;16(1):9552. https://doi.org/10.1038/s41467-025-64567-9.
32. King AE, Blesh J. Crop rotations for increased soil carbon: perenniality as a guiding principle. Ecol Appl. 2018;28(1):249–61. https://doi.org/10.1002/eap.1648.
33. Meng X, Wang B, Zhang X, Liu C, Ji J, Hao X, Yang B, Wang W, Xu D, Zhang S, Wang X, Cao M, Wang Y. Long-term crop rotation revealed the relationship between soil organic carbon physical fraction and bacterial community at aggregate scales. Microorganisms. 2025;13(3):496. https://doi.org/10.3390/microorganisms13030496.
34. Singh AA, Singh AK. Role of bacterial quorum sensing in plant growth promotion. World J Microbiol Biotechnol. 2025;41:18. https://doi.org/10.1007/s11274-024-04232-3.
35. Vidar WS, Baumeister TUH, Caesar LK, Kellogg JJ, Todd DA, Linington RG, Kvalheim O, Cech NB. Interaction metabolomics to discover synergists in natural product mixtures. J Nat Prod. 2023;86(4):655–71. https://doi.org/10.1021/acs.jnatprod.2c00518.
36. Hagan E, Poulin J. Statistics of the early synthetic dye industry. Herit Sci. 2021;9:33. https://doi.org/10.1186/s40494-021-00493-5.
37. Cataldi M, Borriello F, Granata F, Annunziato L, Marone G. Histamine receptors and antihistamines: from discovery to clinical applications. Chem Immunol Allergy. 2014;100:214–26. https://doi.org/10.1159/000358740.
38. Thurmond R. A timeline of histamine and its receptors. Nat Med. 2010;16:1064. https://doi.org/10.1038/nm1010-1064.
39. Oldani M. Pfizer. In: The SAGE encyclopedia of pharmacology and society, vol. 4. London: SAGE Publications; 2016. p. 1063–6. https://doi.org/10.4135/9781483349985.n303.
40. Gaynes R. The discovery of penicillin—new insights after more than 75 years of clinical use. Emerg Infect Dis. 2017;23(5):849–53. https://doi.org/10.3201/eid2305.161556.
41. Schlossman ML, McCarthy JP. Lanolin and its derivatives. J Am Oil Chem Soc. 1978;55:447–50. https://doi.org/10.1007/BF02911911.
42. Turner ME, Beckett JV, Afton B. 'Barley and oats', farm production in England 1700–1914. Oxford: Oxford Academic; 2001. https://doi.org/10.1093/acprof:oso/9780198208044.003.0006.
43. Overton M. Estimating crop yields from probate inventories: an example from East Anglia, 1585-1735. J Econ Hist. 1979;39(2):363–78.
44. Turner M. Agricultural productivity in England in the eighteenth century: evidence from crop yields. Econ Hist Rev. 1982;35(4):489–510.
45. Zhao H, Liu Y, Huang Y, Liang Q, Cai S, Zhang G. Time-course comparative metabolome analysis of different barley varieties during malting. J Agric Food Chem. 2022;70(6):2051–9. https://doi.org/10.1021/acs.jafc.1c08346.
46. Li M, Cai K, Zheng N, Zhang G, Ye L. Identification of the key transcription factors regulating the expression of the genes associated with barley malt quality during malting. J Agric Food Chem. 2023;71(21):8241–51. https://doi.org/10.1021/acs.jafc.3c00988.
47. Bokulich NA, Bamforth CW. The microbiology of malting and brewing. Microbiol Mol Biol Rev. 2013;77(2):157–72. https://doi.org/10.1128/MMBR.00060-12.
48. Bamforth CW. 'Water: and genuine terroir', beer: tap into the art and science of brewing. 4th ed. New York: Oxford Academic; 2023. https://doi.org/10.1093/oso/9780199996742.003.0008.
49. Bamforth CW. Progress in brewing science and beer production. Annu Rev Chem Biomol Eng. 2017;8:161–76. https://doi.org/10.1146/annurev-chembioeng-060816-101450.
50. Antman FM, Flynn J. When beer is safer than water: beer availability and mortality from waterborne illnesses, IZA discussion papers, no. 17164. Institute of Labor Economics (IZA), Bonn; 2024.

51. Osorio-Paz I, Brunauer R, Alavez S. Beer and its non-alcoholic compounds in health and disease. Crit Rev Food Sci Nutr. 2020;60(20):3492–505. https://doi.org/10.1080/10408398.2019.1696278.
52. Marcos A, Serra-Majem L, Pérez-Jiménez F, Pascual V, Tinahones FJ, Estruch R. Moderate consumption of beer and its effects on cardiovascular and metabolic health: an updated review of recent scientific evidence. Nutrients. 2021;13(3):879. https://doi.org/10.3390/nu13030879.
53. Marques C, Dinis L, Barreiros Mota I, Morais J, Ismael S, Pereira-Leal JB, Cardoso J, Ribeiro P, Beato H, Resende M, Espírito Santo C, Cortez AP, Rosário A, Pestana D, Teixeira D, Faria A, Calhau C. Impact of beer and nonalcoholic beer consumption on the gut microbiota: a randomized, double-blind, controlled trial. J Agric Food Chem. 2022;70(41):13062–70. https://doi.org/10.1021/acs.jafc.2c00587.
54. Erhardt N, Martin-Rios C, Bolton J, Luth M. Doing well by creating economic value through social values among craft beer breweries: a case study in responsible innovation and growth. Sustainability. 2022;14:2826. https://doi.org/10.3390/su14052826.
55. Pérez-Bibbins B, Torrado-Agrasar A, Salgado JM, Oliveira RP, Domínguez JM. Potential of lees from wine, beer and cider manufacturing as a source of economic nutrients: an overview. Waste Manag. 2015;40:72–81. https://doi.org/10.1016/j.wasman.2015.03.009.
56. Neylon E, Nyhan L, Zannini E, Sahin AW, Arendt EK. From waste to taste: application of fermented spent rootlet ingredients in a bread system. Foods. 2023;12(7):1549. https://doi.org/10.3390/foods12071549.
57. Lalić A, Jagelavičiūtė J, Trivunović Z, Marić M, Karlović A, Bašinskienė L. Biotechnological valorization of brewer's spent grain from old bread and barley malt: fermentative potential of Saccharomyces cerevisiae. Fermentation. 2025;11:382. https://doi.org/10.3390/fermentation11070382.
58. Sabbatini G, Mari E, Ortore MG, Di Gregorio A, Fattorini D, Di Carlo M, Galeazzi R, Vignaroli C, Simoni S, Giorgini G, Guarrasi V, Chiancone B, Leto L, Cirlini M, Del Vecchio L, Mangione MR, Vilasi S, Minnelli C, Mobbili G. Hop leaves: from waste to a valuable source of bioactive compounds - A multidisciplinary approach to investigating potential applications. Heliyon. 2024;10(18):e37593. https://doi.org/10.1016/j.heliyon.2024.e37593.
59. Perrin CE. A reluctant catalyst: Joseph black and the Edinburgh reception of Lavoisier's chemistry. Ambix. 1982;29(3):141–76. https://doi.org/10.1179/amb.1982.29.3.141.
60. Chang H. Reductionism and the relation between chemistry and physics. In: Arabatzis T, Renn J, Simões A, editors. Relocating the history of science. Boston studies in the philosophy and history of science, vol. 312. Cham: Springer; 2015. https://doi.org/10.1007/978-3-319-14553-2_13.
61. Tang C, Stuyver T, Lu T, et al. Voltage-driven control of single-molecule keto-enol equilibrium in a two-terminal junction system. Nat Commun. 2023;14:3657. https://doi.org/10.1038/s41467-023-39198-7.
62. Kaplan G, Seferoğlu Z. The synthetic approaches for preparation of indigo and applications in denim industry. Curr Org Synth. 2023;20(4):361–4. https://doi.org/10.2174/1570179419666220830091956.
63. van Bommel R, M., de Keijzer, M. Copying nature: synthetic alizarin. In: Bright colours from the past. Cultural heritage science. Cham: Springer; 2025. https://doi.org/10.1007/978-3-031-88664-5_20.
64. Ozdemir MB, Karadag R, Koseoglu I. Agricultural production and international trade analysis of competitiveness between natural and conventional dyes in the textile industry. Sci Rep. 2025;15(1):21596. https://doi.org/10.1038/s41598-025-05032-x.
65. Pribyl K. Farming in Norfolk around 1800. In: Farming, famine and plague. Cham: Springer; 2017. https://doi.org/10.1007/978-3-319-55953-7_4.
66. Overton M. Agricultural revolution in England: the transformation of the agrarian economy 1500–1850. Cambridge: Cambridge University Press; 1996.
67. Curtis HA. Fertilizers: the world supply. Foreign Aff. 1924;2(3):436–45. https://doi.org/10.2307/20028312.

68. Allen RC. Technical change, globalisation and the labour market: British and American experience since 1620. Oxford Open Econ. 2021;3:178.
69. Anomaly J. What's wrong with factory farming? Public Health Ethics. 2015;8(3):246–54. https://doi.org/10.1093/phe/phu001.
70. Magurno F, Sasvári Z, Posta K. From monoculture to the Norfolk system: assessment of arbuscular mycorrhizal fungi communities associated with different crop rotation systems. Symbiosis. 2014;64:115–25. https://doi.org/10.1007/s13199-014-0309-8.
71. García-Llorente M, Rubio-Olivar R, Gutierrez-Briceño I. Farming for life quality and sustainability: a literature review of green care research trends in Europe. Int J Environ Res Public Health. 2018;15(6):1282. https://doi.org/10.3390/ijerph15061282.
72. Sugden K. Clapham revisited: the decline of the Norwich worsted industry (c. 1700–1820). Contin Chang. 2018;33(2):203–24. https://doi.org/10.1017/S0268416018000140.
73. Clark P. Norwich since 1550. Hist Rev. 2006;491:544–6. https://doi.org/10.1093/ehr/cel024.
74. Uluisik S, Chapman NH, Smith R, Poole M, Adams G, Gillis RB, Besong TM, Sheldon J, Stiegelmeyer S, Perez L, Samsulrizal N, Wang D, Fisk ID, Yang N, Baxter C, Rickett D, Fray R, Blanco-Ulate B, Powell AL, Harding SE, Craigon J, Rose JK, Fich EA, Sun L, Domozych DS, Fraser PD, Tucker GA, Grierson D, Seymour GB. Genetic improvement of tomato by targeted control of fruit softening. Nat Biotechnol. 2016;34(9):950–2. https://doi.org/10.1038/nbt.3602.
75. Arvanitoyannis IS. Health risks of genetically modified foods. Crit Rev Food Sci Nutr. 2009;49(2):164–75. https://doi.org/10.1080/10408390701855993.
76. Shen C, Yin XC, Jiao BY, et al. Evaluation of adverse effects/events of genetically modified food consumption: a systematic review of animal and human studies. Environ Sci Eur. 2022;34:8. https://doi.org/10.1186/s12302-021-00578-9.
77. Palmer AC. Golden rice: a quarter-century of innovation, challenges, and the promise of better nutrition. J Nutr. 2025;25:00414–6. https://doi.org/10.1016/j.tjnut.2025.06.025.
78. Hodge C, Taylor C. Vitamin A deficiency. In: StatPearls. Treasure Island: StatPearls Publishing; 2025.
79. Moghissi AA, Pei S, Liu Y. Golden rice: scientific, regulatory and public information processes of a genetically modified organism. Crit Rev Biotechnol. 2016;36(3):535–41. https://doi.org/10.3109/07388551.2014.993586.
80. Martin C, Butelli E. The purple tomato story; from laboratory bench to the consumer. ACS Food Sci Technol. 2024;5(1):19–28. https://doi.org/10.1021/acsfoodscitech.4c00692.
81. Gonçalves AC, Nunes AR, Falcão A, Alves G, Silva LR. Dietary effects of anthocyanins in human health: A comprehensive review. Pharmaceuticals. 2021;14(7):690. https://doi.org/10.3390/ph14070690.
82. Butelli E, Titta L, Giorgio M, Mock HP, Matros A, Peterek S, Schijlen EG, Hall RD, Bovy AG, Luo J, Martin C. Enrichment of tomato fruit with health-promoting anthocyanins by expression of select transcription factors. Nat Biotechnol. 2008;26(11):1301–8. https://doi.org/10.1038/nbt.1506.
83. Gonzali S, Perata P. Anthocyanins from purple tomatoes as novel antioxidants to promote human health. Antioxidants. 2020;9(10):1017. https://doi.org/10.3390/antiox9101017.

MIX
Papier aus verantwortungsvollen Quellen
Paper from responsible sources
FSC® C105338
FSC
www.fsc.org

If you have any concerns about our products,
you can contact us on
ProductSafety@springernature.com

In case Publisher is established outside the EU,
the EU authorized representative is:
Springer Nature Customer Service Center GmbH
Europaplatz 3, 69115 Heidelberg, Germany

Printed by Libri Plureos GmbH
in Hamburg, Germany